Protein Function
Protein Purification Methods
Protein Sequencing
Protein Structure
Proteolytic Enzymes
Solid phase Peptide Synthesis
Spectrophotometry and Spectrofluorimetry
Steroid Hormones
Teratocarcinomas and Embryonic Stem Cells
Transcription and Translation
Virology
Yeast

Ribosomes and Protein Synthesis

A Practical Approach

Edited by

G. SPEDDING
Department of Chemistry and Biochemistry
Southern Illinois University
Illinois, USA

Oxford University Press
Walton Street, Oxford OX2 6DP

Oxford is a trade mark of Oxford University Press

Published in the United States
by Oxford University Press, New York

British Library Cataloguing in Publication Data
Ribosomes and protein synthesis.
1. Organisms. Ribosomes
I. Spedding, G. II. Series
574.87'34

ISBN 0–19–963104–2
ISBN 0–19–963105–0 (pbk)

Library of Congress Cataloging in Publication Data
Ribosomes and protein synthesis: a practical approach/edited by G. Spedding.
(Practical approach series)
Includes bibliographical references.
1. Ribosomes. 2. Proteins—Synthesis. I. Spedding, G. (Gary) II. Series.
QH603.R5R53 1990 574.87'34—dc20 89–26458

ISBN 0–19–963104–2
ISBN 0–19–963105–0 (pbk)

Typeset by Cotswold Typesetting Ltd, Gloucester
Printed by Information Press Ltd, Oxford, England

Preface

This book, in the *Practical Approach* series, has been designed to provide an up-to-date, concise, and self-contained introductory guide to the methodology and many of the techniques currenty being utilized in the analysis of the events of protein biosynthesis and in the elucidation of the structure and function of ribosomes.

Ribosomes and the events of protein synthesis have provided a stimulating challenge to biochemists, geneticists, molecular biologists, and biophysical chemists for over three decades, and many fundamental questions still remain to be answered. In recent years, an increasing emphasis on the potential capability of RNA molecules to act as molecular catalysts has led to a major resurgence of interest in the ribosome. It is therefore hoped that this book will provide an important and timely introduction to many of the tools now required, to enable both the newcomer and the established scientist alike, to enter this exciting and challenging field.

Accurate and faithful cell-free protein-synthesizing systems, utilizing highly purified components, have become powerful assets for studying the mechanisms and controls of protein synthesis and gene regulation. Such systems and their constituent components are detailed in the first part of the book. Subsequent chapters have been designed with a view towards creating a thorough understanding of many of the exciting and powerful new molecular biological methods which are now being used to elucidate the structural and functional features of the ribosome.

Each chapter has been contributed from laboratories at the forefront of research into the field and I should like to express my gratitude to all of the authors for their dedicated efforts in providing outstanding contributions, which are representative of expert, first-hand practical knowledge of each topic.

I should, finally, also like to thank the publishers and staff at Oxford University Press for their patience, hard work, and understanding during this project.

Baltimore
November 1989

GARY SPEDDING

Contents

Contributors

RICHARD H. ARCHER
Department of Biology, University of Rochester, Rochester, NY 14627, USA.

ROSS M. BAXTER
Faculty of Pharmacy, University of Toronto, Ontario, Canada.

NEŞE BILGIN
Department of Molecular Biology, BMC Box 590, S-751 24 Uppsala, Sweden.

MILOSLAV BOUBLIK
Department of Biochemistry, The Roche Institute of Molecular Biology, Roche Research Centre, Nutley, New Jersey 07110, USA.

RICHARD BRIMACOMBE
Max-Planck-Institut für Molekulare Genetik, Abteilung Wittman, D-1000 Berlin-Dahlem, Germany.

C. THOMAS CASKEY
Institute for Molecular Genetics, Baylor College of Medicine, One Baylor Plaza, Houston, Texas, 77030, USA.

JAN CHRISTIANSEN
Department of Clinical Chemistry, Bispebjerg Hospital, DK-2400 Copenhagen NV, Denmark.

DAE-GYUN CHUNG
The Banting and Best Department of Medical Research and the Faculty of Medicine, University of Toronto, Ontario, Canada.

ALBERT E. DAHLBERG
Division of Biology and Medicine, Brown University, Providence, RI 02912, USA.

ELIZABETH A. DE STASIO
Lawrence University, Appleton, WI 54912, USA.

JAYDEV N. DHOLAKIA
Department of Biochemistry, The University of Mississippi Medical Center, 2500 North State Street, Jackson, Mississippi 39216-4505, USA.

B. CAMERON DONLY
Department of Biochemistry, University of Otago, PO Box 56, Dunedin, New Zealand.

JAN EGEBJERG
Biostructural Chemistry, Kemisk Institut, Aarhus Universitet, DK-8000 Aarhus C, Denmark.

MÅNS EHRENBERG
Department of Molecular Biology, BMC Box 590, S-751 24 Uppsala, Sweden.

M. CLELIA GANOZA
C. H. Best Institute, 112 College Street, Toronto, Ontario M5G 1L6, Canada.

ROGER A. GARRETT
Institute for Biological Chemistry B, Copenhagen University, Sølvgade 83, DK-1307 Copenhagen K, Denmark.

H. U. GÖRINGER
Max-Planck-Institut für Molekulare Genetik, Abteilung Wittmann, D-1000 Berlin-Dahlem, Germany.

BARBARA GREUER
Max-Planck-Institut für Molekulare Genetik, Abteilung Wittmann, D-1000 Berlin-Dahlem, Germany.

HEINZ GULLE
Max-Planck-Institut für Molekulare Genetik, Abteilung Wittmann, D-1000 Berlin-Dahlem, Germany.

GEORGE M. C. JANSSEN
Department of Medical Biochemistry, Sylvius Laboratories, University of Leiden, PO Box 9503, 2300 RA, Leiden, The Netherlands.

MICHAEL KOSACK
Max-Planck-Institut für Molekulare Genetik, Abteilung Wittmann, D-1000 Berlin-Dahlem, Germany.

CHARLES G. KURLAND
Department of Molecular Biology, BMC Box 590, S-751 24 Uppsala, Sweden.

NIELS LARSEN
Biostructural Chemistry, Kemisk Institut, Aarhus Universitet, DK-8000 Aarhus C, Denmark.

LASSE LINDAHL
Department of Biology, University of Rochester, Rochester, NY 14627, USA.

J. ANTONIE MAASSEN
Department of Medical Biochemistry, Sylvius Laboratories, University of Leiden, PO Box 9503, 2300 RA, Leiden, The Netherlands.

GEORGE A. MACKIE
Department of Biochemistry, University of Western Ontario, London, Ontario N6A 5C1, Canada.

JOSEPH R. McCORMICK
Department of Biology, University of Rochester, Rochester, NY 14627, USA.

PHILIP MITCHELL
Max-Planck-Institut für Molekulare Genetik, Abteilung Wittmann, D-1000 Berlin Dahlem, Germany.

WIM MÖLLER
Department of Medical Biochemistry, Sylvius Laboratories, University of Leiden, PO Box 9503, 2300 RA, Leiden, The Netherlands.

KNUD H. NIERHAUS
Max-Planck-Institut für Molekulare Genetik, Abt. Wittmann, Ihnestrasse 73, D-1000 Berlin-Dahlem, Germany.

MONIKA OßWALD
Max-Planck-Institut für Molekulare Genetik, Abteilung Wittmann, D-1000 Berlin-Dahlem, Germany.

GARY SPEDDING
Department of Chemistry, The Johns Hopkins University, 3400 Charles Street, Baltimore, MD 21218, USA.

KATRIN STADE
Max-Planck-Institut für Molekulare Genetik, Abteilung Wittmann, D-1000 Berlin-Dahlem, Germany.

WOLFGANG STIEGE
Max-Planck-Institut für Molekulare Genetik, Abteilung Wittmann, D-1000 Berlin-Dahlem, Germany.

WILLIAM E. TAPPRICH
DBS, University of Montana, Missoula, Montana 59812, USA.

WARREN P. TATE
Department of Biochemistry, University of Otago, PO Box 56, Dunedin, New Zealand.

ALBERT J. WAHBA
Department of Biochemistry, The University of Mississippi Medical Center, 2500 North State Street, Jackson, Mississippi 39216-4505, USA.

PHILIP C. WONG
Department of Biological Chemistry, The Johns Hopkins University, Baltimore, MD 21218, USA.

CHARLES L. WOODLEY
Department of Biochemistry, The University of Mississippi Medical Center, 2500 North State Street, Jackson, Mississippi 39216-4505, USA.

NASIR D. ZAHID
Faculty of Pharmacy, University of Toronto, Ontario, Canada.

JANICE M. ZENGEL
Department of Biology, University of Rochester, Rochester, NY 14627, USA.

Abbreviations

A_{260}	absorption at 260 nm
aa-tRNA	aminoacyl-tRNA
ADP	adenosine diphosphate
AMV	avian myeloblastosis virus
ATP	adenosine triphosphate
AUG	initiation codon
BSA	bovine serum albumin
Ci	Curie
cDNA	complementary DNA
CHAPS	3-[(3-cholamidopropyl)di-methylammonio]-1-propane sulphonate
CMCT	1-cyclohexyl-3-[2-morpholinoethyl]carbodiimide metho-*p*-toluene sulphonate
c.p.m.	counts per minute
CTEM	conventional transmission electron microscopy
ddH_2O	double-distilled water
ddNTP	dideoxynucleotide triphosphate
DEAE	diethylaminoethyl
DEPC	diethylpyrocarbonate
DMS	dimethylsulphate
DMSO	dimethylsulphoxide
DNase	deoxyribonuclease
dNTP	deoxynucleotide triphosphate
d.p.m.	disintegrations per minute
DTE	dithioerythritol
DTT	dithiothreitol
EDTA	ethylenediaminetetraacetate
EELS	electron energy loss spectroscopy
EGTA	ethyleneglycolbis-(β-aminoethyl)ether *N*,*N*,*N*′,*N*′ tetra acetic acid
EF-	elongation factor
ELISA	type of sensitive immunological assay
EM	electron microscopy
FAD	flavin adenine dinucleotide
FPLC	fast performance liquid chromatography
fMet-$tRNA_f$	formyl-methionyl-$tRNA_f$ (prokaryotic initiator tRNA)
GDP	guanosine diphosphate
GEF	guanine nucleotide exchange factor
GMP-PCP	(β,γ-methylene guanosine 5′-triphosphate)
Gp_4G	P_1, P_2-diguanosine 5′-tetraphosphate
GTP	guanosine triphosphate
Hepes	*N*-2-hydroxyethylpiperazine *N*′-2-ethanesulphonic acid
HPLC	high performance liquid chromatography
IEM	immuno-electron microscopy

IF-	initiation factor
IPTG	isopropyl-β-D-thiogalactoside
K_M	Michaelis constant
Leu	leucine
M	Molarity
Met	methionine
Met-tRNA$_f$	methionyl-tRNA$_f$ (eukaryotic initiator tRNA)
m.o.i	multiplicity of infection
mRNA	messenger RNA
NAc	*N*-acetyl-
NAD	nicotinamide adenine dinucleotide
NADP	nicotinamide adenine dinucleotide phosphate
NP-40	Nonidet P-40
PAGE	polyacrylamide gel electrophoresis
PEG	polyethylene gycol
PEI	polyethyleneimine
PEP	phosphoenolpyruvate
p.f.u.	plaque forming unit
pI	isoelectric point
PMSF	phenylmethylsulphonyl fluoride
Poly(Phe)	polyphenylalanine
Poly(U)	polyuridylic acid (5′)
PPO	2,5-diphenyloxazole
p.s.i.	pounds per square inch (lb/in^2)
RF-	release factor
RNase	ribonuclease
r-protein	ribosomal protein
rRNA	ribosomal RNA
SDS	sodium dodecyl-sulphate
SSC	standard saline citrate
STEM	scanning transmission electron microscopy
TCA	trichloroacetic acid
TEA	triethanolamine
TEMED	*N*,*N*,*N′*,*N′*-tetramethylethylenediamine
TLC	thin layer chromatography
TMK	Tris-Mg^{2+} buffer
TMV	tobacco mosaic virus
TP70, 50, 30	total proteins derived from 70S ribosomes or 50S or 30S ribosomal subunits
Tris	tris(hydroxy methyl)amino methane
tRNA	transfer RNA
UV	ultraviolet
X-gal	5-bromo-4-chloro-3-indolyl-β-D-galactopyranoside

1

Isolation and analysis of ribosomes from prokaryotes, eukaryotes, and organelles

GARY SPEDDING

1. Introduction

There are many variations upon the theme of isolating ribosomes and ribosomal components. This chapter can therefore only provide a brief introduction to some of the essential principles involved. A few key examples will serve to illustrate some of the routine methodology appropriate for the preparation of ribosomes and their constituent components from eukaryotes, prokaryotes, and organelles. It has had to be assumed, however, that the reader is familiar with the growth of cells and the preparation of starting materials, etc., prior to the isolation of ribosomes. Other books in *The Practical Approach* series may provide an excellent starting point for these procedures.

The efficient breakage or lysis of cells usually represents the first important step in the isolation of intact ribosomal components. Cells may be broken by many of the traditional methods including: grinding with alumina, quartz sand, glassbeads, etc.; by sonication, passage through a French Press; by lysozyme or other forms of enzymatic or detergent lysis, as appropriate. An excellent discussion of these methods may be found elsewhere (1). Brief details for the breakage of bacterial cells by grinding with alumina or by passage through the French press are also presented below.

1.1 Some general notes of guidance

1.1.1 General precautions

Certain precautions and steps need to be taken for the successful isolation of ribosomal components. These precautions are usually based upon the need to eliminate protease and ribonuclease contamination or to minimize such endogenous activities. Some general precautions and steps to be taken are outlined in *Protocol 1*.

Protocol 1. Precautions to be taken when isolating ribosomes

1. Sterilize all plasticware and glassware by autoclaving or baking at 150 °C overnight. Plasticware may also be sterilized by soaking in a 0.1% (v/v) solution of diethylpyrocarbonate (DEPC), followed by extensive rinsing with sterile double-distilled water prior to use. DEPC inactivates ribonucleases.
2. Where possible use pre-wrapped sterile disposable plasticware.
3. All water used and most buffers should be autoclaved or filter-sterilized. Water and many buffer solutions (but not ones containing Tris) may be treated with 0.1% (v/v) final concentration DEPC for 1 h, with stirring, prior to being autoclaved. Autoclaving removes all traces of DEPC. Tris-containing solutions may be made with DEPC-treated autoclaved water and then autoclaved. Direct treatment with DEPC is not possible because DEPC reacts with primary amines.
4. Disposable rubber gloves should be worn to prevent contamination by 'finger ribonucleases'.
5. All dialysis tubing should be boiled or autoclaved in 0.1 M sodium bicarbonate, 10 mM EDTA, washed extensively with, and finally stored in, autoclaved or DEPC-treated double-distilled water at 4 °C.
6. Reagents used should be of the highest quality available and where possible RNase free (e.g. sucrose).

1.1.2 Prevention of nuclease and protease activity

Where calcium-dependent nucleases are prevalent, the use of EGTA, which chelates calcium ions, may be employed. Occasionally, calcium may replace magnesium (which stabilizes ribosomes) in buffers used for the preparation of ribosomal components in order to inactivate magnesium ion dependent nucleases. Polyvinylsulphate or heparin (caution—potential protein synthesis inhibitor!), and particularly placental ribonuclease inhibitor, may be employed for use as ribonuclease inhibitors. Further details concerning ribonucleases and their inhibitors have been presented previously in this series (2). Bentonite and macaloid clays may be used as non-specific inhibitors of ribonuclease activity. Their preparation and use has been detailed recently (3).

As precautions against proteases the use of phenylmethylsulphonylfluoride (PMSF) at a final working concentration of 0.1 mM and/or benzamidine at 0.1–5 mM final concentration is occasionally recommended. See the methods outlined in this chapter for specific uses.

1.1.3 Buffers

Generally, buffers used for the isolation and preparation of ribosomes contain either Tris or Hepes. Whilst Tris has been employed for many years, the use of Hepes may now be preferable in some systems. The pH adjustment of Tris-containing solutions should, ideally, be done at the final working temperature. That this has been the case cannot always be assumed when interpreting the published literature. In many instances workers prepare solutions at room temperature but this is not always obvious. Some caution must be entertained, therefore, when preparing buffers, in order to achieve optimum results.

Generally, the acetate and chloride forms of salts may be used interchangeably in buffers used for preparing ribosomes etc. Again, however, one form may be more advantageous depending upon the particular system under investigation.

Reducing agents such as 2-mercaptoethanol and dithiothreitol (DTT), etc., should always be added freshly to buffers. Buffers containing 2-mercaptoethanol should not, however, be used to determine A_{260} absorbance values for calculation of the concentration of ribosome preparations since this compound absorbs strongly at 260 nm.

1.1.4 Structure of ribosomes—an overview

A brief overview concerning the basic structure of ribosomes will hopefully serve to set the stage for an understanding of the methodology that follows.

Bacterial ribosomes contain three species of rRNA, with sedimentation coefficients of 23S, 16S, and 5S, and about 55 different proteins. The 70S ribosome is composed of two unequal size subunits with sedimentation values of 30S and 50S.

Chloroplast ribosomes are very similar to their bacterial counterparts. The three typical species of rRNA are present together with a small 4.5 S rRNA species in plant (but not in algal) chloroplast ribosomes. The precise number of proteins in chloroplast ribosomes has not yet been determined but appears to be similar to that for bacterial ribosomes.

Higher plant mitochondrial ribosomes (75S) possess rRNA of distinctly larger size (18S and 26S) than those of their prokaryotic counterparts whilst also possessing a 5S rRNA species. Mammalian mitochondrial ribosomes, on the other hand, have very small sedimentation values (55S) and are thus unique. They possess two fairly small ribosomal RNAs (12S and 16S) and no 5S rRNA. Typical mammalian mitochondrial ribosomes contain about 90 proteins.

Cytoplasmic (80S) ribosomes, composed of a 40S and a 60S subunit, from a wide range of organisms including animals, plants, fungi, and unicellular eukaryotes, are all very similar in both their rRNA and protein composition. All contain four rRNA species (17–18S and 26–28S together with a 5.8S and a 5S rRNA species). The precise number of ribosomal proteins in any 80S ribosome has not yet been established. The range is between 75 and 90 proteins.

Several recent reviews may be consulted for more detailed background information concerning this vast subject (4–6).

2. Preparation of ribosomes and ribosomal subunits

2.1 Prokaryotic Ribosomes

2.1.1 Tight-couple ribosomes

The most active prokaryotic ribosomes (as assayed in *in vitro* protein synthesis assays) are those generally prepared according to Noll *et al.* (7) and Jelenc (8). Modifications of both of these methods, which have found widespread use, are presented in *Protocols 2* and *4*. The methods of Noll *et al.* (7), as modified (9), produce 70S ribosomes in a form known as 'vacant couples', with the type known as 'tight couples' generally employed in most studies of protein synthesis. The term tight couple refers to 70S ribosomes that are resistant (unlike loose-couple ribosomes) to dissociation into subunits during low-speed centrifugation in the presence of 5–6 mM magnesium ions. The ability of subsequently derived subunits to form vacant 70S couples (devoid of tRNA, mRNA, etc.) at a magnesium ion concentration of 10 mM is a stringent condition for activity in the translation of natural mRNA (e.g. phage R17 RNA). The method for the preparation of tight-couple ribosomes and ribosomal subunits is outlined in *Protocol 2*. Such preparations typically yield 60–80% active ribosomes as determined by the binding of N-acetylphenylalanyl-tRNA (9). Some reports have indicated that the alumina-grinding method (details below) is to be preferred over the use of the French press for the preparation of cell extracts ready for the isolation of tight-couple ribosomes. The method of choice, however, usually depends upon the physical quantity of cells to be processed.

Large-scale preparations of ribosomes and subunits have been detailed elsewhere (11–13). This usually involves zonal ultracentrifugation with hyperbolic or equivolumetric sucrose density gradients. These systems are capable of handling up to 10 000 A_{260} units of crude ribosomes. Further details concerning large scale zonal ultracentrifugation have also been presented previously in this series (14).

For appropriate modifications to methods for the preparation of archaebacterial ribosomes the reader is referred to reference 15.

Protocol 2. The preparation of tight-couple ribosomes and ribosomal subunits from prokaryotes

1. Grow bacteria in rich media to mid-log phase. Slowly cool the culture to 15 °C (to produce run-off ribosomes). Harvest and wash the cells with buffer A[a]. (Cells may be stored frozen.) Perform all subsequent operations at 4–8 °C.

2. Grind the cells with alumina or break open by passage through a chilled French press at 10 000–12 000 p.s.i. Extract with buffer A. Add DNase (see Section 2.1.2. in the text for details).
3. Centrifuge the extract at 30 000 *g* for 45 min to clear cell debris and alumina. The supernatant obtained at this step is designated 'S30'.
4. Recover the top three-fourths of the S30 supernatant.[b] Layer portions of the supernatant over equal volume 1.1 M sucrose cushions made up in buffer B.[c] Centrifuge at 100 000 *g* for 15 h. (Produces sucrose/salt-washed ribosomes.)
5. Decant the supernatant.[d] Remove the brownish flocculent material on top of the clear ribosome pellet by washing gently with buffer A.
6. Resuspend the ribosomal pellet in buffer A by gentle teasing with a glass rod or by gentle agitation for 1 h at 4 °C.
7. Repeat the sucrose/salt-washing procedure in step 4.
8. (Optional) Wash the ribosomal pellet twice by resuspension in buffer C[e] and sedimentation by centrifugation at 100 000 *g* for 16 h. (These are salt-washing steps in the absence of sucrose.)
9. Resuspend the ribosomes in buffer D[f] (for further purification, step 10) or buffer E[g], and dialyse against that buffer (e.g. 3 changes of 300 volumes over several hours at 4 °C). Store the ribosomes as small aliquots in buffer E at 500 A_{260} units/ml at −80 °C.[h]
10. Obtain tight couple ribosomes from salt-washed ribosomes in buffer D by isolation on 10–30% (w/v) linear sucrose gradients made up in buffer D. Centrifuge at 48 000 *g* for 15 h (e.g. at 19 000 r.p.m. in a Beckman SW28 rotor).
11. Fractionate the gradients by upward displacement using 60% (v/v) glycerol and by pumping through an absorbance monitor and fraction collector. Monitor the absorbance at 260 nm.
12. Recover the 70S ribosomes from appropriately pooled fractions (30S and 50S subunits present in the gradient are derived from loose-couple ribosomes, i.e. those ribosomes that dissociate at a magnesium concentration of 5.25 mM). Ribosomes may then be recovered by either.
    (a) Raising the magnesium ion concentration to 10 mM followed by the addition of 0.65 vol ice cold ethanol. Leave at −20 °C for 2 h then recover the precipitated ribosomes by centrifugation at 16 000 *g* for 15 min, *or*
    (b) Addition of an equal volume of buffer E followed by centrifugation at 56 000 *g* for 24 h.
13. Resuspend the ribosome pellet in buffer E (optional dialysis to remove traces of sucrose) and store at a concentration of 600–1000 A_{260} units/ml in small aliquots at −80 °C[h] (1 A_{260} unit is equivalent to 23 pmol ribosomes when determined in 25 mM Tris–HCl, 10 mM Mg^{2+}, and 100 mM KCl (or NH_4Cl)). The A_{260}/A_{280} ratio will be ≥ 1.8 for clean ribosomes.

14. Determine the purity of the ribosomes by loading 2–5 A_{260} units on a small analytical sucrose gradient containing 10 mM Tris–HCl (pH 7.5), 5 mM magnesium, and 50 mM potassium chloride. Centrifuge at 190 000 *g* for 100 min at 4 °C (e.g. 45 000 r.p.m. in a Beckman SW50.1 rotor). Pump the gradients through a UV analyser set at 260 nm. The ribosomes should now be completely free of ribosomal subunits.
15. *Subunits:* Obtain ribosomal subunits by resuspending tight-couple ribosomes from step 9 in buffer F[i] or by dialysing preparations in buffer E against 2 × 300 ml volumes of buffer F for 6 h at 4 °C.[j]
16. Layer 50–60 A_{260} units of the subunit suspension onto 34 ml 10–30% (w/v) sucrose gradients made up in buffer F. Centrifuge at 43 000 *g* for 16 h at 4 °C (e.g. 18 000 r.p.m. using a Beckman SW27 rotor).
17. Fractionate the gradients, monitoring absorbance at 260 nm. Pool the respective 30S and 50S peak fractions. Raise the magnesium ion concentration to 10mM. Recover subunits by ethanol precipitation (see step 12) or by pelleting at 200 000 *g* for 12 h (e.g. 45 000 r.p.m. in a Beckman Ti60 rotor).
18. Resuspend the ribosomal subunits in buffer E (optional dialysis to remove sucrose). Clarify by centrifugation at 15 000 *g* for 15 min. Store aliquots at −80 °C.[h, k]
19. Test the homogeneity of subunit preparations by loading one A_{260} unit onto 5 ml 10–25% (w/v) sucrose gradients made up in buffer G.[l] Centrifuge at 190 000 *g* for 100 min at 4 °C. Analyse the gradients (see step 11). The percentage cross contamination of subunits should be no more than 2–3%. Repurify on a second sucrose gradient if necessary.
20. With quantitative recovery of ribosomal subunits the peak height (UV absorbance) of 50S to 30S subunits should be 2:1. The concentration of subunits is determined by measuring absorbance at 260 nm. (1 A_{260} unit is equivalent to about 69 pmol and 34.5 pmol, for 30S and 50S subunits, respectively; see step 13).

[a] Buffer A: 20 mM Tris–HCl (pH 7.5 at 4 °C), 10.5 mM magnesium acetate, 100 mM NH_4Cl, 0.5 mM EDTA, and 3 mM 2-mercaptoethanol. The pH of all buffers should be adjusted at the respective experimental temperature.

[b] The S30 may be centrifuged directly at 100 000 *g* for 4 h at 4 °C to yield 'crude' ribosomes and post-ribosomal supernatant. (See note *d*.) The ribosomes may then be salt-washed.

[c] Buffer B: as for buffer A except 0.5 M NH_4Cl.

[d] The S100 post-ribosomal supernatant from crude or salt-washed ribosomes may be reserved as a source of soluble protein synthesis factors. Dialyse against buffer E (see note *g*) and store at −80 °C.

[e] Buffer C: 10 mM Tris–HCl (pH 7.5), 10.5 mM magnesium acetate, 500 mM NH_4Cl, 0.5 mM EDTA, and 7 mM 2-mercaptoethanol.

[f] Buffer D: 10 mM Tris–HCl (pH 7.5), 5.25 mM magnesium acetate, 60 mM NH_4Cl, 0.25 mM EDTA, and 3 mM 2-mercaptoethanol.

[g] Buffer E: as buffer D except 10 mM magnesium acetate and no EDTA. This is a typical ribosome storage buffer.

[h] Ribosomes and subunits should be aliquoted and, preferably, rapidly frozen in a dry ice-methanol (or acetone) bath prior to storage at −80 °C or, alternatively, under liquid nitrogen.

[i] Buffer F: 10 mM Tris–HCl (pH 7.5), 1.1 mM magnesium acetate, 60 mM NH_4Cl, 0.1 mM EDTA, and 2 mM 2-mercaptoethanol.

[j] It is occasionally possible to load ribosomes (from some organisms) directly on to sucrose gradients without prior dialysis against a low magnesium buffer.

[k] For many assays the activity of 30S ribosomal subunits is lost if the magnesium ion concentration has fallen below 2 mM. Thermal reactivation of subunits occurs following incubation at 40 °C for 15–30 min after the magnesium concentration has been raised to 10 mM or greater. Subunits may then be stored at −80 °C and retain full activity. This reactivation step may also be performed after storage of subunits, immediately prior to use (10).

[l] Buffer G: 10 mM Tris–HCl (pH 7.5), 5 mM magnesium acetate, 100 mM NaCl and 3 mM 2-mercaptoethanol. Sodium chloride is used here to prevent aggregation of subunits.

2.1.2 Preparation of cell extracts

Bacterial cells may be broken by grinding with autoclaved, levigated alumina (2–2.5 grams per gram weight of cell paste).

Protocol 3. Preparation of cell extracts

1. Grind in a mortar (cooled at −20 °C) until a smooth, thick, and sticky paste is obtained and 'popping' sounds are heard.
2. Extract the ground paste by slowly adding a buffered salt solution (2.0–2.5 ml per gram wet weight of cells) and continue grinding. See *Protocol 2* for a suitable buffer.
3. After a few minutes, add RNase-free DNase at 3–5 μg/ml and mix by grinding for a few more minutes until a reduction in viscosity is observed; this will yield a cleared supernatant as outlined in step 3 of *Protocol 2*.

If a French press is to be used (e.g. for larger quantities of cells) resuspend each gram wet weight of bacteria in 2.0–2.5 ml of the buffer used. Break the cells (see *Protocol 2*) by passing them twice through the press at 10000–18000 p.s.i. Deoxyribonuclease at 5 μg/ml should then be added followed by a 5 min incubation at 0–4 °C before clarifying the extract (*Protocol 2*, step 3).

2.1.3 Isolation of ribosomes by chromatography on Sephacryl S300

An alternative method for preparing highly active ribosomes from prokaryotes has been developed which avoids repeated high-speed centrifugation (8). The slightly modified procedure, presented in *Protocol 4*, involves the gentle isolation of highly active salt-washed ribosomes by gel filtration through Sephacryl S300 in a buffer which is optimal for high translation rates. Ribosomes prepared in this manner are essentially free of soluble protein synthesis factors such as elongation factor G (EF-G) and typically about 70% of the ribosomes are active as judged by aminoacyl-tRNA binding (8). The procedure outlined in *Protocol 4* may be scaled up or down proportionately.

The preparation and analysis of prokaryotic polysomes, which cannot be detailed here, has been described admirably elsewhere (16–18). For a more comprehensive account of the separation of polysomes by gradient centrifugation the reader is also referred to two previous volumes in this series (14, 19).

Protocol 4. Isolation of prokaryotic ribosomes by sephacryl chromatography

Perform all operations at 4 °C.

1. Obtain bacterial cells and wash the pellets in buffer A.[a]
2. Resuspend 50 g washed cells in 100 ml polymix buffer[b] containing 100 μg DNase.
3. Break the cells by two passages through the French press (or by grinding with alumina—see text). Clear cell debris etc., by centrifugation twice at 28 000 *g* for 30 min (e.g. 13 000 r.p.m. with the Sorvall GSA rotor).
4. Obtain the supernatant. Add 210 mg of solid ammonium sulphate/ml (e.g. enzyme reagent grade BRL) with constant stirring. Adjust the pH to 7.5 with ammonium hydroxide. Continue stirring for 30 min. Centrifuge at 28 000 *g* for 30 min to remove the protein precipitate.
5. Collect the clear yellow supernatant avoiding the floating lipid layer (filter, if necessary, through a plug of glass wool).
6. Load the supernatant (flow rate 30–100 ml/h) onto a column of Sephacryl S300 (9 × 135 cm, Pharmacia) pre-equilibrated with buffer B.[c] Elute at the same flow rate with buffer B.[d] Collect fractions. Ribosomes elute in the excluded volume as a distinct peak primarily detectable by eluent turbidity.
7. Monitor the absorbance of fractions at 260 nm. Pool those containing the majority of A_{260} absorbing material.
8. Measure the volume of the pooled fractions. Add 100 mg PEG_{6000} per ml with stirring for 30 min at 0 °C. Recover ribosomes by centrifugation at 16 000 *g* for 15 min.
9. Rinse the centrifuge tube walls with polymix buffer and then redissolve the pellet in 20 ml polymix buffer (i.e. approximately 400 A_{260} units/ml). Add methanol to give a final concentration of 30% (v/v) and dialyse for several hours against polymix buffer containing 30% (v/v) methanol. Store samples at −20 °C.[e]
10. Prepare ribosomal subunits by sucrose density gradient centrifugation in polymix buffer containing 0.4 M NaCl (see protocol in Secion 2.1.1, step 16). Store purified subunits in the polymix–methanol buffer. Subunits may precipitate upon storage but are easy to resuspend.[e]

[a] Buffer A: 10 mM Hepes–KOH (pH 7.5), 70 mM KCl, 10 mM $MgCl_2$, and 3 mM 2-mercaptoethanol.

[b] Polymix buffer: 20 mM Hepes-KOH (pH 7.5), 5 mM $MgCl_2$, 0.5 mM $CaCl_2$, 8 mM putrescine, 1 mM spermidine, 5 mM NH_4Cl, 95 mM KCl, 1 mM dithioerythritol (DTE) (or 3 mM

2-mercaptoethanol). This polymix buffer has been optimized for *E. coli*. The buffer should be carefully optimized for other organisms. It is best to make 10-fold stock concentrations of polymix and Hepes and mix in correct proportions immediately prior to use. To make 10 × polymix, dissolve magnesium, calcium, ammonium, and potassium chlorides and 1 M spermidine trihydrochloride (Sigma) first, then add 1 M putrescine-free base (Sigma) to final concentrations, whilst closely monitoring and adjusting the pH with 10 M HCl. Do not allow the pH to rise above 9.0. The final pH should be 7.5. Filter through a millipore filter (pore size 0.45 μm). The 10 × polymix can be stored at 4 °C for several months. The addition to all working concentration buffers (except those for storage of ribosomes) of PMSF to 0.1 mM and benzamidine hydrochloride to 0.1 mM final concentrations is strongly recommended. Dithioerythritol or 2-mercaptoethanol should be added to buffers immediately prior to use.

[c] Buffer B: As polymix buffer made 1 M in NH_4Cl and with the pH readjusted to 7.5. (The current trend now, however, is to use 0.5 M NH_4Cl instead of 1 M for salt-washing of ribosomes to avoid the removal of some ribosomal proteins. See p. 164 in Chapter 8).

[d] A slow flow rate is recommended to allow for efficient salt-washing of the ribosomes. No insoluble material should be loaded on to the column.

[e] Storage in polymix buffer without methanol should be at −80 °C (see footnote *h* in *Protocol 2*). The lack of glycerol in storage buffers allows for easier pipetting of samples.

2.2 Eukaryotic (animal) ribosomes

The method chosen to illustrate the preparation of 80S ribosomes and ribosomal subunits from animal sources is based on the use of liver tissues as a source and from details adapted from references 20 and 21. The preparation involves the isolation of a post-mitochondrial supernatant and then the subsequent isolation of microsomes. The microsomes are then extracted to yield ribosomes which are stripped of endogenous peptidyl-tRNA by incubation in the presence of puromycin. These ribosomes may then be used to isolate ribosomal subunits. A reversible dissociation of ribosomes into subunits occurs in buffers containing a high concentration of potassium, following the puromycin treatment. Eukaryotic ribosomes, unlike their prokaryotic counterparts, do not readily dissociate into subunits by a simple lowering of the magnesium ion concentration.

The method outlined in *Protocol 5* yields active ribosomal subunits which are capable of association to form intact 80S ribosomes. Activity of such 80S ribosomes for incorporation of amino acids into protein is often lower than for polysomes, however, owing to degradation of mRNA by ribonucleases present in the microsomal membrane fraction (22). Ribosomes from other animal sources may be prepared in a similar fashion. Special precautions unique to other systems may, however, need to be taken. As an example, contaminating actomyosin has to be removed from ribosome pellets prepared from muscle tissues (23).

The methods of preparing ribosomes from tissue culture cells are essentially similar to those presented in *Protocol 5*. Generally, cells are lysed in a solution buffered with Tris or Hepes at physiological pH. Such buffers also typically contain 5 mM magnesium ions, 100–250 mM potassium (or sodium) chloride, 0.5% (v/v) Nonidet-P-40 (NP-40) and up to 20 mM 2-mercaptoethanol. Occasionally, 0.1 mM PMSF is employed as a protease inhibitor and the use of placental ribonuclease inhibitor (5–20 units/ml) may also be considered.

Protocol 5. Isolation and preparation of mammalian liver ribosomes

1. Starve rats (e.g. Wistar strain or Sprague–Dawley weighing 200–250 g) overnight. This depletes the glycogen level in the liver.
2. Sacrifice rats by decapitation. Remove the livers and place them in several volumes of ice-cold homogenization buffer.[a] Perform all subsequent operations at 0–4 °C, unless stated otherwise.
3. Blot the livers to remove excess buffer, weigh, mince finely, and homogenize with 2 ml homogenization buffer per gram of tissue. This may be done in three strokes using a loosely fitted teflon-glass homogenizer, followed by 5–6 strokes of a motor driven teflon pestle at 1000 r.p.m.
4. Centrifuge the homogenate at 13 000 *g* for 30 min to remove mitochondria.
5. Remove the top two-thirds of the post-mitochondrial supernatant, avoiding the turbid layer, and put on one side in ice.
6. Re-extract the pellet and turbid zone by adding one volume of homogenization buffer and re-homogenize (3 strokes). Re-centrifuge as above and pool the resultant supernatant with that obtained at step 5.
7. Filter the pooled supernatants through four layers of cheesecloth. Centrifuge at 145 000 *g* for 2 h. Discard the supernatant or reserve the cell sap as a source of soluble protein synthesis factors.
8. Remove any loose material above the clear microsomal pellet by swirling with a small amount of resuspension buffer.[b] Resuspend the pellet in resuspension buffer (one volume of original weight of tissue).
9. Measure the volume. Adjust the concentration of potassium chloride to 0.5 M (make allowance for the volume of sodium deoxycholate to be added subsequently) by adding a stock solution containing 2.5 M KCl and 10 mM $MgCl_2$. Add one-tenth the volume of a freshly prepared solution of 10% sodium deoxycholate. Mix the solution for several minutes. (This releases ribosomes from microsomal membranes.)
10. Layer the resultant suspension over an equal volume 0.3 M sucrose pad made up in buffer A.[c] Centrifuge at 176 000 *g* for 90 min.
11. Carefully decant the supernatant and discard. Rinse the ribosomal pellet with buffer B[d] then gently resuspend the pellet in the same buffer at 20–30 mg rRNA/ml. Manually resuspend using a glass rod or a teflon pestle[e]. (About 3 mg rRNA is obtained/gram of liver. A_{260} values multiplied by 50 will give the concentration of rRNA in μg/ml.)
12. Clarify by centrifugation at 20 000 *g* for 10 min to remove aggregates and debris. Store, if desired, following addition of DTT to 1 mM final concentration.
13. Raise the concentration of KCl in the ribosome suspension to 0.5 M. Add a freshly prepared solution of puromycin in water to give 0.5 mM final

concentration. Incubate at 37 °C for 15 min. (This step eliminates peptidyl-tRNA and soluble protein synthesis factors.) Proceed to step 14 for further purification of 80S ribosomes or directly to step 16 for preparation of ribosomal subunits.

14. Pellet the stripped 80S ribosomes by centrifugation at 105 000 *g* for 4 h through a 0.5 M sucrose cushion made up in buffer C.[f]

15. Resuspend the 80S ribosomes at 300 A_{260} units/ml in buffer D.[g] Store at −80 °C (liquid nitrogen is better for long-term storage) or go to step 16.

16. *Subunits*: Add sufficient 2.5 M KCl and 1 M $MgCl_2$ stock solutions to raise the concentrations to 1 M and 10 mM, respectively. Add 0.1 M stock solution of 2-mercaptoethanol to raise the concentration to 20 mM. Mix. (This treatment causes dissociation of ribosomes into subunits.)

17. Layer the suspension of dissociated ribosomes (e.g. 10–12.5 mg in rRNA) onto 15–30% (w/v) linear sucrose gradients made up in buffer E.[h] Centrifuge at 95 000 *g* for 5 h at 26 °C (e.g. using a Beckman SW27 rotor).

18. Fractionate the gradients, monitoring absorbance at 260 nm. Pool respective fractions (40S and 60S subunits) from the two well-resolved peaks. Avoid material in the leading and trailing edges of the peaks (see step 21).

19. Recover the subunits by centrifugation at 130 000 *g* for 12 h. (Note direct ethanol precipitation is not effective due to high concentrations of sucrose and KC1.)

20. Resuspend the subunit pellets in buffer D at 100–150 A_{260} units/ml and clarify as in step 12. Store at −80 °C (or under liquid nitrogen) in small aliquots. (See footnote *h* in protocol in Scction 2.1.1.)

21. The 60S ribosomal subunits are often contaminated with dimerized 40S subunits which may be removed by dialysing the pooled 60S subunit fractions against buffer F[i] overnight.

22. Layer portions of the dialysed suspension on to 15–30% (w/v) sucrose gradients made up in buffer F. Centrifuge at 238 000 *g* for $4\frac{1}{4}$ h at 26 °C.[j]

23. Fractionate the gradients and collect the 60S subunits. Recover by centrifugation at 80 000 *g* for 12 h.

24. Resuspend the 60S subunits in buffer D and clarify as in step 12. Store at a concentration of 100–150 A_{260} units/ml under liquid nitrogen or at −80 °C.[k, l]

25. Analyse for subunit purity by loading a small quantity of subunits on 5.2 ml sucrose gradients made up in buffer G.[m] Centrifuge 240 000 *g* for 50 min at 26 °C.

[a] Homogenization buffer: 50 mM Tris–HCl (pH 7.6 at 20 °C), 25 mM KCl, 5 mM $MgCl_2$, 0.25 M sucrose.

[b] Resuspension buffer: 35 mM Tris–HCl (pH 7.8), 25 mM KCl, 10 mM $MgCl_2$, 0.15 M sucrose, 6 mM 2-mercaptoethanol (or 1 mM DTT).

[c] Buffer A: 35 mM Tris–HCl (pH 7.8), 600 mM KCl, 10 mM $MgCl_2$, 6 mM 2-mercaptoethanol.
[d] Buffer B: 50 mM Tris–HCl (pH 7.8), 50 mM KCl, 5 mM $MgCl_2$, 10 mM potassium bicarbonate, 0.25 M sucrose, 6 mM 2-mercaptoethanol.
[e] Clean ribosome preparations will have absorbance ratios of A_{260}/A_{280} and A_{260}/A_{235} of 1.80–1.85 and 1.60 respectively. 1 A_{260} unit is equivalent to 18 pmol of 80S ribosomes.
[f] Buffer C: 20 mM Tris–HCl (pH 7.6), 3 mM $MgCl_2$, 300 mM KCl, 10 mM 2-mercaptoethanol.
[g] Buffer D: 20 mM Tris–HCl (pH 7.6), 100 mM NH_4Cl (or KCl), 5 mM $MgCl_2$, 10 mM 2-mercaptoethanol (or 1 mM DTT), and 0.25 M sucrose.
[h] Buffer E; 50 mM Tris–HCl (pH 7.6), 850 mM KCl, 10 mM $MgCl_2$, 10 mM 2-mercaptoethanol.
[i] Buffer F: 20 mM Tris–HCl (pH 7.6), 50 mM KCl, 2 mM $MgCl_2$, 10 mM 2-mercaptoethanol.
[j] The contaminating 40S subunits in the 60S subunit preparation associate with 60S subunits to form 80S ribosomes. This allows for the separation of excess pure 60S subunits. Note that ribosomal subunits from other mammalian sources do not tend to dimerize as much as do rat liver ribosomal subunits and so therefore do not pose as much of a problem.
[k] Subunit preparations should have an A_{260}/A_{235} ratio of 1.45 or above and an A_{260}/A_{280} ratio of 1.80–1.85 or above. One A_{260} unit is equivalent to 50 pmol and 25 pmol of 40S and 60S ribosomal subunits, respectively (22).
[l] Eukaryotic subunits may be heat activated for 5 min at 40 °C in a buffer containing 20 mM Tris–HCl (pH 7.4), 200 mM NH_4Cl, 20 mM $MgCl_2$ and 1 mM EDTA prior to use (22).
[m] Buffer G: 10 mM Tris–HCl (pH 7.6), 500 mM KCl, 5 mM $MgCl_2$.

Buffers containing 0.25 M sucrose, 0.2% (v/v) Triton X-100, 0.05% (w/v) sodium deoxycholate and 0.2 mg/ml heparin have also been used for lysis of cultured cells. Following lysis, the preparations are cleared of mitochondria and the post-mitochondrial supernatants are passed through a 30% (w/v) sucrose cushion in order to prepare crude ribosomes. Procedures for further purification and the preparation of subunits may then be based upon those presented in *Protocol* 5. Preparations of ribosomes from other common eukaryotic sources are detailed elsewhere, e.g. from *Drosophila melanogaster* (25), rabbit reticulocytes (26), and yeast (27, 28). The preparation of ribosomes from *Artemia salina* is covered in Chapter 3 of this volume. Large-scale isolation of eukaryotic ribosomes and subunits may be performed using a zonal rotor as described in detail elsewhere (29).

The subject of mammalian eukaryotic polysomes has received extensive coverage previously in this series (2). The interested reader may consult that work for full details concerning the purification and analysis of such eukaryotic polysomes.

2.3 Chloroplast ribosomes

Two examples will be given to illustrate the preparation of chloroplast ribosomes. The first example, presented in *Protocol 6*, has been adapted from references 30 and 31, and details a rapid preparation of ribosomes from large quantities of fresh spinach (*Spinacia oleracea*) leaves. The method allows the preparation of 400 mg (6400 A_{260} units) of ribosomes from washed chloroplasts isolated from 10 kg of leaves. Such preparations have been used for the purification of chloroplast ribosomal proteins (31). The details concerning the zonal sucrose density gradient centrifugation have been taken from ribosome and subunit profiles kindly provided by A. Subramanian (unpublished observations, personal communication). Consult references 11–14 for further details of large-scale purification of ribosomes and ribosomal subunits by zonal centrifugation.

Protocol 6. Isolation of spinach chloroplast ribosomes

Perform all operations at 0–4 °C.

1. Obtain fresh leaves, wash thoroughly with cold deionized water, and then remove the veins.
2. Homogenize (two 10 sec bursts) in a Waring Blender using 2 litres of homogenization buffer[a] per kg of leaves.
3. Filter the homogenate through several layers of cheesecloth then through one layer of miracloth. Centrifuge the filtrate at 1200 *g* for 15 min.
4. Decant the supernatant.[b] Resuspend the chloroplast pellet and wash once with buffer A.[c] Recentrifuge at 1200 *g* for 15 min.
5. Resuspend the washed chloroplast pellet in lysis buffer[d] and incubate for 30 min. Clarify the lysed suspension by centrifugation at 26 000 *g* for 30 min.
6. Layer the supernatant over a 1 M sucrose cushion made up in buffer B.[e] Centrifuge at 86 000 *g* for 17 h.
7. Dissolve the greenish ribosomal pellet in a small volume of buffer B containing 10% (v/v) glycerol.
8. Clarify by centrifugation at 26 000 *g* for 15 min.
9. Repurify the ribosomes (if desired) by a second pelleting through a sucrose cushion as in step 6. (Purified ribosomes typically have A_{260}/A_{280} ratios of 1.78–1.88.)
10. Obtain highly purified ribosomes on a large scale by zonal centrifugation on 0–40% (w/v) sucrose gradients made up in buffer B.[e] Load up to 4600 A_{260} units of ribosomes. Perform the centrifugation using a Beckman type Ti15 rotor or its equivalent, centrifuging at 22 000 r.p.m. for 16 h.
11. Fractionate the gradients. Pool those fractions containing 70S ribosomes.[f]
12. Recover the ribosomes by centrifugation at 150 000 *g* for 16 h. Resuspend and store at −80 °C in buffer B containing 10% (v/v) glycerol. (See footnote *h* in *Protocol 2*)
13. Obtain ribosomal subunits by loading up to 1700 A_{260} units of ribosomes on to 0–40% (w/v) sucrose gradients made up in buffer C.[g] Separation of subunits may be achieved using the Beckman Ti15 zonal rotor. Centrifuge at 25 000 r.p.m. for 16 h.
14. Fractionate the gradients. Pool the respective subunit fractions[h] and recover subunits after raising the Mg^{2+} concentration to 10 mM by pelleting at 150 000 *g* for 24 h. Store as aliquots after resuspension in buffer B (see step 12).

[a] Homogenization buffer: 10 mM Tris–HCl (pH 7.6 at 20 °C), 50 mM KCl, 10 mM magnesium acetate, 0.7 M sorbitol, 7 mM 2-mercaptoethanol.

[b] Cytoplasmic 80S ribosomes may be obtained from the post-chloroplast supernatant by centrifugation at 26 000 *g* for 30 min to remove mitochondria, proplastids, and chloroplast fragments. Then pellet the ribosomes at 120 000 *g*.

[c] Buffer A; as homogenization buffer with sorbitol reduced to 0.4 M.

[d] Lysis buffer; as homogenization buffer without sorbitol and containing 2% (v/v) Triton-X100.

[e] Buffer B as homogenization buffer but without sorbitol.

[f] With appropriate displacement of gradients, the first peak of A_{260} absorbing material located near the top of the gradient is ribulose bisphosphate carboxylase. This is usually followed by small 30S and 50S ribosomal subunit peaks. The main peak should consist of 70S ribosomes.

[g] Buffer C: 10 mM potassium dihydrogen phosphate (pH 7.0), 100 mM KCl, 1 mM magnesium acetate, 7 mM 2-mercaptoethanol.

[h] When pooling fractions containing the 50S subunits, avoid those fractions corresponding to the trailing edge of the absorbance peak as these may contain some undissociated 70S ribosomes.

A more involved procedure producing good 'clean' ribosomes and ribosomal subunits from tobacco chloroplasts, is also presented in *Protocol* 7. The methodology outlined in *Protocol* 7 is essentially that developed by Bourque *et al.* (32, 33) who have obtained chloroplast ribosomes from either isolated chloroplasts or from total-leaf homogenates. Whole leaf homogenates have generally proved better than intact chloroplasts for overall ribosome yield (32, 33). This method of purification of chloroplast ribosomes relies on the differential susceptibility for dissociation of 70S and 80S ribosomes under appropriate ionic conditions in order to yield chloroplast ribosomes essentially free of cytoplasmic 80S ribosomes. A carefully chosen magnesium ion concentration together with appropriate concentrations of monovalent ions leads to a clean separation, upon sucrose gradients, of the 70S ribosomes from contaminating 80S ribosomes, the latter sedimenting ahead of the 70S form.

Bourque (33) has generally made use of isokinetic sucrose gradients for the isolation of chloroplast ribosomes and the separation of ribosomal subunits. Such systems are not detailed here. The reader is referred to the original work (33) for further details. For the sake of simplicity, and as chloroplast 70S ribosomes and their constituent 30S and 50S subunits are similar to their bacterial equivalents, simple gradient conditions for the separation of these components may be based on those conditions used for *E. coli* ribosomes (see *Protocol* 2) using the buffers devised by Bourque (33, *Protocol* 7). Profiles of the sucrose gradient separation of chloroplast ribosomes and subunits have been presented elsewhere (34). Details for large-scale isolation of chloroplast ribosomes and subunits have also been detailed elsewhere (32, 33). Other methods dealing with the preparation of chloroplasts and ribosomes (including those from *Euglena*), and with aspects of protein synthesis, may be found in a comprehensive monograph (35).

Protocol 7. Preparation of chloroplast ribosomes from tobacco leaves (*Nicotiana tabacum*)

1. Place tobacco plants in the dark for 24 h to reduce starch levels. Perform all subsequent steps at 4 °C.

2. Harvest young leaves less than 10 cm long. Sterilize the leaf surfaces with a 1.3% solution of sodium hypochlorite and rinse with water. Remove the ribs from all leaves longer than 5 cm. Chill on ice.
3. Homogenize the leaves (e.g. 100 g with 100 ml of buffer A[a]) in a chilled Waring Blender for 2–3 min.
4. Filter the homogenate through four layers of cheesecloth and one layer of miracloth (Calbiochem) and put on one side in ice.
5. Rinse the remaining pulp with 20–50 ml buffer A and filter again.
6. Pool the filtered homogenates and measure the volume. Add solid $MgCl_2$ to 5 mg/ml of fluid in excess of the total original buffer A volume. The final concentration of $MgCl_2$ will be 25 mM.
7. Clarify the suspension by centrifugation at 46 000 *g* for 1 h.
8. Decant the supernatant, carefully avoiding the pellet. Pass through one layer of miracloth.
9. Pellet the ribosomes through 3–5 ml of a 1 M sucrose cushion made up in buffer B.[b] Centrifuge at 105 000 *g* for 14 h.
10. Discard the supernatant and any loose greenish material on top of the pellet of ribosomes. Invert the tube and drain well.
11. Resuspend the straw coloured pellet in a minimum volume of either buffer B (ready for step13) or C[c] (for step 18) (use about 2 ml for a preparation from 100 grams of leaves). Resuspension should be done without agitation by leaving the solution on ice for several hours.
12. Clarify the ribosome preparation by centrifugation at 16 000 *g* for 15 min. Decant the cleared supernatant. Store at 4 °C ready for further purification of ribosomes (step 13) or direct preparation of subunits (step 18).
13. Dialyse ribosomes in buffer B against 200 volumes of that buffer for several hours.
14. Separate ribosomes on sucrose gradients (made up in buffer B) similar to those for bacterial ribosomes (see protocol for the preparation of tight-couple ribosomes and ribosomal subunits from prokaryotes, step 10).
15. Collect fractions and pool those containing the 70S ribosomes.[d]
16. Dialyse the 70S ribosomes against buffer B and recover through a sucrose cushion as described in step 9.
17. Resuspend the pellet in either buffer B and store aliquots at −80 °C (see footnote *h*, in *Protocol 2*) or resuspend in buffer C ready for subunit isolation (step 18).
18. *Subunits.* Dialyse ribosomes resuspended in buffer C against three changes of 200 volumes of buffer C over 24 h.[e] (Chloroplast ribosomes dissociate into subunits whilst contaminating 80S ribosomes remain intact.)
19. Clarify the dialysed preparation by centrifugation at 10 000 *g* for 15 min.

20. Fractionate the subunits in sucrose gradients made up in buffer C (see *Protocol 2*, step 16).

21. Recover the subunits from appropriately pooled fractions, after dialysis against buffer B, by centrifugation (see *Protocol 2*, step 17).

22. Repurify on a second sucrose gradient immediately, if necessary, to eliminate cross-contaminating subunits. Store in buffer B.[f]

[a] Buffer A; 100 mM Tris–HCl (pH 8.0 at 20 °C), 25 mM $MgCl_2$, 25 mM KCl, 58 mM 2-mercaptoethanol.
[b] Buffer B; 25 mM Tris–HCl (pH 7.5), 25 mM $MgCl_2$, 25 mM KCl, 8.5 mM 2-mercaptoethanol.
[c] Buffer C; as buffer B but with 2.5 mM $MgCl_2$.
[d] Use caution in pooling fractions. Avoid the contaminating cytoplasmic 80S ribosomes which should sediment closer to the bottom of the tube. Vary the sucrose density gradient conditions as appropriate.
[e] During dialysis of crude ribosomes to allow dissociation into subunits a precipitate may be observed. This is very often due to ribulose bisphosphate carboxylase/oxygenase. This will separate from subunits during gradient centrifugation and will appear near the top of the gradient.
[f] Criteria of purity. The ratio of areas under the 50S and 30S subunit peaks should be 2:1. The ratio of absorbance at 260 nm to that at 280 nm should be 1.7–1.9.

2.4 Mitochondrial ribosomes

Two methods will be outlined to describe the requirements for, and the preparation of, low abundance mitochondrial ribosomes which are free from contaminating cytosolic ribosomes. A method for the preparation of liver mitochondrial ribosomes is derived from reference 36 (and references contained therein) and is presented in detail in *Protocol 8*. Details are then provided in *Protocol 9* for a typical preparation of well-characterized mitochondrial ribosomes from *Neurospora* (37). This latter method serves to illustrate a situation which in plagued by the presence of contaminating endogenous nucleases. The essentials of the techniques, which may be readily adapted for the preparation of ribosomes from similar sources, include the purification of mitochondria by flotation gradient centrifugation and the substitution of calcium for magnesium during mitochondrial lysis. This latter condition is essential in order to minimize endogenous ribonuclease activities (see *Protocol 9*).

Protocol 8. Preparation of mammalian mitochondrial ribosomes

1. Obtain livers from freshly sacrificed animals. Small animals should be fasted overnight to reduce glycogen levels. Place the livers into isolation medium[a] at 4 °C. Perform all subsequent operations at 0–4 °C.

2. Weigh the liver tissue and mince with scissors or pass through a meat grinder. Homogenize with six volumes of isolation medium (6 ml per gram). Use glass homogenizers and motor-driven teflon pestles.

3. Centrifuge the homogenate at 970 *g* for 10 min (sediments nuclei, red cells, whole cells, and debris). Collect the supernatant, carefully avoiding any pelleted material, and put on one side in ice.

4. Re-homogenize the pellet in 3 volumes of isolation medium. Re-centrifuge, collect the supernatant, and discard the pellet.
5. Pool the supernatants and pellet the mitochondria by centrifugation at 5200 *g* for 10 min.
6. Resuspend the mitochondria in isolation medium by homogenization (3–5 strokes in an homogenizer with 0.004–0.006 inch clearance). Re-centrifuge at 670 *g* for 10 min to remove final traces of nuclei and red blood cells. (Purity may be verified by phase microscopy.)
7. Decant the supernatant and recover mitochondria by centrifugation at 5200 *g* for 10 min. Resuspend the mitochondria to a concentration of 20 mg protein/ml in isolation medium and incubate for 15 min at 4 °C with constant stirring in the presence of digitonin (100 μg/ml final concentration).[b]
8. Dilute the digitonin-treated preparation of mitochondria five-fold with isolation medium and recover 'stripped' mitochondria by centrifugation at 10 000 *g* for 10 min. Invert the tubes several times before decanting the supernatant to remove the loose packed layer of material above the mitochondrial pellet. Decant the supernatant, invert the tubes, and drain well.
9. Resuspend and wash the pellet one more time in isolation medium followed by recovery as in step 8. Store the mitochondria as a concentrated suspension in buffer A[c] at −80 °C.
10. Quickly thaw the mitochondrial preparations and adjust with buffer A to a concentration of 20 mg protein/ml. Lyse the mitochondria by adding Triton-X100 to 1.6% (v/v). Clarify by centrifugation at 16 000 *g* for 45 min.
11. Recover the mitochondrial ribosomes from the cleared supernatant by layering it over a 20 ml 34% (w/v) sucrose cushion prepared in buffer B[d] containing 1% (v/v) Triton X-100. Centrifuge at 100 000 *g* for 17 h.
12. Resuspend the ribosomes in buffer B containing 1 mM puromycin to discharge nascent polypeptides.
13. Purify mitochondrial ribosomes by layering onto 10–30% (w/v) linear sucrose density gradients made up in buffer C.[e] Centrifuge at 96 000 *g* for 5 h.
14. Recover the 55S ribosomes from gradients and concentrate by high speed centrifugation at 100 000 *g* for 17 h.
15. Dissociate the ribosomes by resuspending pellets in buffer D.[f] Separate the ribosomal subunits by sedimentation through 10–30% (w/v) sucrose gradients made up in buffer D. Centrifuge at 47 000 *g* for 14 h (e.g. 19 000 r.p.m. in a Beckman SW27.1 rotor).

16. Pool the subunit particles from fractionated gradients and recover by centrifugation at 230 000 *g* for 15 h.

17. Resuspend the subunit pellets in buffer E.[g] Store at −80 °C (one A_{260} unit corresponds to 32 pmol 55S ribosomes, 55 pmol or 77 pmol of 39S or 28S ribosomal subunits, respectively).[h]

[a] Isolation medium: 5 mM Tris–HCl (pH 7.5 at 20 °C), 1 mM EDTA, and 340 mM sucrose.

[b] Digitonin is used to strip off the outer mitochondrial membrane to yield 'mitoplasts'. Digitonin as a 2% stock solution may be prepared by addition of warm 0.25 M sucrose to powder, mixing, and then sonicating for 1–2 min in an ultrasonic bath.

[c] Buffer A: 14 mM Tris–HCl (pH 7.5), 40 mM KCl, 15 mM $MgCl_2$, 0.8 mM EDTA, 0.26 mM sucrose, 5 mM 2-mercaptoethanol, 50 μM spermidine, and 50 μM spermine.

[d] Buffer B: 20 mM Triethanolamine (pH 7.5), 100 mM KCl, 20 mM $MgCl_2$, 5 mM 2-mercaptoethanol.

[e] Buffer C: 10 mM Triethanolamine (pH 7.5), 100 mM KCl, 10 mM $MgCl_2$ and 5 mM 2-mercaptoethanol.

[f] Buffer D: 10 mM Tris–HCl (pH 7.5), 300 mM KCl, 5 mM $MgCl_2$ and 5 mM 2-mercaptoethanol.

[g] Buffer E: 10 mM Tris–HCl or triethanolamine (pH 7.5), 25 mM KCl, 2.5 mM $MgCl_2$ and 5 mM 2-mercaptoethanol.

[h] Calculated for Bovine mitochondrial ribosomes (36). The value for 28S ribosomal subunits has been revised.

The preparation and analysis of both cytoplasmic and mitochondrial ribosomes from *Xenopus laevis*, together with procedures for removing contaminating ferritin by DEAE cellulose chromatography, have also been detailed previously (38).

Protocol 9. Preparation of mitochondrial ribosomes from *Neurospora*

Perform all steps at 0–4 °C.

1. Disrupt *Neurospora* cells, e.g. by grinding with sand in the presence of buffer A.[a] Pellet nuclei and cell debris by centrifugation at 1000 *g* for 10 min.

2. Recover mitochondria from the supernatant by centrifugation at 15 000 *g* for 30 min. Resuspend the pellet in 5–10 ml of buffer B[bc] and pellet again.

3. Carefully remove all excess buffer from the pellet. Resuspend in 4–5 ml of buffer C[d] using a teflon pestle.

4. Perform flotation gradient centrifugation (e.g. in a Beckman SW41 rotor), as follows: prepare linear gradients containing 44–55% sucrose in 10 mM Tricine–KOH pH 7.5. Underlay with 0.5 ml packed mitochondrial suspension using a Pasteur pipette. (Each gradient may accept an amount of mitochondria obtained from 2–5 g wet weight of cells.) Centrifuge at 198 000 *g* for 90 min.[e]

5. Recover mitochondria from the flotation gradients by using a Pasteur pipette. Dilute the recovered mitochondrial suspension with 3–4 volumes of buffer D.[f] Centrifuge at 56 500 *g* for 10 min to pellet the mitochondria and to remove gradient media.

6. Discard the supernatant and evenly resuspend the packed pellet. Take 0.5 ml packed mitochondria in 3.8 ml buffer D. Dissolve all lumps. Clarify by low speed centrifugation if necessary. Lyse the mitochondria by addition of 0.2 ml 20% (v/v) NP-40. (This lysis buffer must be carefully adjusted to pH 7.5 at 4 °C.)
7. Purify mitochondrial ribosomes from contaminating membrane fragments by layering the lysate over a 5.5 ml 1.85 M sucrose cushion made up in buffer E.[g] Centrifuge at 264 000 *g* for 17 h at 4 °C.
8. Remove the top portion of the lysate and sucrose layer. Wash the sides of the tube several times with distilled water to remove contaminating nucleases. Finally, remove the remaining 1.85 M sucrose. Rinse the translucent ribosome pellet with 1 ml ice cold distilled water.
9. Dissolve the high salt-washed ribosomal pellets in buffer F[h] and remove any remaining membrane fragments by centrifugation at 15 000 *g* for 15 min.[i]
10. Allow the ribosomes to dissociate into subunits by the addition, to 1 mM final concentration, of puromycin-KOH (pH 7.5). Incubate at 35 °C for 15 min.
11. Layer 0.3–0.4 ml of the suspension of subunits containing up to five A_{260} units over linear 5–20% sucrose gradients made up in buffer F. Centrifuge at 197 000 *g* for 3 h at 4 °C (e.g., 40 000 r.p.m. in a Beckman SW41 rotor).
12. Fractionate the gradients and pool appropriate subunit fractions. Dilute with buffer F and pellet the subunits at 226 000 *g* for 14 h at 4 °C.
13. Resuspend the subunits in an appropriate buffer, e.g., buffer F, and store −80 °C. (See footnote *h* in *Protocol 2*.)

[a] Buffer A: 10 mM Tricine–KOH (pH 7.5), 0.2 mM EDTA, 15% (w/v) sucrose. May include 1 mM PMSF during initial homogenization of cells. Generally all buffers may be pH adjusted at 20 °C, but see step 6.

[b] Buffer B: as for buffer A but with 0.1 mM EDTA and with 20% (w/v) sucrose.

[c] The choice of the concentration of EDTA in early procedural steps is important to prevent damage to mitochondria and to maximize recovery of mitochondrial ribosomes in the absence of contaminating cytoplasmic 80S ribosomes. No magnesium ions should be present during mitochondrial purification.

[d] Buffer C: 10 mM Tricine-KOH (pH 7.5), 0.1 mM EDTA, 60% (w/v) sucrose.

[e] Step gradients may also be used here. Place the mitochondrial suspension in a centrifuge tube. Overlay with 4 ml 55% (w/v) sucrose solution followed by 2–3 ml 44% (w/v) sucrose solution. The mitochondria will be found as a tight layer between the 44% and 55% sucrose layers following centrifugation.

[f] Buffer D: 25 mM Tris–HCl (pH 7.5), 500 mM KCl, 50 mM $CaCl_2$, 5 mM DTT.

[g] Buffer E: as buffer D with $CaCl_2$ concentration reduced to 25 mM.

[h] Buffer F: 25 mM Tris–HCl (pH 7.5), 500 mM KCl, 25 mM $MgCl_2$ and 5 mM DTT. Once the initial steps are completed and the nuclease activities have been eliminated, magnesium containing buffers may be employed.

[i] Ribosome monomers may be analysed by resuspension in 25 mM Tris–HCl (pH 7.5), 50 mM KCl, and 10 mM $MgCl_2$, and layering over linear gradients of 5–20% sucrose in the same buffer. Centrifuge at 197 400 *g* for 3 h at 4 °C. (The A_{260}/A_{280} ratio for mitochondrial ribosomes should be ~1.47.)

2.5 Plant ribosomes

The preparation of ribosomes from plant sources is often hampered by the presence of endogenous ribonucleases. It has been a common practice to use very high concentrations of Tris at a somewhat higher than physiological pH in order to minimize the nuclease activities. A method for the preparation of ribosomes and subunits from wheat seedlings is presented in *Protocol 10*. It has been derived from references 39–41. Ribosomes have also been isolated from leaf tissues by grinding leaves in a chilled mortar using 10 grams of quartz sand and 60 ml of extraction buffer (see *Protocol 10*) per 20 grams of tissue. Other details for the preparation of ribosomes are then similar to those outlined in *Protocol 10*. It is usual, however, to use either 300 mM Tris–HCl (pH 8.5) or 100 mM potassium bicine buffer (pH 8.0) when extracting ribosomes from leaves (42). PMSF at 1 mM final concentration is sometimes employed just after homogenization of tissues.

Protocol 10. Preparation of ribosomes from wheat shoots

Perform all steps at 2–4 °C unless stated otherwise.

1. Freeze wheat shoots in liquid nitrogen. Pulverize and extract with further grinding in 4–5 volumes (ml/gram of tissue) of extraction buffer.[a]
2. Strain the slurry through four layers of cheesecloth then centrifuge at 21 000 *g* for 15 min to remove mitochondria.
3. Remove any floating lipid, then filter the supernatant through a layer of miracloth. Recentrifuge as in step 2.
4. Recover the upper seven-eighths of the supernatant. Layer26 ml of the post-mitochondrial supernatant over an 8 ml cushion of 1 M sucrose made up in extraction buffer. Centrifuge at 105 000 *g* for 4 h.[b] Decant the supernatant.
5. Rinse the ribosomal pellet once, then resuspend in dissociation buffer.[c] Leave on ice for 45–60 min.
6. Clarify from non-dissolved material by centrifugation at 8700 *g* for 2 min.
7. Adjust the solution of ribosomes to 25 A_{260} units/ml with dissociation buffer and incubate for 30 min at 37 °C in the presence of 1 mM GTP and 1 mM puromycin.
8. Separate the dissociated ribosomal subunits by loading 30 A_{260} units onto 8–38% (w/v) linear sucrose gradients made up in dissociation buffer. Centrifuge at 82 000 *g* for 15 h (e.g., using a Beckman SW27 rotor).

9. Fractionate the gradients and monitor the absorbance at 260 nm. Pool the 40S and 60S subunit peak fractions, respectively.
10. Recover the subunits by centrifugation at 147 000 *g* for 5 h at 4 °C.
11. Resuspend the pelleted subunits in buffer A[d] or B.[e] Store at −80 °C at approximately 800 A_{260} units/ml. (See footnote *h*, *Protocol 2*.)
12. Analyse for subunit cross-contamination on small analytical sucrose gradients.[f, g] [Characteristic absorbance values for ribosomal subunit preparations from plants are A_{260}/A_{280}—1.67 and A_{260}/A_{235}—1.20. Concentration determinations may be made in buffer B (lacking 2-mercaptoethanol).[e] 1 A_{260} unit is equivalent to 20 pmol 80S ribosomes or to 50 or 30 pmol of 40S and 60S subunits, respectively.]

[a] Extraction buffer: 50 mM Tris–acetate (pH 8.2), 50 mM KCl, 5 mM magnesium acetate, 250 mM sucrose, and 5 mM 2-mercaptoethanol.
[b] The upper portion of the supernatant may be reserved. Store portions up to one week only at −20 °C. Dialyse against 150 volumes of buffer B (see note *e*) without glycerol for 2 h and use immediately as a source of soluble protein synthesis factors.
[c] Dissociation buffer: 20 mM Tris–HCl (pH 7.6), 400 mM KCl, 3 mM $MgCl_2$, 5 mM 2-mercaptoethanol and 5% (w/v) sucrose.
[d] Buffer A: 20 mM Hepes–KOH (pH 7.6), 125 mM potassium acetate, 5 mM magnesium acetate, and 6 mM 2-mercaptoethanol.
[e] Buffer B: 50 mM Tris–acetate (pH 7.6), 50 mM KCl, 5 mM magnesium acetate, 3 mM DTT (or 5 mM 2-mercaptoethanol), and 10% (v/v) glycerol.
[f] Typical recovery—300 A_{260} units of 40S ribosomal subunits and 580 A_{260} units of 60S subunits from 1500 A_{260} units of 80S ribosomes. Subunits should associate readily to yield active ribosomes.
[g] Isolated subunits should sediment as single symmetric peaks. If gradients show broad subunit peaks increase the magnesium ion concentration.

The preparation of plant polysomes is generally hampered by the presence of endogenous ribonucleases. Whilst such preparations have not been dealt with here, it should be pointed out that some of the most successful ones are those based on techniques developed by Larkins and Davies (43, 44). Modifications to these methods have been used to prepare membrane-bound and free polysomes, which have been used for *in vitro* protein synthesis assays (45). The reader is also advised to consult references 46–49 before embarking upon a program for the isolation of polysomes from plant sources. All of these references address the potential problems encountered with endogenous ribonuclease activity.

3. Preparation of ribosomal RNA and proteins

It has not been possible to cover many common methods for the preparation of ribosomal RNA and ribosomal proteins in this chapter. Some of these methods are, of course, presented or referred to elsewhere in this volume (see also Appendix 1) and are often reviewed in the literature. A collection of important methods can be found in reference 50. The reader may also consult references 51–53 for an introduction to, or guidance in, the preparation of ribosomal RNA from prokaryotes, and references 54–56 for the preparation of eukaryotic ribosomal RNA.

Total ribosomal proteins from purified ribosomes, or from whole cells (in particular, bacterial and lysed yeast cells) for two-dimensional gel electrophoretic analysis are usually prepared by the method originally described (57) or as later modified (58). The 2D gel electrophoretic analysis of ribosomal proteins prepared from whole cells is possible because most ribosomal proteins are small and basic and can be clearly resolved from non-ribosomal proteins on appropriate two-dimensional gel systems (see Section 4). The purification of individual ribosomal proteins by HPLC has been discussed recently in a series of articles within reference 75. See also Appendix 1.

4. Two-dimensional gel electrophoresis of ribosomal proteins

A number of two-dimensional gel electrophoretic systems have been described for the separation and analysis of eukaryotic and prokaryotic ribosomal proteins. These systems were developed initially due to the fact that it is impossible to analyse ribosomal proteins by means of specific assays. The proteins do not have any individual activities when separated from the ribosome. The gel systems have proved to be very powerful tools in the analysis of mutational alterations in ribosomal proteins, and for the comparison of patterns of and numbers of ribosomal proteins from many different organisms and cell organelles. The measurement of the rates of synthesis of ribosomal proteins has been facilitated by means of double radioactive-labelling of ribosomal proteins and subsequent resolution and analysis of the radioactive ribosomal proteins.

No one electrophoretic system alone has proved capable of resolving all of the many ribosomal proteins from any single source. Each method possesses its own distinct advantages and disadvantages. A combination of methods must still be used if the resolution of all proteins is desired.

Many features, properties, and parameters of two-dimensional gel electrophoretic systems have been discussed previously in this series (59). The reader is referred to this work and to the original references cited in *Table 1* for further basic information concerning the recipes and manipulations needed to achieve success in performing two-dimensional gel electrophoretic analysis of ribosomal proteins. Although such details cannot be presented here, some pointers to recent developments in this area are given below.

Two original systems, modified on many occasions, have provided the basis for the two-dimensional gel electrophoretic analysis of ribosomal proteins. One system (60) subsequently modified (61) consists of a discontinuous gel in the first dimension, followed by a continuous one in the second dimension. Both dimensions employ high concentrations of urea (see *Table 1*). The second major system (62) consists of a urea-containing acrylamide gel in which the basis of

Table 1. Two-dimensional gel systems for separation of ribosomal proteins

Gel System		References
A.	1st Dimension Rod Gel: 6 M urea, 4% acrylamide, pH 8.6	Kaltschmidt and Wittmann (60)
	2nd Dimension Slab Gel: 6.2 M urea, 18.6% acrylamide, pH 4.5	modified by Howard and Traut (61)
B.	1st Dimension Rod Gel: 8 M urea 4% acrylamide, pH 5.0	Mets and Bogorad (62)
	2nd Dimension Slab Gel: 10% acrylamide, containing SDS, pH 6.5	
C.	1st Dimension Rod Gel: 6 M urea, 4% acrylamide, pH 5.0	Geyl, Böck, and Isono (63)
	2nd Dimension Slab Gel: 6.2 M urea, 18.6% acrylamide, pH 4.5	
D.	1st Dimension Rod Gel: Stacking gel and resolving gel. Resolving gel: 8 M urea, 11.4% acrylamide, pH 4.5	Datta, Changchien, Nierras, Strycharz, and Craven (64)
	2nd Dimension Slab Gel: 6 M urea, 10% acrylamide, 0.1% SDS, pH 7.1	

protein separation in the first dimension is the mobility of the proteins at an acidic pH. Proteins are further resolved, in the second dimension, on the basis of molecular weight by incorporating SDS into the acrylamide slab gel.

A more modern approach developed by Geyl *et al.* (63) incorporates the procedures for the first dimension of one system (62) and the second dimension of another (60) with several other modifications (see *Table 1*). In the opinion of this author, this latter development has proved to be a very useful one. A typical gel protein profile, obtained with this system, for ribosomal proteins extracted directly from *E. coli* cells is presented in *Figure 1*.

The most recent development worthy of mention is that of Datta *et al.* (64). These authors have developed a new system for identifying *E. coli* ribosomal proteins which should be applicable to other prokaryotic and eukaryotic systems. It incorporates a number of distinct advantages. The system employs a microscale approach and exhibits good resolution and high sensitivity. See *Table 1* and the original reference (64) for other details. The reader is also encouraged to consult other references cited in this chapter, covering the preparation of ribosomes, for details concerning the two-dimensional gel electrophoretic analysis of ribosomal proteins from various sources, together with the appropriate nomenclature for ribosomal proteins.

5. Protein synthesis

Ribosomes, polysomes and microsome preparations (as detailed or referenced in this chapter) may be routinely utilized in cell-free protein-synthesizing systems. These systems are used to determine many parameters of protein synthesis (see

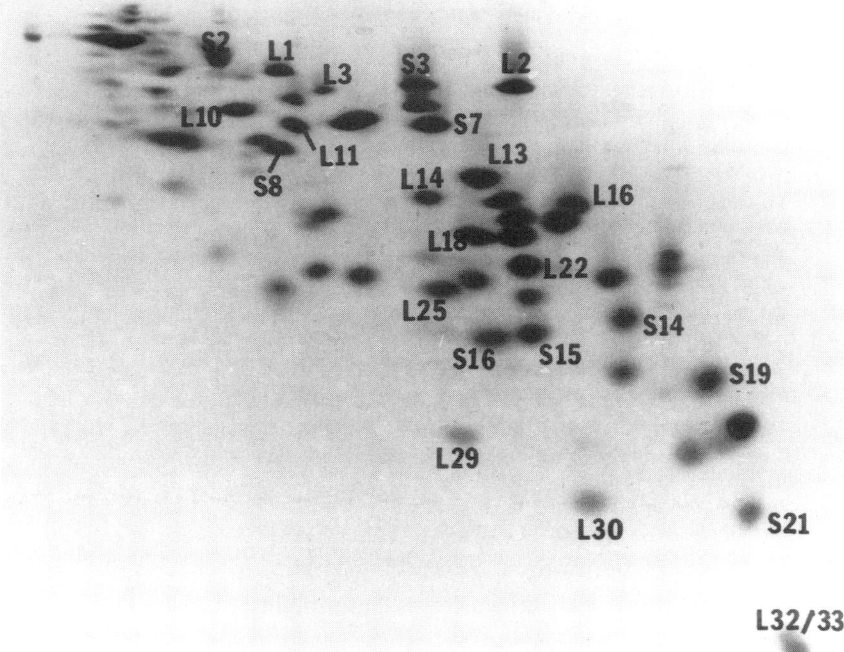

Figure 1. Two-dimensional gel electrophoretic separation of ribosomal proteins. Total proteins were extracted from *E. coli* by acetic acid treatment and acetone precipitation. Ribosomal proteins were resolved by using the gel system of Geyl *et al.* (63). See the text (Section 4) for further details. A representative set of proteins are numbered. Over 44 out of 52 individual ribosomal proteins have been clearly resolved. L refers to proteins from the large ribosomal subunit and S refers to proteins derived from the small subunit. For the full numbering scheme of the ribosomal proteins, see Geyl *et al.*(63).

also chapter 6) or to study effects of ribosome-targeted antibiotics (65), etc. Such reactions are relatively easy to perform and may make use of artificial mRNA's [usually polyuridylic acid; poly(U)] or natural mRNA's (e.g., phage R17 or MS2 RNA). The use of poly(U) as mRNA bypasses physiological initiation and so cannot be used for studies of initiation events. However, it does offer certain advantages compared to natural mRNA when interest is focused on translation rates rather than translation products. By using poly(U), saturating conditions with respect to mRNA can be more easily and economically established in the test systems, thus improving reproducibility. Typical requirements for both bacterial and eukaryotic systems are outlined in *Table 2*. An extensive list of requirements for various pro- and eukaryotic systems has also been presented elsewhere (66). Cell-free protein-synthesizing systems typically employ S30 extracts (i.e. $30\,000 \times g$ cleared bacterial extracts) or purified polysomes, ribosomes or microsomes together with $100\,000 \times g$ post-ribosomal supernatants, i.e. S100s. Post-ribosomal supernatants from the preparation of crude or salt-washed ribosomes are dialysed against a suitable buffer [e.g. 10 mM Tris–HCl (pH 7.5), 10 mM $MgCl_2$, 60 mM NH_4Cl, and 3–5 mM DTT or 2-mercaptoethanol] and stored frozen. The S100 supernatant supplies the soluble factors required for protein synthesis and is employed in the assays at pre-determined optimum amounts (e.g., as mg protein content/ml of assay mixture). The S100 supernatants may also be concentrated by ammonium sulphate fractionation with the cut between 30 and 70% saturation being retained. Protein synthesis reactions are very often initiated by addition of the S100. If desired, the S100's may be passed through a Sephadex G–25 column (5 ml bed volume/ml of supernatant equilibrated in the final desired buffer at 4 °C) in order to lower the concentration of nucleotides and amino acids and to standardize the ionic conditions before storage. The removal of endogenous amino acids either by dialysis or by Sephadex G–25 chromatography can facilitate the design of systems to gain maximal incorporation of radioactive amino acids.

In addition to the requirements of *Table 2*, some systems including those programmed with a natural mRNA are often supplemented with spermidine (1 mM final concentration) and putrescine (8 mM final concentration) (67, 68).

Table 2. Typical requirements for polyuridylic acid-dependent cell-free protein synthesis assays

Component	Typical Concentration in Assay	
Tris–HCl or Hepes–KOH (pH 7.0–8.0)	20–60	mM
Mg^{2+} (chloride or acetate)	10–15	mM
NH_4Cl or KCl	10–100	mM
2-mercaptoethanol or DTT	1–10	mM
ATP	1–2	mM
GTP	0.1–0.5	mM
Phosphoenolpyruvate	5–15	mM[a]
Pyruvate kinase	2.0–50	μg/ml[a]
19 non-radioactive amino acids (minus phenylalanine)	50–170	μM each
Radioactive amino acid (phenylalanine)	5–100	μM[b]
tRNA	100–1000	μg/ml[c]
Poly(U)	100–800	μg/ml
Ribosomes or subunits	2–50	A_{260} units/ml
S100	up to 2 mg protein/ml	

[a] Phosphoenolpyruvate and pyruvate kinase are added as a nucleoside triphosphate regeneration system in prokaryotic systems. Eukaryotic systems are often supplemented with creatine phosphate (4–16 mM) and creatine phosphokinase (30–80 μg/ml) instead.

[b] e.g., L [U–^{14}C] phenylalanine 15 μM at 10–100 μCi/μmol or L [2, 6 – ^{3}H] phenylalanine, usually at higher specific activity. Other radiolabelled amino acids may be used, e.g., with different synthetic templates or with phage mRNA's. The cold amino acids mixture is modified accordingly.

[c] Prokaryotic systems are supplemented with either phenylalanine specific tRNA or crude tRNA from *E. coli* or occasionally with homologous source tRNA when the protein-synthesizing system is derived from a different organism. Eukaryotic systems often incorporate brewers yeast tRNA or homologous source tRNA.

Other systems have included 4 mg/ml PEG_{6000}. The final ionic requirements of the system depend on the particular organism and must be optimized accordingly. Systems should be designed to ensure linear incorporation of amino acids over at least 30 min.

Following incubation at appropriate temperatures and incubation times, assay samples or whole reaction mixtures (typically 100–500 µl) are processed as in *Protocol 11*.

Protocol 11. Processing protein synthesis assay samples

1. Withdraw the sample into 2–5 ml ice cold 10% (w/v) trichloroacetic acid (TCA) to terminate the reaction and precipitate proteins. (Occasionally 1% casein amino acids or 0.5% of non-labelled amino acid, corresponding to the labelled amino acid used in the reaction, is added to the TCA during precipitation in order to reduce background effects.)
2. Heat the terminated reaction mixtures at 90 °C for 20 min in order to deacylate tRNAs, again to reduce background effects.
3. After cooling, collect the acid-precipitable material by filtration on to Whatman GF/C filters [presoaked in 5% (w/v) TCA]. Wash the filters extensively with 5% (w/v) TCA and, optionally, rinse with ethanol, then dry under infrared lamps.
4. The radioactivity retained on the filters is a measure of the phenylalanine incorporation into polyphenylalanine. Estimate this by liquid scintillation spectrometry. Knowing the specific activity of the radioactive phenylalanine used and the counting efficency of the counting system for the isotope employed, the level of incorporation in pmol of phenylalanine residues per ribosome per hour may now be determined.

In vitro protein synthesizing systems measuring the accuracy of the process are detailed in Chapter 6. Protein synthesis may also be readily measured in intact bacteria (69), in isolated chloroplasts (70), and in isolated mitochondria (71).

A number of commercially available protein synthesis systems are on the market. These include a prokaryotic DNA-directed translation system, rabbit reticulocyte lysates, wheat-germ translation systems, and canine microsomal membranes. These systems may be obtained from Amersham, BRL, or Promega, and are usually supplied with excellent technical protocols. Some of these types of systems have, together with the details of electrophoretic analysis of the protein translation products, also been detailed in the book, *Transcription and translation: a practical approach*, in this series (2). The preparation and characterization of a cell-free system from *Saccharomyces cerevisiae*, capable of translation of natural mRNA, has been described (72). The system has been modified recently to enable efficient synthesis at 37 °C (73). The reader should also consult another volume in this series for further information concerning *in vitro* translation in yeast cell-free systems (74). Coupled transcription and translation systems derived from *E. coli*, together with their uses, are also covered in Chapter 9 of this volume.

Acknowledgements

Thanks are due to J. D. Bewley, Don P. Bourque, Michael Calcutt, Prasun Datta, Ramesh Gupta, Donna M. Janes, Michael Madigan, Kikuo Ogata, and Alap R. Subramanian for comments, support, and advice. Thank you also to Terri Schwalb for typing the manuscript.

References

1. Schleif, R. and Wensink, P. C. (1981). *Practical Methods in Molecular Biology*. Springer-Verlag, NY.
2. Hames, B. D. and Higgins, S. J. (ed.) (1984). *Transcription and Translation: A Practical Approach*. IRL Press, Oxford.
3. Blumberg, D. D. (1987). In *Methods in Enzymology*, vol. 152, (ed. S. L. Berger and A. R. Kimmel), p. 20. Academic Press, Orlando, Florida.
4. Subramanian, A. R. (1985). In *Essays in Biochemistry*, vol. 21, (ed. R. D. Marshall and K. Tipton), p. 45. Academic Press, London.
5. Spirin, A. S. (1986). *Ribosome Structure and Protein Biosynthesis*. Benjamin-Cummings Publishing Co., California.
6. Hardesty, B. and Kramer, G. (ed.) (1986). *Structure, Function and Genetics of Ribosomes*. Springer-Verlag, NY.
7. Noll, M., Hapke, B., Schreier, M. H., and Noll, H. (1973). *J. Mol. Biol.*, **75**, 281.
8. Jelenc, P. C. (1980). *Anal. Biochem.*, **105**, 369.
9. Robertson, J. M. and Wintermeyer, W. (1981). *J. Mol. Biol.*, **151**, 57.
10. Zamir, A., Miskin, R., Vogel, Z., and Elson, D. (1974). In *Methods in Enzymology*, Vol. 30, (ed. K. Moldave and L. Grossman), p. 406. Academic Press, NY.
11. Eikenberry, E. F., Bickle, T. A., Traut, R. R., and Price, C. A. (1970). *Eur. J. Biochem.*, **12**, 113.
12. Sypherd, P. S. and Wireman, J. W. (1974). In *Methods in Enzymology*, Vol. 30, (ed. K. Moldave and L. Grossman), p. 349. Academic Press, NY.
13. Saruyama, H. and Nierhaus, K. H. (1986). *Mol. Gen. Genet.*, **204**, 221.
14. Rickwood, D. (ed.) (1984). *Centrifugation: A Practical Approach*. IRL Press, Oxford.
15. Teichner, A., Londei, P., and Cammarano, P. (1986). *J. Mol. Evol.*, **23**, 343.
16. Sirdeshmukh, R., Krych, M., and Schlessinger, D. (1985). *Nucleic Acids Res.*, **13**, 1185.
17. Gourse, R. L., Takebe, Y., Sharrock, R. A., and Nomura, M. (1985). *Proc. Nat. Acad. Sci. USA*, **82**, 1069.
18. Angeloni, S. V. and Potts, M. (1987). *J. Microbiol. Methods*, **6**, 61.
19. Rickwood, D. (ed.) (1983). *Iodinated Density Gradient Media: A Practical Approach*. IRL Press, Oxford.
20. Ogata, K. and Terao, K. (1979). In *Methods in Enzymology*, Vol. 59, (ed. K. Moldave and L. Grossman), p. 502 Academic Press, NY.

21. Moldave, K., Sadnik, I., and Sabo, W. (1979). ibid, p. 402.
22. Sugano, H., Watanabe, I., and Ogata, K. (1967). *J. Biochem.*, **61**, 778.
23. Ashford, A. J. and Pain, V. M. (1986). *J. Biol. Chem.*, **261**, 4059.
24. Semenkov, Yu P., Kirillov, S. V., and Stahl, J. (1985). *FEBS Lett.*, **193**, 105.
25. Santon, J. B. and Pellegrini, M. (1980). *Proc. Nat. Acad. Sci. USA*, **77**, 5649.
26. St. Clair, D. K., Rybak, J. F., Riordan, J. F., and Vallee, B. L. (1988). *J. Biochem.*, **27**, 7263.
27. Adoutte-Panvier, A., Davies, J. E., Gritz, L. R., and Littlewood, B. S. (1980). *Mol. Gen. Genet.*, **179**, 273.
28. Lhoest, J., Lobet, Y., Costers, E., and Colson, C. (1984). *Eur. J. Biochem.*, **141**, 585.
29. Sherton, C. C., DiCamelli, R. F., and Wool, I. G. (1974). In *Methods in Enzymology*, Vol. 30 (ed. K. Moldave and L. Grossman), p. 354. Academic Press, NY.
30. Bartsch, M., Kimura, M., and Subramanian, A. R. (1982). *Proc. Nat. Acad. Sci. USA*, **79**, 6871.
31. Kamp, R. M., Srinivasa, B. R., von Knoblauch, K., and Subramanian, A. R. (1987). *Biochemistry*, **26**, 5866.
32. Hewlett, N. G. and Bourque, D. P. (1986). In *Methods in Enzymology*, Vol. 118 (ed. A. Weissbach and H. Weissbach), p. 201 Academic Press, Florida.
33. Capel, M. and Bourque, D. P. (1982). In *Methods in Chloroplast Molecular Biology*, (ed. Edelman, M., Hallick, R. B., and Chua, N.-H.), p. 1029. Elsevier Biomedical Press, Amsterdam.
34. Bourque, D. P. and Wildman, S. G. (1973). *Biochem. Biophys. Res. Commun.*, **50**, 532.
35. Edelman, M., Hallick, R. B., and Chua, N.-H. (ed.) (1982). *Methods in Chloroplast Molecular Biology*. Elsevier Biomedical Press, Amsterdam.
36. Matthews, D. E., Hessler, R. A., Denslow, N. D., Edwards, J. S., and O'Brien, T. W. (1982). *J. Biol. Chem.*, **257**, 8788.
37. Lambowitz, A. M. (1979). In *Methods in Enzymology*, Vol. 59, (ed. K. Moldave and L. Grossman), p. 421. Academic Press, NY.
38. Leister, D. E. and Dawid, I. B. (1974). *J. Biol. Chem.*, **249**, 5108.
39. Fehling, E. and Weidner, M. (1988). *Plant Physiol.*, **87**, 562.
40. Sikorski, M. M., Pryzybyl, D., Legocki, A. B., Kudlicki, W., Gasior, E., Zalac, J., and Borkowski, T. (1979). *Plant Sci. Lett.*, **15**, 387.
41. Sikorski, M. M., Przybyl, D., Legocki, A. B., and Nierhaus, K. H. (1983). *Plant Sci. Lett.*, **30**, 303.
42. Battelli, M. G., Lorenzoni, E., Stirpe, F., Cella, R., and Parisi, B. (1984). *J. Exp. Botany*, **35**, 882.
43. Larkins, B. A. and Davies, E. (1973). *Plant Physiol.*, **52**, 655.
44. Larkins, B. A. and Davies, E. (1975). *Plant Physiol.*, **55**, 749.
45. Bewley, J. D. and Larsen, K. M. (1982). *J. Exp. Botany*, **33**, 406.
46. Akalehiywot, T., Gedamu, L., and Bewley, J. D. (1977). *Can. J. Biochem.*, **55**, 901.
47. Drouet, A., Nivet, C., and Hartmann, C. (1983). *Plant Physiol.*, **73**, 754.
48. Fourcroy, P. (1980). *Phytochem.*, **19**, 7.
49. Lord, J. M. (1987). In *Methods in Enzymology*, Vol. 148, (ed. L. Packer and R. Douce), p. 576. Academic Press, California.
50. Moldave, K. and Grossman, L. (ed.) (1979). *Methods in Enzymology*, Vol. 59. Academic Press, NY.
51. Hochkeppel, H., Spicer, E., and Craven, G. R. (1976). *J. Mol. Biol.*, **101**, 155.
52. Moazed, D., Stern, S., and Noller, H. F. (1986). *J. Mol. Biol.*, **187**, 399.

53. Kime, M. J. and Moore, P. B. (1983). *Biochemistry*, **22**, 2615.
54. El-Baradi, T. T. A. L., Raué, H. A., DeRegt, V. C. H. F., and Planta, R. J. (1984). *Eur. J. Biochem.*, **144**, 393.
55. Ulbrich, N. and Wool, I. G. (1978). *J. Biol. Chem.*, **253**, 9049.
56. Lee, J. C., Henry, B., and Yeh, Y. (1983). *J. Biol. Chem.*, **258**, 854.
57. Hardy, S. J. S., Kurland, C. G., Voynow, P., and Mora, G. (1969). *Biochemistry*, **8**, 2897.
58. Barritault, D., Expert-Bezançon, A., Guérin, M., and Hayes, D. (1976). *Eur. J. Biochem.*, **63**, 131.
59. Hames, B. D. and Rickwood, D. (ed.) (1981). In *Gel Electrophoresis of Proteins: A Practical Approach*. IRL Press, Oxford.
60. Kaltschmidt, E. and Wittmann, H. G. (1970). *Anal. Biochem.*, **36**, 401.
61. Howard, G. A. and Traut, R. R. (1973). *FEBS Lett.*, **29**, 177.
62. Mets, L. J. and Bogorad, L. (1974). *Anal. Biochem.*, **57**, 200.
63. Geyl, D., Böck, A., and Isono, K. (1981). *Mol. Gen. Genet.*, **181**, 309.
64. Datta, D. B., Changchien, L., Nierras, C. R., Strycharz, W. A., and Craven, G. R. (1988). *Anal. Biochem.*, **173**, 241.
65. Spedding, G. and Cundliffe, E. (1984). *Eur. J. Biochem.*, **140**, 453.
66. Cammarano, P., Teichner, A., Londei, P., Acca, M., Nicolaus, B., Sanz, J. L., and Amils, R. (1985). *EMBO J.*, **4**, 811.
67. Jelenc, P. C. and Kurland, C. G. (1979). *Proc. Nat. Acad. Sci. USA*, **76**, 3174.
68. Moreau, N., Jaxel, C., and LeGoffic, F. (1984). *Antimicrob. Ag. Chemother*, **26**, 857.
69. Davis, B. D., Luger, S. M., and Tai, P. C. (1986). *J. Bacteriol.*, **166**, 439.
70. Fromm, H., Edelman, M., Aviv, D., and Galun, E. (1987). *EMBO J.*, **6**, 3233.
71. Hanson, M. R., Boeshore, M. L., McClean, P. E., O'Connell, M. A., and Nivison, N. T. (1986). In *Methods in Enzymology*, Vol. 118, (ed. A. Weissbach and H. Weissbach, H., p. 437. Academic Press, Florida.
72. Gasior, E., Herrera, F., Sadnik, I., McLaughlin, C. S., and Moldave, K. (1979). *J. Biol. Chem.*, **254**, 3965.
73. Mandel, T. and Trachsel, H. (1989). *Biochim. Biophys. Acta*, **1007**, 80.
74. Campbell, I. and Duffus, J. H. (ed.) (1988). *Yeast: A Practical Approach*. IRL Press, Oxford.
75. Noller, H. F. and Moldave, K. (ed.) (1988). *Methods in Enzymology*, Vol. 164, Section VII. Academic Press, San Diego.

2

Initiation of protein synthesis

ALBERT J. WAHBA,
CHARLES L. WOODLEY, and JAYDEV N. DHOLAKIA

1. Polypeptide chain initiation in *Eschericia coli*

The first step in protein synthesis in *E. coli* is the formation of a polypeptide chain initiation complex at the peptidyl site of the ribosome. This complex contains fMet–tRNA$_f$ associated with an AUG in mRNA. Three protein factors, IF-1, IF-2, and IF-3, are required for the formation of a stable initiation complex and for maximal rates of amino acid incorporation with natural mRNA (1, 2).

2. Initiation factor assays

2.1 Determination of IF-2 activity

Initiation factor 2 (IF-2) stimulates the binding of fMet–tRNA$_f$ to *E. coli* 70S ribosomes in the presence of IF-1 and IF-3 with either an AUG triplet or natural mRNA as a template. The 70S initiation complex may be detected either by filtration through nitrocellulose membranes or by sucrose density gradient centrifugation. The former procedure is readily adaptable to the routine assay of a large number of samples, whereas the latter is primarily used for detection of IF-2 activity when phage RNA is used as the template. With IF-2, a 16-fold stimulation of fMet–tRNA$_f$ binding to ribosomes is observed by the addition of purified preparations of IF-1 and IF-3. The binding of fMet–tRNA$_f$ increases linearly with added IF-2, usually in the range of 0.02–0.2 μg/50 μl reaction mixture.

Protocol 1. AUG-dependent binding of f [^{14}C]Met–tRNA$_f$ to ribosomes

The standard assay contains, in a volume of 50 μl: 100 mM NH_4Cl, 50 mM Tris–HCl (pH 7.2), 5 mM Mg acetate, 1.0 mM dithiothreitol, 0.2 mM GTP, 1 mmol of AUG, 18 pmol of f[^{14}C]Met–tRNA$_f$ (221 Ci/mol), 1 A_{260} unit of ribosomes, 2.0 μg of IF-1, 0.03–0.3 μg of IF-2, and 0.65–1.04 μg of IF-3.

1. Samples of IF-2 should be diluted in buffer A (1.0 M NH_4Cl, 20 mM

Tris–HCl (pH 7.6), 0.2 mM Mg acetate, 1.0 mM dithiothreitol, 5% glycerol) containing 1 mg/ml bovine serum albumin and the solution then incubated at 25 °C for 5 min before addition to the assay mixture.

2. Incubate for 15 min at 25 °C, and then terminate the reaction by the addition of 1 ml of ice-cold buffer containing 1.0 M NH_4Cl, 50 mM Tris–HCl (pH 7.2), and 15 mM Mg acetate.
3. Filter the solution through a nitrocellulose membrane (Millipore, 0.45 μm); and wash the filter three times with 1 ml aliquots of the same buffer, dry, and count for radioactivity in a liquid scintillation counter. One unit of IF-2 activity is defined as 1 nmol f[^{14}C]Met–$tRNA_f$ bound/mg of protein under standard assay conditions (2–4).

Protocol 2. R17 RNA-dependent binding of f [^{14}C]Met–$tRNA_f$ to ribosomes

The standard assay contains, in a volume of 125 μl: 60 mM NH_4Cl, 50 mM Tris–HCl buffer (pH 7.8), 5 mM Mg acetate, 1 mM dithiothreitol, 0.2 mM GTP, 64 pmol of R17 RNA, 34 pmol of f[^{14}C]Met–$tRNA_f$, 5 A_{260} units of ribosomes, 2.0 μg of IF-1, 0.25–0.6 μg of IF-2, and 2.6 μg of IF-3.

1. Incubate for 15 min at 37 °C then layer a 0.1 ml aliquot of each sample over a 5.1 ml linear 5–20% (w/v) sucrose gradient in buffer containing 60 mM NH_4Cl, 50 mM Tris–HCl (pH 7.8), and 5 mM Mg acetate.
2. Centrifuge at 150 000 × g for 130 min at 5 °C.
3. Collect aliquots (0.16 ml) and dilute with 1 ml of water.
4. Measure the absorbance of each aliquot at 260 nm, and determine the amount of radioactivity in a scintillation counter using a scintillation fluid suitable for aqueous samples (2–4).

2.2 Determination of IF-1 activity

The assay for IF-1 utilizes the same components as in *Protocol 1* for IF-2 with the following differences. Each reaction mixture contains approximately 1.5–3.0 μg of crude IF-2 (Section 3.3.i). Under these conditions the amount of f Met–$tRNA_f$ bound with IF-2 alone is quite low and five to ten-fold stimulation of f Met–$tRNA_f$ binding is observed upon addition of IF-1. Incubation is carried out for 5 min at 25 °C, and complex formation is detected as described in *Protocol 1* for IF-2.

2.3 Determination of IF-3 activity

Initiation factor 3 (IF-3) exhibits a stimulatory effect on the following reactions:

(a) AUG-dependent binding of fMet–tRNA$_f$ to ribosomes (2–4).

(b) R17 RNA-dependent binding of fMet–tRNA$_f$ to ribosomes (2–4).

(c) Dissociation of 70S ribosomal particles into 30S and 50S subunits (2, 3, 5, 6).

(d) Translation of natural mRNA such as R17, MS2, or T4 RNA (2, 4, 7).

(e) poly(U)-directed polyphenylalanine synthesis (2, 4, 7).

During purification, reaction (e) is used for the routine assay of column fractions. Reaction (d) may be also used for detection of IF-3 activity with phage RNA as a template.

Protocol 3. Poly(U)-dependent phenylalanine incorporation

The assay contains in a volume of 125 μl: 84 mM NH_4Cl, 63 mM Tris–HCl (pH 7.8), 14 mM Mg acetate, 16 mM 2-mercaptoethanol, 1.3 mM ATP, 0.3 mM GTP, 17 mM creatine phosphate, 8 μg of creatine kinase, 0.2–0.3 mg of *E. coli* S-150 fraction (dialysed), 2 μg poly(U), 8 A_{260} units of ribosomes, 0.3–1.3 μg of IF-3, 770 μg *E. coli* W tRNA, and 0.1 mM [^{14}C]phenylalanine (10 Ci/mol).

1. Incubate for 20 min at 37 °C, then stop the reaction by the addition of 3 ml of 5% trichloroacetic acid.
2. Heat the samples for 15 min at 90 °C.
3. Collect precipitated material on nitrocellulose membranes and wash with approximately 6 ml of 5% trichloroacetic acid.
4. Dry the membranes and count for radioactivity in a scintillation counter using a xylene- or toluene-based scintillation fluid.

---

i. R17 RNA-dependent lysine incorporation

A typical assay contains, in a volume of 125 μl: 76 mM NH_4Cl, 50 mM Tris–HCl buffer (pH 7.8), 12 mM Mg acetate, 16 mM 2-mercaptoethanol, 1.3 mM ATP, 0.3 mM, GTP, 17 mM creatine phosphate, 8 μg creatine kinase, 125 μg *E. coli* W tRNA, 40–80 μg R17 RNA, 5 A_{260} units of ribosomes, 1.5 μg of IF-1, 40–80 μg of crude IF-2 (Section 3.3.i), 1.0–5.0 μg of IF-3, 0.3 mg of *E. coli* S-150 fraction (dialysed), 0.1 mM [^{14}C]lysine (10 Ci/mol), and 0.1 mM of each of the remaining (unlabelled) 19 amino acids. After incubation for 20 min at 37 °C, the reaction is stopped and radioactivity incorporated into protein is determined as in *Protocol 3*.

3. Isolation of polypeptide chain initiation factors

3.1 Preparation of initial extracts

Protocol 4. Preparation of 1.0 M NH_4Cl ribosomal wash

All procedures are carried out at 0–5 °C unless otherwise indicated.

1. Grind frozen *E. coli* MRE 600 cells (500 g) with 1 kg of alumina and extract the paste with buffer containing 30 mM NH_4Cl, 20 mM Tris–HCl (pH 7.8), 10 mM Mg acetate, and 10 mM 2-mercaptoethanol.
2. Remove cell debris and alumina by centrifugation at 24 000 × *g* for 40 min.
3. Treat the viscous supernatant with DNase I (final concentration 3 μg/ml) for 15 min at 5 °C, and clarify the solution by centrifugation at 24 000 × *g* for 40 min.
4. Recover ribosomes by centrifugation of the extract at 340 000 × g for 2 h.
5. Resuspend the pellets and then gently stir overnight in a buffer containing 1.0 M NH_4Cl, 20 mM Tris–HCl, pH 7.8, 10 mM Mg acetate, and 10 mM 2-mercaptoethanol.
6. Sediment the ribosomes by centrifugation at 340 000 × *g* for 2 h and retain the supernatant (ribosomal salt wash) as the source of initiation factors.

---

i. Ammonium sulphate fractionation of the ribosomal salt wash

The ribosomal salt wash is fractionated step-wise by the addition of solid ammonium sulphate. The protein precipitating between 35% and 45% saturated ammonium sulphate contains IF-2, and that between 55% and 70% contains IF-1 and IF-3.

3.2 Purification of IF-1

Protocol 5. DEAE-cellulose chromatography

From 2 kg of *E. coli* a total of 5.0 g of protein is obtained from the 55–70% saturated ammonium sulphate fraction of the ribosomal high–salt wash (Section 3.1.i).

1. Resuspend and dialyse this fraction against buffer containing 5 mM phosphate-Tris (pH 7.5), and 5% glycerol, dilute in the same buffer to contain 17 mg/ml protein and load on to a column (2.2 × 66 cm) of DEAE-cellulose equilibrated with the same buffer.

2. Wash the column with buffer and collect 20 ml fractions. IF-1 is in the fraction that is not adsorbed to the column and the yield is c. 195 mg of protein.

Protocol 6. Heating of IF-1

1. Heat the solution from step 2 in *Protocol 5* with shaking at 65 °C for 5 min and then rapidly cool to 0 °C.
2. Remove precipitated proteins by centrifugation and retain the supernatant which contains IF-1 in a yield of about 155 mg protein. This step inactivates about 60% of an enzymatic activity which hydrolyses the ester linkage of fMet–$tRNA_f$.

Protocol 7. Carboxymethyl cellulose chromatography

1. Load the supernatant from *Protocol 6*, step 2 on to a column (1.5 × 36 cm) of carboxymethyl cellulose equilibrated with 10 mM Tris–HCl (pH 7.4).
2. Wash the column with this buffer to remove protein not adsorbed to the resin, and then apply a linear gradient from 0–350 mM NH_4Cl in a total volume of 1500 ml to the column.
3. Collect and pool those fractions containing IF-1 activity, yielding about 8.05 mg of protein. The IF-1 is eluted from carboxymethyl cellulose at approximately 100 mM NH_4Cl.

Protocol 8. Phosphocellulose chromatography

1. Raise the concentration of NH_4Cl in the solution from *Protocol 7*, step 3 to 200 mM and load the sample on to a column (0.9 × 23 cm) of phosphocellulose equilibrated with buffer containing 200 mM NH_4Cl and 10 mM Tris–HCl (pH 7.4).
2. Wash the column to remove protein that is not adsorbed and then apply a linear gradient of 200–700 mM NH_4Cl in a total volume of 200 ml. IF-1 activity is eluted at approximately 350 mM NH_4Cl. At this step, 5 mg of pure IF-1 may be recovered (8).

---

i. Concentration of IF-1

IF-1 may be concentrated up to 0.80 mg/ml by step-wise elution from a small phosphocellulose column equilibrated with buffer containing 0.2 M NH_4Cl and

10 mM Tris–HCl (pH 7.4). Adjust the preparation to 200 mM NH_4Cl, load on to the column and elute with the same buffer containing 0.7 M NH_4Cl. Concentrated solutions of IF-1 may be stored at 4 °C for several months with no loss of activity. At this stage, IF-1 has an $A_{280}:A_{260}$ ratio of 1.8. The protein concentration determined by the Lowry method (9) is approximately three times higher than that estimated by ultraviolet absorption. The average molecular weight of IF-1, as determined by the meniscus depletion method of Yphantis, is 9207 (10).

3.3 Purification of IF-2

i. Preparation of ammonium sulphate fraction

Dissolve the 35–45% saturated ammonium sulphate fraction of the ribosomal salt wash (3.1.i) in buffer A (20 mM Tris–HCl (pH 7.6), 0.2 mM Mg acetate, 1 mM dithiothreitol and 5% glycerol) containing 20 mM NH_4Cl, and dialyse overnight against the same buffer.

Protocol 9. DEAE-Cellulose chromatography

1. Load the dialysed 35–45% $(NH_4)_2SO_4$ fraction from 1.5 kg of cells (513 mg of protein) on to a column (2.5 × 100 cm) of DEAE-cellulose equilibrated with buffer A.
2. Wash the column with buffer A and then with buffer A containing 6 M urea. No IF-2 activity is eluted at this stage.
3. Apply a linear gradient of 20–100 mM NH_4Cl to the column. IF-2 elutes in two peaks. The first, containing IF-2α, elutes at 44 mM NH_4Cl and the second, containing IF-2β and an unresolved mixture of IF-2α and β, elutes at 55 mM NH_4Cl. Each pool of IF-2 is further purified by chromatography on phosphocellulose.

Protocol 10. Phosphocellulose chromatography

1. Dilute each of the two pools from the DEAE-cellulose column 2-fold with buffer A containing 6 M urea (without NH_4Cl) to reduce the concentration of NH_4Cl to approximately 20 mM.
2. Apply each pool of IF-2 to a column (0.9 × 30 cm) of phosphocellulose equilibrated with the same buffer.
3. Wash the column with the application buffer and then elute IF-2α by using a 200 ml linear gradient of 20–200 mM NH_4Cl in buffer A containing 6 M urea. IF-2β and a mixture of IF-2α and β (from the second DEAE-cellulose pool) are eluted by a linear gradient of 20–300 mM NH_4Cl in buffer A containing

6 M urea. IF-2α elutes at 135 mM NH_4Cl and IF-2β at 130 mM NH_4Cl. The unresolved mixture of IF-2α and β elutes at 157 mM NH_4Cl. Preparations of IF-2 obtained at this stage may be concentrated on a 2 ml phosphocellulose column.

Phosphocellulose fractions of IF-2, as well as concentrated IF-2, should be stored in 50% glycerol after dialysis against buffer A containing 1.0 M NH_4Cl. Factor activity is rapidly lost if samples are diluted in a buffer with a concentration of NH_4Cl less than 250 mM. In this procedure $(NH_4)_2SO_4$ precipitation of IF-2 is avoided since this invariably leads to loss of activity. An overall yield of 26% is obtained at this stage. Of the total IF-2, approximately 75% is present as IF-2α and 25% as IF-2β. Both species may be stored at −80 °C without appreciable loss of activity. Purified IF-2α and β have an $A_{280}:A_{260}$ ratio of 1.28 and 1.39, respectively. The protein concentration determined by the Lowry method (9) is 2–3 times higher than that estimated by ultraviolet absorption for both species of IF-2. The molecular weight, as measured by dodecylsulfate/polyacrylamide gel electrophoresis, of IF-2α is 98 000 and that of IF-2β is 83 000. In the presence of IF-1 and IF-3, both IF-2α and IF-2β promote the binding of fMet–$tRNA_f$ to ribosomes with either AUG, GUG, or R17 RNA as messenger, and the activities of both species are additive (3). They are, however, inactive in promoting the translation of natural mRNA.

i. Preparation of partially purified IF-2

A partially purified preparation of IF-2 may be obtained from the 35–80% saturated ammonium sulphate fraction of the ribosomal salt wash by chromatography on DEAE-cellulose in buffer containing 5 mM phosphate-Tris (pH 7.5), and 5% glycerol (buffer C). The IF-2 activity which elutes at approximately 200 mM NH_4Cl will stimulate phage RNA-dependent amino acid incorporation in the presence of IF-1 and IF-3 (4, 7, 8).

3.4 Purification of IF-3

Protocol 11. DEAE-cellulose chromatography

1. Dilute the dialysed 55–70% $(NH_4)_2SO_4$ fraction (600–800 mg) (Section 3.1.i) with buffer C to a protein concentration of 8.5 mg/ml and load on to a column (2.6 × 100 cm) of DEAE-cellulose equilibrated with the same buffer.
2. Wash the column until all protein that is not adsorbed is removed.
3. Elute the IF-3 with buffer C containing 150 mM phosphate–Tris (pH 7.5).

Protocol 12. Phosphocellulose chromatography

1. Pool the fractions from *Protocol 11*, step 3 containing IF-3 activity and load on to a column (1.5 × 30 cm) of phosphocellulose equilibrated with buffer C containing 150 mM phosphate–Tris (pH 7.5).
2. Wash the column with the same buffer to remove protein that is not retained, and elute IF-3 using buffer C containing 750 mM phosphate–Tris (pH 7.5).
3. Pool the fractions containing IF-3 activity, dialyse against buffer C containing 0.15 M phosphate–Tris (pH 7.5) and then load on to a second column (0.8 × 6 cm) of phosphocellulose equilibrated with the same buffer.
4. Elute IF-3 with buffer C containing 750 mM phosphate–Tris (pH 7.5), combine the fractions containing activity, then store at 4 °C.

Thc factor may be further purified by chromatography on a Sephadex G-150 Superfine column (2.5 × 150 cm) in 20 mM Tris–HCl (pH 7.5), 0.5 M NH_4Cl and 5% glycerol. Throughout the purification procedure, IF-3 activity in column fractions may be determined by the poly(U) or R17 RNA assays (7). A sample of purified IF-3 has an $A_{280}:A_{260}$ ratio of 1.7, and has a molecular weight of 23 000 as estimated by dodecyl sulphate electrophoresis.

3.5 Preparation of reagents

(a) *E. coli* W tRNA is aminoacylated with [^{14}C]methionine, specific activity 221 Ci/mol, in the presence of N^{10}-formyltetrahydrofolic acid and a dialysed *E. coli* MRE 600 S-150 fraction (4, 7).

(b) The RNA of R17 coliphage is prepared by phenol extraction and ethanol precipitation (7).

(c) Stock solutions of 10 M urea are prepared and deionized immediately before use.

(d) For the preparation of phosphate-Tris stock solution, 1 M orthophosphoric acid is adjusted to pH 7.5 at 0 °C by the addition of solid Tris base.

(e) Ribosomes and S-150 fractions are prepared from freshly harvested MRE 600 cells as previously described (3, 7). Pellet the ribosomes at 4 °C by centrifugation overnight at 40 000 r.p.m. Resuspend the pellets in the same buffer (as in Section 3.1), wash again for 4 h at 4 °C, and concentrate by centrifugation at 365 000 × *g* for 2 h at −4 °C. Resuspend ribosomes in a buffer containing 500 mM NH_4Cl, 20 mM Tris–HCl (pH 7.8), 10 mM Mg acetate, 1.0 mM dithiothreitol, and 50% glycerol and store at −80 °C at a concentration of 1000 A_{260} units/ml. Preparations of ribosomes stored in this manner are stable for 4 months.

4. Polypeptide chain initiation in eukaryotes

The mechanism of polypeptide chain initiation in eukaryotes is similar in many ways to that in prokaryotes. The eukaryotic small ribosomal subunit (40S) first forms a pre-initiation complex with the initiator tRNA, GTP, and mRNA. GTP is hydrolysed only upon joining of the large ribosomal subunit (60S) to form an 80S initiation complex. However, in contrast to the three well-defined initiation factors in *E. coli*, the eukaryotic initiation process involves a multitude of factors (designated eIF's), and their interplay is more intricate.

The first step in eukaryotic polypeptide chain initiation is the formation of a ternary complex between the initiator tRNA, GTP, and eIF-2 (*Figure 1*). One of the major differences between prokaryotic and eukaryotic initiation is the regulation of eIF-2 activity in eukaryotes (26, 27). Upon formation of the 80S initiation complex, GTP is hydrolysed and eIF-2 is released as the eIF-2·GDP binary complex. In mammalian systems, this binary complex is stable in the presence of Mg^{2+} and a second protein, the guanine nucleotide exchange factor (GEF), is required for the exchange of bound GDP for GTP. It is at this point in the eIF-2 cycle that regulation appears to occur. Phosphorylation of the α-subunit of eIF-2 by a heme-controlled repressor or dsRNA-dependent kinase during viral infection results in the cessation of protein synthesis and is due to the inability of GEF to catalyse the GTP/GDP nucleotide exchange reaction. Recently, GEF was shown to catalyse the GTP/GDP nucleotide exchange

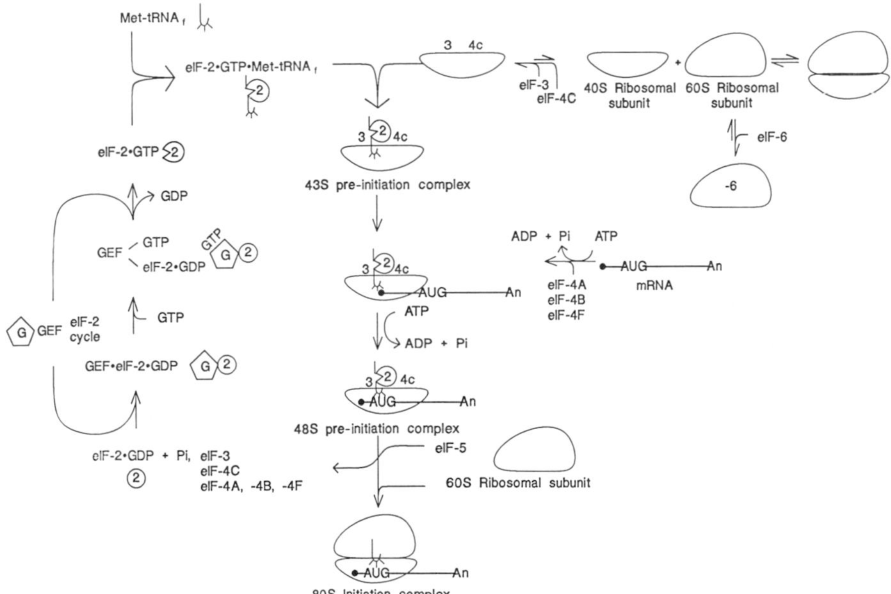

Figure 1. Reaction scheme for the initiation of protein synthesis in eukaryotes.

reaction by a sequential mechanism (15). The activity of GEF may also be influenced directly by the phosphorylation of its M_r 82 000 subunit (12) and the redox state of the cell (13).

Like its prokaryotic counterpart, eIF-3 binds to the small ribosomal subunit and has ribosome dissociation activity. As illustrated in *Figure 1*, the ternary complex is transferred to the 40S ribosomal subunit to form the 43S pre-initiation complex.

The eIF-4E component of the eIF-4F complex recognizes the 'cap' structure of eukaryotic mRNA. eIF-4A, eIF-4B, and eIF-4F bind and transfer the mRNA in an ATP-dependent reaction to the 43S complex forming a 48S pre-initiation complex in which the initiator codon AUG is aligned with the anticodon loop of the initiator tRNA. The exact sequence of events involved in mRNA selection and the interaction between various factors involved in mRNA binding are presently poorly understood. For a recent review of the RNA-binding proteins see reference 30.

The factor eIF-5 facilitates the binding of the 60S ribosomal subunit to the 48S initiation complex. At this point eIF-2-bound GTP is hydrolysed, resulting in the removal of initiation factors from the ribosomal complex and the formation of the 80S initiation complex.

5. Functional assays

The eukaryotic initiation factors for protein synthesis may be assayed by partial reactions. The procedures for performing many of these assays are given in Sections 5.1 to 5.7.

5.1 Ternary (eIF-2–GTP–Met–tRNA$_f$) complex formation

The eIF-2 activity may be estimated by GTP-dependent retention of the initiator tRNA on a nitrocellulose membrane (11, 21).

Protocol 13. Ternary complex formation

1. Each reaction mixture contains in 75 μl: 20 mM Tris–HCl (pH 7.5), 1 mM dithiothreitol, 100 mM KCl, 10 μg of bovine serum albumin, 0.2 mM GTP, 5 pmol [^{35}S]Met–tRNA$_f$ (approximately 15 000 c.p.m./pmol) and eIF-2 (1–3 pmol).
2. Incubate for 10 min at 30 °C, then stop the reactions by the addition of 1 ml cold wash buffer (20 mM Tris–HCl (pH 7.8), 100 mM KCl).
3. Bound Met–tRNA$_f$ is measured by its retention on nitrocellulose membranes (Millipore, type HA, 0.45 μm) after filtration of the diluted reaction mixture.
4. Rinse the tube and the membrane with three-additional 2 ml aliquots of cold wash buffer.

5. Dry the membranes under an infrared lamp and determine radioactivity by counting the membranes in 8 ml of a toluene-based liquid scintillation fluid.

Since GDP is a potent inhibitor of ternary complex formation (22), the use of HPLC-purified GTP (11) or the incorporation of a phosphoenolpyruvate/pyruvate kinase-GTP regenerating system (1 mM phosphoenol pyruvate, 0.1 IU pyruvate kinase) in the reaction mixture is recommended (23). Since eIF-2 binds to glass surfaces reactions should be carried out in plastic or silicone-treated glassware. In the absence of GEF, Mg^{2+} inhibits ternary complex formation. [^{35}S]Met–$tRNA_f$ can be prepared as described (21). The specific activity of eIF-2 is defined as the number of pmol [^{35}S]Met–$tRNA_f$ bound/mg protein. Non-specific binding of Met–$tRNA_f$ in crude extracts may be corrected by subtracting a blank value when the assay is carried out in the absence of GTP.

5.2 Nucleotide binding to initiation factors

The binding of radioactively labelled nucleotides to initiation factors may be determined by nitrocellulose filtration assays. Only protein-bound nucleotides are adsorbed to nitrocellulose membranes. A typical assay mixture for [^{3}H]GDP binding to eIF-2 contains, in a volume of 75 μl, 20 mM Tris–HCl (pH 7.5), 100 mM KCl, 1 mM dithiothreitol, 10 μg bovine serum albumin, 4 μM [^{3}H]GDP (6000 c.p.m./pmol) and eIF-2. Incubate at 30 °C for 10 min and then stop the reaction by transferring the reaction tubes to an ice bath followed by the addition of 1 ml of wash buffer (as in *Protocol 13*) containing 1 mM $MgCl_2$ to each sample. The reaction mixtures may either be processed as described for ternary complex formation (with wash buffer containing 1 mM $MgCl_2$) or by chromatography on Mono S (11) or phosphocellulose (21) columns in a buffer containing 20 mM Tris–HCl (pH 7.5), 2 mM DTT, 50 μM EDTA, 10% (v/v) glycerol, and 1 mM $MgCl_2$ to isolate a binary complex of eIF-2–[^{3}H]GDP which is free of unbound GDP.

5.3 GEF assay

GEF catalyses the exchange of eIF-2-bound GDP for GTP and may be assayed by monitoring the release of [^{3}H]GDP from eIF-2–[^{3}H]GDP in the presence of added guanine nucleotides (12, 13, 15).

Protocol 14. GEF assay

1. Each reaction mixture contains, in 75 μl: 20 mM Tris–HCl (pH 7.5), 1 mM dithiothreitol, 1 mM $MgCl_2$, 4 μM GDP or 20 μM GTP, 100 mM KCl, 10 μg bovine serum albumin, 3 pmol eIF-2–[^{3}H]GDP complex (6000 c.p.m./pmol) and GEF.

2. Incubate for 10 min at 30 °C.
3. Terminate the reaction by the addition of cold-wash buffer containing 1 mM $MgCl_2$ and process as described for the ternary complex assay.

Nucleotide exchange is dependent on GEF only in the presence of Mg^{2+}. GEF activity is expressed as pmol of [^{3}H]GDP released/mg protein.

5.4 Assay of eIF-4A, eIF-4B, and eIF-4F

These factors are most conveniently assayed by their cooperative activity in the release of [^{32}P]orthophosphate from [γ-^{32}P]ATP (29). This hydrolysis is increased in the presence of natural or synthetic mRNA. These proteins may also be assayed by measuring the ATP-dependent binding of labelled mRNA to a nitrocellulose membrane (29).

5.4.1 ATP hydrolysis

The mRNA-dependent hydrolysis of [γ-^{32}P]ATP by eIF-4A, eIF-4B, and eIF-4F is assayed at 30 °C for 15 min. The phosphate reagent used for complexing with [^{32}P]orthophosphate released during the reaction should be prepared immediately before use. It contains one volume 10 mM NaH_2PO_4, one volume 1 M $HClO_4$ and two volumes 0.3 M Na_2MoO_4 in 1 M H_2SO_4.

Protocol 15. ATP hydrolysis

1. Each assay, in a final volume of 38 μl, contains 100 mM KCl, 2 mM dithiothreitol, 30 mM Hepes-KOH (pH 7.4), 1 mg/ml bovine serum albumin, 2 mM Mg acetate, and, as required eIF-4A (1–2 μg), eIF-4B (1–2 μg), eIF-4F (1–3 μg) in buffer B, and 0.125 A_{260} unit poly (A). Buffer B contains 20 mM Tris–HCl (pH 7.5), 2 mM dithiothreitol, 50 μM EDTA, 10% (v/v) glycerol and KCl as indicated in subsequent protocols.
2. Stop the reactions by the addition of 0.4 ml of the phosphate reagent and 0.75 ml N-butyl acetate and agitate the contents thoroughly with a vortex mixer. The tubes should be centrifuged briefly to ensure complete separation of the lower aqueous phase from the upper organic layer. The complete extraction of the ^{32}P-labelled complex into the organic phase may be monitored by the transfer of the yellow phosphate complex into the upper layer. Place 0.5 ml aliquot from the top layer of each tube into a vial with 5 ml of a scintillation fluid suitable for aqueous samples and count for radioactivity. 'Crude' preparations of [γ-^{32}P]ATP will produce a very high background.

5.4.2 mRNA binding

The binding of labelled mRNA to nitrocellulose membranes is assayed at 30 °C for 5 min. Synthetic polynucleotides may be labelled with [γ-^{32}P]ATP by using polynucleotide kinase, whereas natural mRNA may be labelled with [α-^{32}P]ATP by using RNA ligase or with (α-[^{35}S]thio)ATP by using poly (A) polymerase (28, 29). ^{3}H-labelled mRNA with an adequate specific activity (30–50 000 c.p.m./μg) may also be obtained by incubating rat ascites tumor cells in tissue culture with [^{3}H]uridine and isolating the poly (A^{+}) RNA from these cells.

Protocol 16. mRNA binding

1. Each reaction mixture contains in a volume of 75 μl, 100 mM KCl, 5 mM Mg acetate, 30 mM Hepes–KOH (pH 7.4), 2 mM dithiothreitol, 1 mg/ml bovine serum albumin, 1 mM ATP, 4 mM phosphoenolpyruvate, 0.4 IU pyruvate kinase, 5–20 μg/ml labelled mRNA. The reactions also contain 15 μl buffer B (as in *Protocol 15*) with 100 mM KCl or initiation factors (0.8–1.2 μg eIF–4F, 2–2.5 μg eIF-4B, and 1.5–2 μg eIF-4A) as required.
2. Terminate the reactions by the addition of 1 ml cold buffer B with 100 mM KCl and 5 mM $MgCl_2$.
3. Process the samples as described in the ternary complex assay (*Protocol 13*).

5.5 eIF-5-dependent GTP hydrolysis

eIF-5 is assayed either by its requirement in the Met–puromycin reaction (see below) or by ribosome-dependent hydrolysis of GTP (32). This reaction is the same as the ATPase assay described above (*Protocol 15*) for eIF-4A, eIF-4B, and eIF-4F, with the following modifications. The reactions contain 0.5–1 μg eIF-5, 0.8–1 A_{260} unit 80S ribosomes, and 20 μM [γ-^{32}P]GTP at 5000 c.p.m./pmol. The reactions are stopped, processed, and counted as described for the ATPase reaction.

5.6 Met-puromycin synthesis

The transfer of [^{35}S]Met from Met–$tRNA_f$ to puromycin may be used to assay several factors including eIF-2, eIF-4C, eIF-4D, and eIF-5 (31).

Protocol 17. Met-puromycin synthesis

1. Each reaction contains, in a final volume of 75 μl, 40 mM KCl, 90 mM potassium acetate, 30 mM Hepes-KOH (pH 7.4), 2 mM dithiothreitol, 8 mM creatine phosphate, 15 μg creatine kinase, 1 mM ATP, 0.25 mM GTP,

30 μM spermine, 2.5 mM Mg acetate, 1.2 mM puromycin, 0.075 A_{260} unit AUG, 20 μl buffer B (*Protocol 15*) with 100 mM KCl or initiation factors in this buffer, 3–5 pmol [^{35}S]Met–tRNA$_f$ (15 000 c.p.m./pmol) and 0.8–1 A_{260} unit of run-off rabbit reticulocyte 80S ribosomes. Each reaction may contain as required 1 μg eIF-2, 0.6 μg eIF-5, 0.5–1 μg eIF-4C, 1–1.5 μg eIF-3, 0.05 μg GEF, 0.65 μg eIF-4B, and 1–1.5 μg eIF-4D.

2. Incubate for 20 min at 30 °C.
3. Terminate reactions by the addition of 0.15 ml 0.5 M Tris–glycine (pH 9.0) and 1 ml ethyl acetate.
4. Agitate the samples on a vortex mixer and process as described for the ATPase and GTPase assays.

5.7 40S initiation complex formation

This assay may be used to measure the initiation factor-dependent transfer of the initiator tRNA, [^{35}S]Met–tRNA$_f$, to a 40S ribosomal subunit in the presence of a template [AUG triplet, poly (A, U, G), or natural mRNA]. If the mRNA is radioactively labelled, then its binding to the ribosomal subunit may also be monitored.

Protocol 18. 40S initiation complex formation

1. Each reaction contains, in a final volume of 75 μl, the same components used for the assay of Met–puromycin synthesis (see *Protocol 17*) with the following changes: the puromycin is deleted and approximately 0.5 A_{260} unit of 40S ribosomal subunits is substituted for the 80S ribosomes.
2. Incubate the reaction mixtures for 10 min at 30 °C.
3. At the end of the incubation remove a 50 μl aliquot for analysis. Analyse the samples by sedimenting the initiation complex at 380 000 × g for 2 h at 4 °C through a 5.1 ml, 5–30% (w/v) sucrose gradient made up in 20 mM Tris–HCl (pH 7.5), 100 mM KCl, 3 mM $MgCl_2$, and 1 mM dithiothreitol.
4. Fractionate the gradients and count aliquots for radioactivity in a scintillation cocktail suitable for aqueous samples.

The background produced from free [^{35}S]Met–tRNA$_f$ may be reduced by filtering the fractions through a nitrocellulose membrane as described for eIF-4F-dependent message binding. Alternatively, the 50 μl aliquot may be analysed by sedimenting the initiation complex through a 10–40% sucrose gradient and into an 80% sucrose cushion. Collect the lower 50 μl with a blunt-tip microsyringe, and count the sample in a liquid scintillation cocktail. This assay may be performed in a Beckman Airfuge (40 min at 28 p.s.i. and room temperature) or the TL-100 ultracentrifuge (45 min at 60 000 r.p.m. and 25 °C in a TLA-100 rotor). The gradient layers consist of: 20 μl 80% sucrose, 40 μl 40% sucrose, 35 μl 30% sucrose, 30 μl 20% sucrose, and 25 μl 10% sucrose.

6. Purification and properties of the eukaryotic initiation factors

The purification procedure for the initiation factors from rabbit reticulocyte lysates is outlined in *Figure 2* and in *Protocol 19*.

Protocol 19. Purification procedure

1. All procedures are carried out at 4 °C except for HPLC chromatography which is performed at room temperature.
2. Fractions from FPLC columns are collected in tubes immersed in ice.
3. The buffer used to equilibrate gel filtration and ion-exchange columns is buffer B (as in *Protocol 15*) with KCl as indicated in the individual procedure.
4. Fractions containing GEF are dialysed against buffer B containing 50% (v/v) glycerol and are stored in liquid nitrogen.
5. DEAE-cellulose (DE-52) and phosphocellulose (P-11) are obtained from Whatman and m^7GTP-Sepharose from Pharmacia.
6. Mono S, Mono Q, and Mono P FPLC columns are obtained from Pharmacia LKB Biotechnology Inc., and they may be used with either Pharmacia FPLC or ISCO HPLC systems.
7. The ionic strength of the buffers used for column chromatography is shown in the flow diagram (Figure 2).
8. Protein samples for sucrose gradient fractionation are applied to 8–24% linear sucrose gradients (prepared in buffer B containing 500 mM KCl and lacking glycerol) and centrifuged at 120 000 × g for 26 h.
9. Mono S, Mono Q, and Mono P columns are equilibrated with buffer containing 100 mM KCl or as indicated. The columns are first washed with 2 ml of the same buffer, and the proteins are then eluted with three consecutive linear KCl gradients in buffer B, 5 ml gradient, 100–200 mM KCl, 20 ml gradient, 200–500 mM KCl, and 10 ml gradient, 500–1000 mM KCl. Fractions of 0.5 ml are collected.

The purification of the individual factors, eIF-2 (11), GEF (12, 13), eIF-3, and eIF-4A, eIF-4B and eIF-4F (14) have been previously described and their properties are summarized in Table I and in references 11–20. The β-subunit of eIF-2 migrates anomalously in dodecylsulphate/polyacrylamide gels. Depending

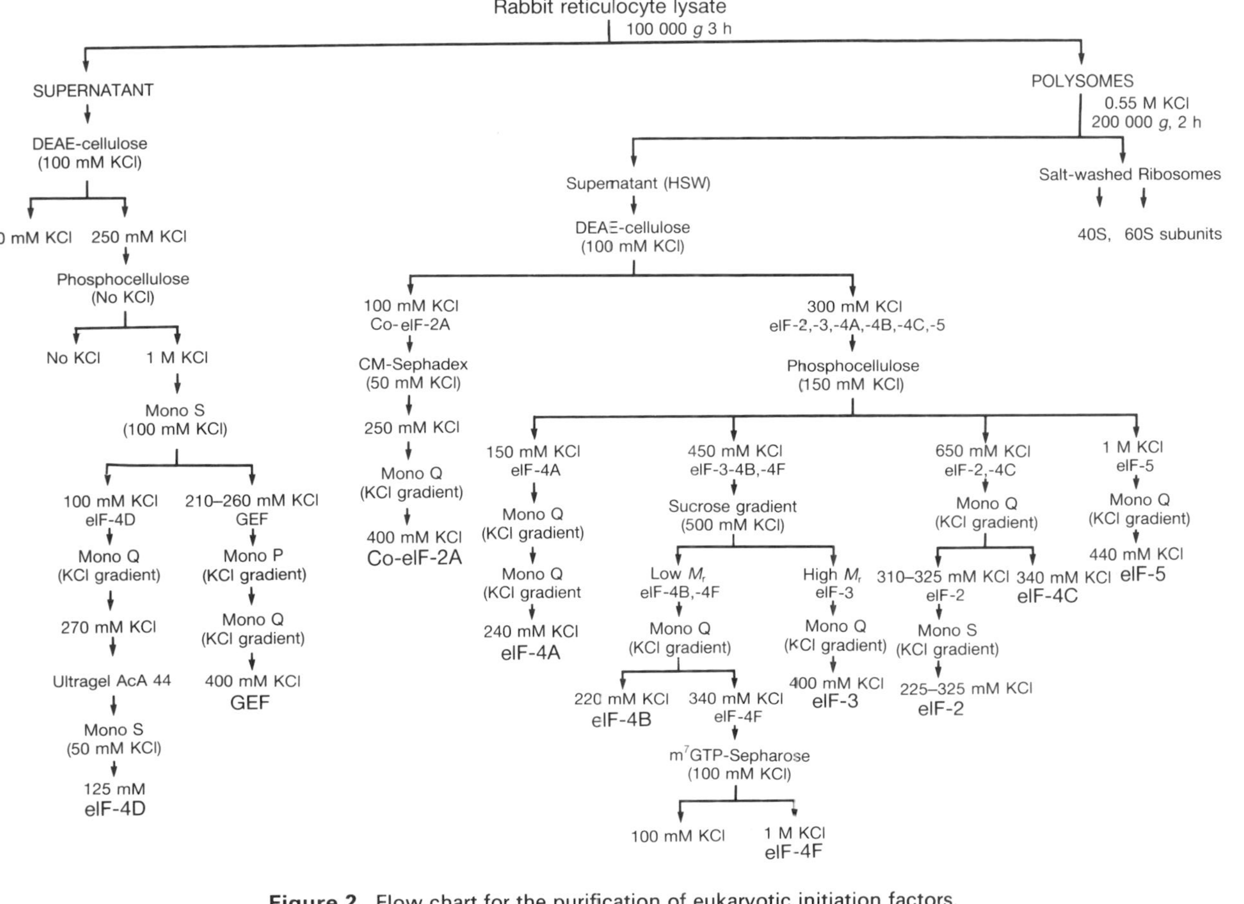

Figure 2. Flow chart for the purification of eukaryotic initiation factors. (HSW = high salt washate)

Table 1. Properties of eukaryotic protein synthesis initiation factors.

Factor	Activities	Native protein composition		Subunit composition		Comments
		M_r	pI	M_r	pI	
eIF-2	GTP-dependent binding of Met–tRNA$_f$ to 40S ribosomal subunits	122 700	6.4	α-36 100	5.1	Phosphorylation by HCR and DAI inhibits protein synthesis
				β-38 400	5.4	GTP binding (Guanine moeity)
				γ-55 000	8.9	GTP binding (γ-phosphate moeity)
GEF	Exchange of eIF-2-bound GDP for GTP	274 000		32 000		
				40 000		GTP binding
				55 000		NADPH/ATP binding
				65 000		NADPH/ATP binding
				82 000		Phosphorylation by CK II stimulates GEF activity
Co-eIF-2A	Protection of eIF-2–GTP–Met-tRNA$_f$ against inhibition by mRNA	90 000				
eIF-3	Anti-association activity against 40S and 60S ribosomal subunits	700 000	6.7	35 000		
				39 000		
				45 000		
				69 000		
				110 000		
				130 000		
eIF-4A	mRNA binding to 43S preinitiation complex. RNA-dependent ATPase	46 000	6.1	46 000	6.1	Also present in eIF-4F complex
eIF-4B	Stimulates mRNA binding to 43S preinitiation complex	80 000	6.0	80 000	6.0	Stimulates eIF-4A and eIF-4F activities
eIF-4C	Ribosome subunit joining	17 600	5.6	17 600	5.6	
eIF-4D	Stimulates Met–puromycin synthesis	16 800	5.1	16 800	5.1	^{50}Lys is modified to hypusine
eIF-4E	Recognition of mRNA 5′ cap structure	26 000	5.6	26 000	5.6	Also present in eIF-4F complex
eIF-4F	mRNA binding to 43S preinitiation complex	290 000		26 000	5.6	eIF-4E or CBP I
				46 000	6.1	eIF-4A
				220 000		Also known as CBP II
eIF-5	Hydrolysis of eIF-2-bound GTP and the release of initiation factors upon 60S ribosomal subunit joining to 43S pre-initiation complex	150 000	6.4	150 000	6.4	
eIF-6	Ribosomal subunit dissociation activity	25 000		25 000		

HCR = heme-controlled repressor. DAI = dsRNA activated inhibitor. CK II = casein kinase II.

upon the acrylamide/bisacrylamide ratio of the gel, the apparent M_r of the β-subunit of eIF-2 is either 48 000 or 52 000 daltons (11). The M_r reported in Table 1 is derived from the corresponding cDNA sequence of the eIF-2 α- and β-polypeptides (24, 25).

Acknowledgements

We are grateful to Anthony Jones and Usha Dholakia for their excellent technical assistance, and to Connie Mascher for expert typing of this manuscript. This work was supported in part by Public Health Service Grant GM-25451.

References

1. Wahba, A. J., Chae, Y.-B., Iwasaki, K., Mazumder, R., Miller, M. J., Sabol, S., and Silero, M. A. G. (1969). *Cold Spring Harbor Symp. Quant. Biol.*, **34**, 285.
2. Wahba, A. J., Iwasaki, K., Miller, M. J., Sabol, S., Sillero, M. A. G., and Vasquez, C. (1969). *Cold Spring Harbor Symp. Quant. Biol.*, **34**, 291.
3. Miller, M. J. and Wahba, A. J. (1973). *J. Biol. Chem.*, **248**, 1084.
4. Sobura, J. E., Chowdhury, M. R., Hawley, D. A., and Wahba, A. J. (1977). *Nucleic Acids Research*, **4**, 17.
5. Subramanian, A. R. and Davis, B. D. (1970). *Nature*, **228**, 1273.
6. Kaempher, R. (1971). *Proc. Nat. Acad. Sci. USA*, **68**, 2458.
7. Schiff, N., Miller, M. J., and Wahba, A. J. (1974). *J. Biol. Sci.*, **249**, 3797.
8. Wahba, A. J. and Miller, M. J. (1974). In *Methods in Enzymology*, vol. 30 (ed. K. Moldave and L. Grossman), p. 3. Academic Press, New York.
9. Lowry, O. H., Rosebrough, N. J., Farr, A. L., and Randall, R. J. (1951). *J. Biol. Chem.*, **193**, 265.
10. Yphantis, D. A. (1964). *Biochemistry*, **3**, 297.
11. Dholakia, J. N. and Wahba, A. J. (1987). *J. Biol. Chem.*, **262**, 10164.
12. Dholakia, J. N. and Wahba, A. J. (1988). *Proc. Nat. Acad. Sci. USA*, **85**, 51.
13. Dholakia, J. N., Mueser, T. C., Woodley, C. L., Parkhurst, L. J., and Wahba, A. J. (1986). *Proc. Nat. Acad. Sci. USA*, **83**, 6746.
14. Colin, A. M., Brown, B. D., Dholakia, J. N., Woodley, C. L., Wahba, A. J., and Hille, M. B. (1987). *Dev. Biol.*, **123**, 354.
15. Dholakia, J. N. and Wahba, A. J. (1989). *J. Biol. Chem.*, **264**, 546.
16. Merrick, W. C. (1979). In *Methods in Enzymology*, vol. 60 (ed. K. Moldave and L. Grossman), p. 108. Academic Press, New York.
17. Voorma, H. O., Thomas, A., Gowmans, H., Amesz, H., and Mast, C. V. D. (1979). In *Methods in Enzymology*, vol. 60 (ed. K. Moldave and L. Grossman), p. 124. Academic Press, New York.
18. Floyd, G. A., Merrick, W. C., and Traugh, J. A. (1979). *Eur. J. Biochem.*, **96**, 277.
19. Safer, B., Jagus, R., and Kemper, W. M. (1979). In *Methods in Enzymology*, vol. 60 (ed. K. Moldave and L. Grossman), p. 61. Academic Press, New York.
20. Benne, R. and Hershey, J. W. B. (1978). *J. Biol. Chem.*, **253**, 3078.
21. McRae, T. H., Roychowdhury, M., Houston, K. J., Woodley, C. L., and Wahba, A. J. (1979). *Eur. J. Biochem.*, **100**, 67.

22. Walton, G. M. and Gill, G. N. (1975). *Biochim. Biophys. Acta*, **390**, 231.
23. Goldstein, J. and Safer, B. (1979). In *Methods in Enzymology*, vol. 60 (ed. K. Moldave and L. Grossman), p. 165. Academic Press, New York.
24. Ernst, H., Duncan, R. F., and Hershey, J. W. B. (1987). *J. Biol. Chem.*, **262**, 1206.
25. Pathak, V. K., Nielsen, P. J., Trachsel, H., and Hershey, J. W. B. (1988). *Cell*, **54**, 633.
26. Pain, V. M. (1986). *Biochem. J.*, **235**, 625.
27. Moldave, K. (1985). *Ann. Rev. Biochem.*, **54**, 1109.
28. Sippel, A. (1973). *Eur. J. Biochem.*, **37**, 31.
29. Abramson, R. D., Dever, T. E., Lawson, T. G., Ray, B. K., Thach, R. E., and Merrick, W. C. (1987). *J. Biol. Chem.*, **262**, 3826.
30. Sonenberg, N. (1988). In *Prog. Nucl. Acids Res. Mol. Biol.* Vol. 35 (ed. W. E. Cohn and K. Moldave), p. 173. Academic Press, New York.
31. Benne, R., Brown-Leudi, M. L., and Hershey, J. W. B. (1977). *J. Biol. Chem.*, **253**, 3070.
32. Merrick, W. C., Kemper, W. M., and Anderson, W. F. (1975). *J. Biol. Chem.*, **250**, 5556.

3

Purification of elongation factors from *Artemia salina*

GEORGE M. C. JANSSEN,
J. ANTONIE MAASSEN, and WIM MÖLLER

1. Introduction

The study of eucaryotic protein synthesis relies on the availability of pure components. Particularly in the area concerning elongation factors, progress in this direction has been slow. Part of the difficulty has been their large complexity, the strong tendency of eucaryotic elongation factors to occur as aggregates, their instability, and the difficulties in defining precisely their mechanism and target on the ribosome. Nevertheless, there is little doubt that the basic biochemistry of elongation factors runs along similar lines throughout all forms of life.

Renewed interest in elongation factors stems from the growing feeling that they may act as translational control elements especially during periods of cell differentiation and cell growth. Another reason is that insight into the mode of transcriptional regulation of an elongation factor may have important ramifications for our understanding of co-ordinately regulated synthesis of the translation apparatus by and within itself. It should be remembered that elongation factors belong to the most abundant proteins in the cell and that their half lives are short compared to most structural proteins, namely 5–6 h (1). This high turnover makes them attractive candidates for translational control.

The evidence that eucaryotic elongation factors are involved in translational control is that in exponentially growing cells the rate of chain elongation is considerably greater than in stationary cells, and that during this transition the activity of elongation factor 1 decreases markedly (2). Another indication for translational control is the arrest of elongation of the nascent polypeptide chain during the period when the ribosome is being transported to the membrane (3). Since elongation factors are critically involved in the traffic of tRNA's in and out of the ribosome (4), it is not surprising that in-depth characterization of the interaction of elongation factors with the eucaryotic ribosome receives long-term interest from those interested in translational control. Moreover, since a sizeable portion of a cell's energy is spent on the synthesis of proteins, the study of elongation factor-dependent hydrolysis of GTP keeps attracting devotees.

Furthermore, a knowledge of the mechanism of guanine nucleotide function during protein synthesis may help us to understand the processes of hormonal signal transduction mediated by G proteins and RAS.

Concerning EF-1, this normally occurs as a series of different molecular weight aggregates of a basic, 50 000 MW protein, EF-1α, that binds to the dimeric elongation factor EF-1βγ. These aggregated forms of EF-1αβγ are designated as $EF\text{-}1_{Heavy}$ (EF-1H). During the pre-emergence period of the brine shrimp *Artemia* this EF-1H form is converted into the low-molecular weight form EF-1α, and presumably EF-1βγ, as judged from functional assays for EF-1α on crude extracts (5). Upon isolation, indeed two separate proteins EF-1α and EF-1βγ are obtained.

EF-1α is responsible for the attachment of aminoacyl-tRNA to the A-site of the ribosome with concomitant hydrolysis of one molecule of GTP. Subsequent to the release of EF-α. GDP from the ribosome, EF-1βγ catalyses the exchange of GDP for GTP on EF-1α, so that another round of EF-1α-dependent transport of aminoacyl tRNA to the ribosome can occur. The actual guanine nucleotide exchange activity resides in EF-1β whereas the function of EF-1γ correlates best with anchoring EF-1β to cell structures like membranes and microtubules (6).

With respect to function, the EF-1α-EF-1βγ complexes resemble the G-protein receptors. Recently, the conservation of a guanine nucleotide binding site in EF-1α and G-proteins has been discussed (7). Less attention has been paid to common motives with respect to the guanine nucleotide exchange mechanism. Phosphorylation of EF-1β, at amino acid residue Ser 89 by an endogenous casein kinase II, decreases its nucleotide exchange activity (8).

The second elongation factor, EF-2, is responsible for the translocation of peptidyl-tRNA and mRNA with respect to the ribosome such that the uncharged tRNA can leave the ribosome and a new codon is exposed for reacting with the incoming aminoacyl tRNA. Under the influence of translocation factor EF-2 another molecule of GTP is hydrolysed in the elongation cycle, but this time no guanine nucleotide exchange factor specific for EF-2 has been reported.

This chapter describes in detail the techniques for purifying elongation factor 1 and elongation factor 2 from *Artemia*. The methods described have been improved over the years and avoid some of the difficulties encountered in earlier procedures (9, 10). For coverage of this vast topic in a short space, we have confined ourselves to one type of eucaryote, namely *Artemia*, although we are confident that the assays as described could be adapted to other organisms, tissues, or even individual types of eucaryotic cells. Those interested in specific problems concerning factor-ribosome coupling and development of *Artemia* should consult references 7, 11–13 and references cited therein. For an up-to-date general review on eucaryotic elongation factors, see (14).

The use of *Artemia* as a source for elongation factors warrants an explanation. The least scientific reason is that cysts can be bought all year round and processed instantly from laboratory stocks. A more respectable reason is that dried *Artemia* cysts develop in sea water within 24 h to emerging pre-nauplii, a marvel happening simultaneously with the turning-on of the transcription and translation apparatus. During this development the dormant cyst embryo, composed of a syncytium of about 4000 nuclei, undergoes a series of morphogenetic changes which lead to a gastrula having the visible features of a nauplius, like head, eyes, claws, and antenna. This occurs without cell division or DNA synthesis until after emergence of the nauplius (15). For references to the isolation of elongation factors from other organisms see (6, 16, 32). Details concerning the purification of elongation factors from *E. coli* are also presented in Chapter 6 of this volume.

2. Purification of elongation factor 1α

2.1 Assays for EF-1α

The activity of EF-1α may be measured by either tRNA binding to ribosomes (17) or polyphenylalanine synthesis (18), the former method being the more usual one.

2.1.1 tRNA binding assay

The principle of this assay is based on the ability of EF-1α.GTP to bind phe-tRNA to ribosomes. In the presence of limiting amounts of EF-1α and no EF-1β, stoichiometric binding of phe-tRNA to the ribosome can be observed. This is because under these conditions the EF-1α.GDP complex as released from the ribosome is slow in recycling and thus there is a slow catalysis of binding another tRNA.

Protocol 1. tRNA binding assay

1. Mix for each assay, in the following order:
   (a) 5 μl poly(U) 2 mg/ml
   (b) 5 μl 2.5 mM GTP (pH 7.5)
   (c) 5 μl 5 × assay buffer[a]
   (d) 20 pmoles 80S ribosomes (*Protocol 2*)
   (e) 20 pmoles [^{3}H]phe–tRNA (1000 c.p.m./pmol) (*Protocol 3*)
   (f) Adjust with water to a total volume of 40μl.
2. Add dilution buffer[b] to a sample of EF-1α (1 to 10 pmoles) to a final volume of 10 μl.
3. Add the mix to the EF-1α sample and incubate for 7 min at 37 °C.
4. Chill on ice.
5. Add 1 ml of ice-cold washing buffer[c] and filter immediately through a 0.45 μm nitrocellulose filter (e.g., BA85 from Whatman). Try to avoid contact of the solution with the filter housing.

6. Wash the filter twice with 2 ml of ice-cold washing buffer.
7. Dry the filters and determine the amount of radioactivity.

[a] 5 × assay buffer: 100 mM Tris–HCl (pH 7.5 at 20 °C), 500 mM KCl, 50 mM 2-mercaptoethanol, 500 μM EDTA, 45 mM magnesium acetate.
[b]dilution buffer: 20 mM Tris–HCl (pH 7.5 at 20 °C), 100 mM KCl, 10 mM 2-mercaptoethanol, 100 μM EDTA, 5 mM magnesium acetate, 25% (v/v) glycerol.
[c] Washing buffer: 20 mM Tris–HCl (pH 7.5), 100 mM KCl, 5 mM magnesium acetate.

2.1.2 Poly(phe) synthesis

EF-1α activity can be measured by using the protocol for EF-2 activity given in *Protocol 8* for polyphenylalanine synthesis. However, when assaying EF-1α, poly(phe) synthesis is measured in the presence of low amounts of EF-1α (1–5 pmol) and ribosomes (1.6 pmoles) with respect to the amount of EF-2 (10 pmol).

For the preparation of salt-washed 80S ribosomes (19) and [^{3}H]phe–tRNA (20, 21) refer to *Protocols 2* and *3*.

Protocol 2. Purification of 80S ribosomes from *Artemia*

1. Suspend 425 g of cysts (e.g., from San Francisco Bay Brand Inc.) In 1 litre of ice-cold 2% NaOCl and leave for 5 min. Dilute with 4 litres of ice-cold water and leave for another 10 min.
2. Pour the resuspended cysts over a filter of fine nylon cheese cloth and wash the cysts on the filter with 10 litres of water and 1 litre of homogenization buffer[a], successively.
3. Grind about 50 g (on dry weight basis) portions of cysts with adherent buffer in a large mortar, preferably a motor-driven type, until essentially all cysts are broken. Usually this takes about 10 to 15 min as checked under a microscope. Dilute with homogenization buffer to a total volume of 750 ml.
4. Centrifuge at 15 000 *g* for 20 min. The supernatant after this centrifugation is called S-15.
5. Centrifuge this supernatant at 100 000 *g* for 30 min and save the supernatant.
6. Centrifuge the latter supernatant at 100 000 *g* for 180 min yielding the post-ribosomal supernatant (S-100) and crude ribosomal pellets.
7. Remove all traces of lipid floating on top of the post-ribosomal supernatant. Remove by vigorous shaking with 5 ml of buffer A[b], the thin, brownish material laying on top of the clear, glassy ribosomal pellet. Next rinse the ribosomal pellet twice with 5 ml of buffer A.
8. Dissolve the pellets in buffer A to an absorbance (260 nm) of about 300 units per ml (approximately 15 ml).

9. Layer the combined solutions of ribosomes onto a cushion of one volume of buffer B[c]. Centrifuge at 225 000 *g* for 120 min.
10. Rinse the pellets with buffer C[d] and dissolve in 5 ml of the same buffer.
11. Measure the absorbance at 260 nm. Dilute to a concentration of 220 A_{260} units per ml and divide in 100 μl-aliquots (sufficient for 20 assays; 1 A_{260} unit = 18 pmoles of 80S ribosomes). Ribosomes can be stored at −80 °C for at least one year without significant loss of activity. *Do not freeze 80S ribosomes more than once.*
12. Using this protocol, the yield of purified ribosomes is about 400 to 500 mg per 425 g of cysts.

[a] Homogenization buffer: 35 mM Tris–HCl (pH 7.4 at 20 °C), 70 mM KCl, 9 mM magnesium acetate, 100 μM EDTA, 10 mM 2-mercaptoethanol, 250 mM sucrose.
[b] Buffer A: 50 mM Tris–HCl (pH 7.5 at 20 °C), 500 mM KCl, 20 mM 2-mercaptoethanol.
[c] Buffer B is homogenization buffer, but containing 1 M sucrose.
[d] Buffer C: 50 mM Tris–HCl (pH 7.5), 25 mM KCl, 100 μM EDTA, 5 mM magnesium acetate, 10 mM 2-mercaptoethanol, 250 mM sucrose, 10% (v/v) glycerol.

Protocol 3. Preparation of [^{3}H]phe-tRNA

1. Mix in the following order:
   (a) (1250 − x) μl water
   (b) 1 μl 2-mercaptoethanol
   (c) 190 μl 80 mM ATP
   (d) 60 μl 1 M magnesium acetate
   (e) 300 μl 1 M Tris–HCl (pH 7.2)
   (f) 218 μl [^{3}H]phenylalanine[a]
   (g) x μl charging enzymes preparation[b]
   (h) 1000 μg tRNA[c], dissolved in 1 ml of water
2. Incubate at 37 °C for the appropriate time.[b] Chill on ice.
3. Add 75 μl of 2 M potassium acetate (pH 5.0) and mix.
4. Extract proteins by shaking for 3 min with 3 ml of phenol, saturated with 0.05 M potassium acetate (pH 5.0). Separation of the phases is achieved by low-speed centrifugation (e.g., 2000 r.p.m. in a clinical centrifuge).
5. Re-extract the (lower) phenol phase with 3 ml of 0.05 M potassium acetate (pH 5.0).
6. Combine the aqueous phases and shake three times with 5 ml of diethylether. Remove residual ether with a gentle stream of nitrogen bubbled through the solution.
7. Add 0.1 volume of 2 M potassium acetate (pH 5.0) and three volumes of

ethanol (−20 °C). Leave for 2 h at −20 °C. Recover the pellet by centrifugation (15 000 *g* for 15 min) and dissolve in a minimal volume (less than 400 μl) of 0.05 M potassium acetate (pH 5.0).

8. Load this onto a column (1 × 25 cm) of Sephadex G-25, equilibrated with 0.05 M potassium acetate (pH 5.0) and elute with the same buffer. Collect fractions of 2 ml. Determine the radioactivity of 10 μl aliquots.
9. Mix the peak fraction(s) with 0.1 volume of 2 M potassium acetate (pH 5.0) and three volumes of ethanol (−20 °C). Leave overnight at −20 °C.
10. Recover the precipitate by centrifugation and dissolve in a minimal volume (less than 500 μl) of water.
11. Determine the amount of radioactivity and the concentration of tRNA, in terms of A_{260} units[d]. Calculate the degree of aminoacylation by using these data, and the value for the specific radioactivity of the phenylalanine used. The degree of aminoacylation should exceed a value of 25%, and is usually in the order of 75–100%. The recovery of tRNA (based on A_{260} units) should be 75% or more.
12. Divide the preparation into aliquots sufficient for 10 assays. Liquid preparations can be stored as such at −80 °C, when used within about two months. For long-term purposes lyophilization is recommended. Since samples of lyophilized tRNA are extremely electrostatic, great care should be exercised during handling.

[a] [^{3}H]phenylalanine: Dilute 200 μCi of [^{3}H]phenylalanine (e.g., L-[2, 3, 4, 5, 6-^{3}H]phenylalanine; 140 Ci/mmol; 1 μCi/μl; Amersham, UK) with 18 μl of 6 mM phenylalanine and determine exactly the specific radioactivity of this mixture.

[b] Prepare a fresh batch of 'charging enzymes' (stable for only two weeks at −80 °C) from *E. coli* MRE 600 post-ribosomal supernatant. Use the alumina-grinding method, followed by the DEAE procedure as described on pages 392 and 405 of ref. (22), respectively. Determine the conditions (concentration and incubation time) for optimal aminoacylation. Incubate at 37 °C, several mixtures (as described in step 1 of this protocol) containing various amounts of charging enzymes (1–5 mg of protein per ml) and determine every 10 min (to a maximum of 60 min) the radioactivity incorporated in 5% TCA-precipitable material (23). Usually, an incubation of 30–50 min in the presence of 3 mg of charging enzymes per ml is found to be optimal.

[c] tRNA: phenylalanyl-specific tRNA ($tRNA^{phe}$) from brewer's yeast, e.g., from Boehringer Mannheim (FRG). Dissolve the tRNA in water just before use and determine the total amount of A_{260} units. Total soluble tRNA (brewer's yeast, Boehringer) may also be used. In this case, increase the amount of tRNA to 20 mg.

[d] 1 mg of $tRNA^{phe}$ corresponds to about 35 nmol of tRNA and yields an A_{260} value of 18 when dissolved in 1 ml of water (i.e. 18 A_{260} units).

2.2 Purification protocol

EF-1α is isolated from post-ribosomal supernatants by a combination of ion-exchange chromatography, hydrophobic interaction chromatography, and molecular sieving (9), as detailed in the following protocol. The estimated yield of EF-1α is approximately 25 mg per 200 g of cysts. The protein thus isolated remains fully active for several years, when stored at −20 °C.

Protocol 4. Purification of EF-1α

1. Perform all operations at 0–4 °C.
2. Prepare a post-ribosomal supernatant as described in the protocol for the purification of 80S ribosomes from *Artemia* (up to and including step 6), starting from 200 g of cysts. Use buffer A[a] instead of homogenization buffer.
3. Dialyse the supernatant for at least 24 h against 9 volumes of buffer B[b]; this results in a 41% ammonium sulphate saturation at equilibrium. Occasionally, mix the contents of the dialysis tubing. After dialysis, centrifuge the dialysate at 25 000 *g* for 30 min.
4. Add 25.6 g of ammonium sulphate per 100 ml of supernatant (80% saturation). Leave for 1 h. Recover precipitated protein by centrifugation at 25 000 *g* for 30 min, and dissolve it in a minimal volume of buffer C[c] (about 50 ml). Dialyse, during 24 h, against several litres of buffer C with one change of buffer. Centrifuge at 25 000 *g* for 30 min.
5. Equilibrate a column (3.5 × 12 cm) of CM-Sephadex C-50 with buffer C. Apply the supernatant of the previous step on the column. Develop in succession with 500 ml of buffer C, 500 ml of buffer D[d] and 500 ml of buffer E[e]. Collect fractions of 15 ml. Analyse aliquots of fractions by SDS-PAGE. EF-1α elutes from the column with buffer E. Pool fractions containing EF-1α and dilute with one volume of buffer F[f].
6. Load onto a column (1.8 × 11 cm) of phenyl-Sepharose, previously equilibrated with buffer G[g]. Wash with one column-volume of the same buffer and develop with a linear gradient of 450 ml of buffer G to 450 ml of buffer H[h]. Collect fractions of 5 ml. Analyse aliquots by SDS-PAGE. EF-1α elutes at about 60% ethylene glycol. Pool appropriate fractions. At this stage of purification the preparation should be essentially composed of EF-1α and a protein of slightly higher molecular weight (55 kDa) in a ratio of about 1:1.
7. Concentrate the pooled sample to a volume of about 10 ml using an ultrafiltration cell.
8. Equilibrate a column (2 × 130 cm) of Sephacryl S-200 with buffer I[i]. Load the ultrafiltrate onto the column and elute with the same buffer. Collect fractions of 4 ml. Analyse by SDS-PAGE. The 55 kDa protein elutes just before EF-1α from the column. Pool fractions containing pure EF-1α and store at −20 °C. The preparation of EF-1α may be concentrated by ultrafiltration (as in step 7) to 10 mg/ml.

[a] Buffer A: 50 mM Tris–HCl (pH 7.5 at 20 °C), 25 mM KCl, 100 μM EDTA, 5 mM magnesium acetate, 10 mM 2-mercaptoethanol, 250 mM sucrose.

[b] Buffer B: 1.78 M ammonium sulphate, 100 μM EDTA, 6 mM 2-mercaptoethanol.

[c] Buffer C: 20 mM Tris–HCl (pH 7.5), 50 mM KCl, 100 μM EDTA, 10 mM 2-mercaptoethanol, 25% (v/v) glycerol.

[d] Buffer D: as buffer C, but with 165 mM KCl.

[e] Buffer E: as buffer C, but with 350 mM KCl.

[f] Buffer F: 1.6 M ammonium sulphate, 20 mM Tris–HCl (pH 7.5), 100 μM EDTA, 10 mM 2-mercaptoethanol.

[g] Buffer G: as buffer F, but with 0.8 M ammonium sulphate.

[h] Buffer H: 20 mM Tris–HCl (pH 7.5), 50 mM KCl, 100 μM EDTA, 10 mM 2-mercaptoethanol, 80% (v/v) ethylene glycol.

[i] Buffer I: 20 mM Tris–HCl (pH 7.5), 100 mM KCl, 100 μM EDTA, 10 mM 2-mercaptoethanol, 25% (v/v) glycerol.

2.3 Comments

2.3.1 Ribosome-elongation factor coupling

Elongation factors act on a restricted region of the 60S ribosomal subunit, at a place where proteins equivalent to *E. coli* proteins L10, L11 and L12 are located (24). The corresponding site within the large rRNA has also been considered as a prime target for factor interaction (25). There are strong indications that structural changes in the factor-binding region of the ribosome induce efficient exit and entry of tRNAs. EF-1α protects the 3′ aminoacylated end of tRNA and positions this part correctly in the A site of the ribosome. Presumably, A-site binding also facilitates the exit of deacylated tRNA from the exit (E) site (Nierhaus, personal communication).

2.3.2 Isoelectric point

In contrast to bacterial EF-Tu, eucaryotic EF-1α is a basic molecule, the isoelectric point being about 8.3 in the case of *Artemia* (9). *In vitro* experiments indicate a strong interaction of the factor with tRNA as well as with other types of RNA. For instance, in *Xenopus laevis* pre-vitellogenic oocytes three abundant proteins are involved in storage of tRNA and 5S RNA, TFIIIA, thesaurin b, and a protein homologous to EF-1α, thesaurin a, (26). This preference of EF-1α to react with several types of nucleic acids may be indicative of a storage function for EF-1α for nucleic acids, besides its role in translation.

2.3.3 Diguanosine tetraphosphate

The encysted gastrulae of *Artemia* contain an unusual guanine nucleotide, P1, P2-diguanosine 5′-tetraphosphate (Gp_4G), as 2.8% of the dry weight. This compound can be considered as two guanosine 5′-diphosphate molecules linked by a pyrophosphate bond. It has been suggested that DNA synthesis is intimately linked to this guanine tetraphosphate (11). It is also an intriguing question as to whether Gp_4G may interfere with the function of EF-1α, which relies upon GTP.

3. Purification of elongation factor 1$\beta\gamma$ and its derived subunits

3.1 Assays for EF-1β

The activity of EF-1β may be measured by using one of the following assay systems (10):

(a) guanine nucleotide exchange assay,

(b) stimulation of EF-1α-dependent binding of phe-tRNA to ribosomes, and

(c) stimulation of poly(U)-directed polyphenylalanine synthesis.

Of these, the guanine nucleotide exchange assay is the most usual and convenient one.

3.1.1 Guanine nucleotide exchange assay

The exchange assay uses labelled EF-1α. [^{3}H]GDP complex as prepared in part A of the following protocol. Exchange of [^{3}H]GDP bound to EF-1α is measured in the presence of free unlabelled GDP and EF-1β, as given in part B of the following protocol. A control experiment without EF-1β should always be included. The decrease of radioactivity with respect to the control is a measure of the amount of EF-1β activity.

Protocol 5. Guanine nucleotide exchange assay

Preparation of EF-1α-[^{3}H]GDP *complex*

1. Mix for each assay in the following order:
   (a) x μl pure EF-1α (40 pmoles; $x \leq 5$ μl)
   (b) 1 μl 0.05 mM [^{3}H]GDP (500 Ci/mol)
   (c) 6 μl complex buffer[a]
   (d) $(5-x)$ μl water
2. Incubate at 37 °C for 5 min.
3. Chill on ice.
4. Dilute the solution with 24 μl of exchange buffer.[b]

Exchange assay

1. Mix in the following order:
   (a) x μl sample containing 1 to 10 pmoles of EF-1β ($x \leq 65$ μl)
   (b) 15 μl 1 mM GDP in exchange buffer
   (c) $(65-x)$ μl exchange buffer
2. At $t = 0$ min, add 30 μl of the preformed EF-1α-[^{3}H]GDP complex (step 4, part A of this protocol) and mix.
3. Incubate at 0 °C for 3 min, exactly.
4. Take a 100 μl aliquot at $t = 3$ min and pass it immediately through a nitrocellulose filter (0.45 μm pore size). Wash the filter twice with 2 ml of ice-cold washing buffer.[c]
5. Dry the filter and determine its radioactivity.

[a] Complex buffer: 80 mM Tris–HCl (pH 7.5 at 20 °C), 400 μM dithiothreitol, 20 mM magnesium acetate, 200 mM NH_4Cl, 2 mg/ml bovine serum albumin (BSA), 50% (v/v) glycerol.

[b] Exchange buffer: 20 mM Tris–HCl (pH 7.5), 10 mM magnesium acetate, 50 mM NH_4Cl, 10% (v/v) glycerol.
[c] Washing buffer: 20 mM Tris–HCl (pH 7.5 at 20 °C), 10 mM magnesium acetate, 100 mM NH_4Cl, 100 μg/ml BSA.

3.1.2 tRNA binding stimulation

The assay is carried out as given in *Protocol 1* for the determination of EF-1α activity. However, when assaying EF-1β, tRNA binding is measured in the presence of a limiting amount of EF-1α (0.5 pmole per assay) and variable amounts of EF-1β (1–10 pmoles). A typical result under these assay conditions is a three-fold stimulation of phe-tRNA binding per 1 pmole of EF-1β.

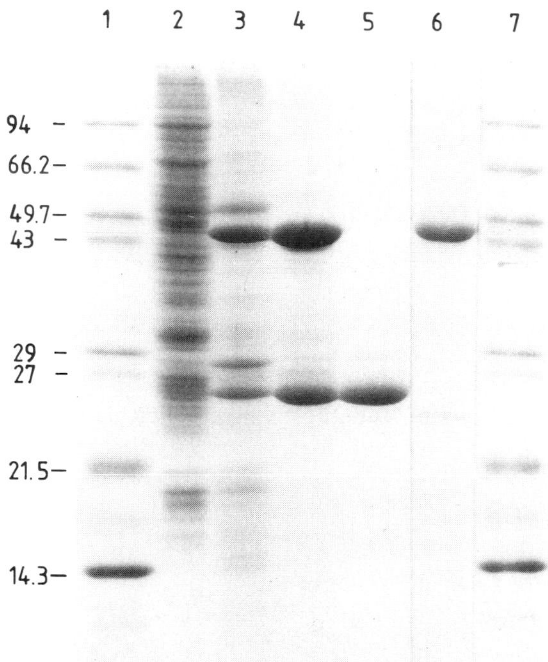

Figure 1. Purification of EF-1β and EF-1γ followed by means of SDS-PAGE. Proteins from individual purification steps were subjected to electrophoresis in a polyacrylamide slab gel (13%) in the presence of SDS according to Laemmli (8). Lanes 1 and 7; marker proteins: phosphorylase *a* (94 000), bovine serum albumin (66 200), EF-1α from *Artemia salina* (49 700), ovalbumin (43 000), carbonic anhydrase (29 000), chymotrypsinogen (27 000), trypsin inhibitor from soy bean (21 500), and lysozyme (14 300): lane 2; 35–50% ammonium sulphate saturation fraction (*Protocol 6*, step 6; 25 μg): lane 3; DEAE-cellulose preparation, after sedimentation of tubulin (step 9; 25 μg): lane 4; EF-1βγ, after phenyl-Sepharose chromatography (step 10; 14 μg): lane 5, EF-1β (7 μg): lane 6; EF-1γ (7 μg), both after DEAE-cellulose chromatography (step 12). Numbers on the left are $M_r \times 10^{-3}$.

3.1.3 Polyphenylalanine synthesis

Factor-dependent poly(phe) synthesis is measured as described in *Protocol 8* for polyphenylalanine synthesis, but in the presence of 2.4 pmol of EF-1α and 1.3 pmol of EF-2. The amount of EF-1β is varied between 0 and 2 pmol, giving typical values of about 3 pmol of phenylalanine incorporated into polyphenylalanine per pmol of EF-1β.

3.2 Purification protocol

EF-1$\beta\gamma$ is purified from post-ribosomal supernatants by ammonium sulphate fractionation, followed by ion-exchange and hydrophobic interaction chromatography. The protein complex is dissociated into its subunits by the addition of urea (6 M), and the proteins are resolved on DEAE-cellulose on the basis of their isoelectric points. Triton X-100 is included in the buffers to prevent precipitation of EF-1γ (6). The purification procedure is given in *Protocol 6*. Typical yields from 425 g of cysts are 10 mg of EF-1$\beta\gamma$, or 3 mg of EF-1β and 5 mg of EF-1γ. Purification of EF-1$\beta\gamma$ and its derived subunits can be monitored by SDS-PAGE and/or activity measurements of the various steps. Typical results are given in *Figure 1* and *Table 1*, respectively.

Table 1. Purification of EF-1$\beta\gamma$ from 425 g of *A. salina* cysts

Purification step	Volume (ml)	Total Protein (mg)	Specific activity (units/mg)		
			exchange[a]	tRNA binding stimulation[b]	poly-(phe) synth[c]
Ammonium sulphate fractionation, step 6	200	3400	1.7	nd[d]	nd
DEAE-cellulose, step 8	840	70	26	12	32
Phenyl-Sepharose, step 10	205	10	49	18	57
DEAE-cellulose, step 12: EF-1β	140	2.8	120	61	150
EF-1γ	480	4.8	0	nd	nd

[a] One unit of exchange activity is defined as the amount which promotes the exchange of 1 pmol [^{3}H]GDP bound to EF-1α with exogenous GDP.

[b] One unit of stimulatory activity with respect to tRNA binding is defined as the amount required to double the amount of [^{3}H]phe–tRNA binding to ribosomes.

[c] One unit of activity of EF-1β with respect to the stimulation of poly(phe) synthesis is defined as the amount which is required to incorporate 1 pmol [^{3}H]phe–tRNA into polyphenylalanine.

[d] nd = not determined.

Protocol 6. Purification of EF-1$\beta\gamma$ and its derived subunits

1. It is crucial to avoid the preparation being degraded or aggregated. Therefore, perform all operations at 0–4 °C, do not interrupt the procedure at any other steps than indicated, and do not freeze partially pure preparations.
2. Prepare post-ribosomal supernatant starting from 425 g of cysts as described in the protocol for the purification of 80S ribosomes from *Artemia*, up to and including step 6.
3. Add to the supernatant 0.5 M sodium pyrophosphate (pH 7.0) and solid GTP to make final concentrations of 50 mM and 0.1 mM, respectively. Check the pH and adjust to 7.0 if necessary. Centrifuge at 20 000 *g* for 30 min and recover the supernatant.
4. Add to the supernatant ammonium sulphate to 35% saturation, stir for 1 h and centrifuge at 25 000 *g* for 20 min.
5. Adjust the supernatant to 50% saturation with ammonium sulphate and stir for 1 h. At this point the proteins can be stored at 0 °C for at least 7 days. Centrifuge at 25 000 *g* for 20 min and recover protein.
6. Dissolve the precipitated proteins in 150 ml of buffer A[a]. Dialyse for at least 1 h against 1 litre of the same buffer and then overnight against 5 litres of fresh buffer. Centrifuge at 20 000 *g* for 30 min and recover the supernatant.
7. Equilibrate a column (5 × 10 cm) of DEAE-cellulose DE-52 (Whatman) with buffer A. Load the supernatant onto the column. Wash the column with 1 litre of buffer A containing 150 mM NaCl. Develop with a linear 2-litre gradient of 150 mM to 300 mM NaCl in buffer A, flow rate 200 ml/h. Analyse the protein composition by SDS-PAGE.
8. Pool fractions containing the purest EF-1$\beta\gamma$. At this stage the preparation is already of high purity and usually contains only α/β tubulin (50 kDa) and a 28 kDa protein as major impurity. Store the pooled fractions overnight at −20 °C.
9. Thaw the preparation, add GTP to a final concentration of 10 μM and centrifuge for 1 h at 100 000 *g*. Add solid ammonium sulphate to a final concentration of 0.4 M.
10. Equilibrate a column (2 × 15 cm) of phenyl-Sepharose CL-4B (Pharmacia) with buffer B[b]. Load the supernatant of the previous step onto the column, wash with 1 litre of buffer B, and elute the proteins with a linear gradient of 1 litre of buffer B to 1 litre of buffer C[a]. Maintain the flow rate at about 100 ml/h. Pool fractions containing pure EF-1$\beta\gamma$ only.
11. Add Triton X-100 and urea to the EF-1$\beta\gamma$ preparation to give final concentrations of 1% and 6 M, respectively. Dialyse overnight against buffer D[d]

12. Load the dissociated subunits onto a column (1 × 5 cm) of DEAE-cellulose DE-52, previously equilibrated with the same buffer. Wash with 25 ml of buffer D and apply a gradient of 100 ml buffer D to 100 ml buffer D containing 0.3 M NaCl, at a flow rate of 20 ml/h. EF-1γ does not bind and is found in a pure form in the flow-through of the column. Pure EF-1β elutes at about 0.2 M NaCl.
13. Preparations of EF-1βγ and EF-1β are stable at −20 °C for at least one year. The preparation of EF-1γ is stable for only several weeks at 0 °C. After this period a precipitate of EF-1γ appears. Preparations can be stored as such, or after dialysis against a suitable buffer e.g., dilution buffer (*Protocol 1*) or exchange buffer (*Protocol 5*). Assays are carried out after dialysis.

[a] Buffer A: 50 mM sodium pyrophosphate (pH 7.4), 2.5 mM $MgCl_2$. Adjust the pH of a 50 mM sodium pyrophosphate solution (20 °C) with 85% H_3PO_4. Then add a solution of 1 M $MgCl_2$. A transient precipitate may be formed.
[b] Buffer B: 20 mM Tris–HCl (pH 7.5), 100 μM EDTA, 10 mM 2-mercaptoethanol, 0.4 M ammonium sulphate.
[c] Buffer C: 20 mM Tris–HCl (pH 7.5), 100 μM EDTA, 10 mM 2-mercaptoethanol, 80% (v/v) ethylene glycol.
[d] Buffer D: 5 mM sodium pyrophosphate (pH 6.7), 2.5 mM $MgCl_2$, 6 M urea, 1% Triton X-100.

3.3 Comments

3.3.1 EF-1βγ, the functional analogue of bacterial EF-Ts

EF-1βγ unquestionably increases the rate of guanine nucleotide exchange on EF-1α. The rate constant for the dissociation of GDP from EF-1α-GDP is 0.7×10^{-3}, while that of GDP from the ternary complex EF-1α-GDP-EF-1βγ is 0.7 per sec (27). However, it would be premature to conclude that the sole function of EF-1βγ is connected with this exchange, especially since there is such a high concentration of EF-1βγ in the cell; in *Artemia*, the molar ratio of ribosomes: EF-1βγ: EF-1α = 1:5:20. It seems probable that EF-1γ, the extra polypeptide of eucaryotic 'eEF-Ts', is responsible for a transient interaction of EF-1βγ with the cytoskeleton and membrane structures (6).

3.3.2 Translational control through phosphorylation

Phosphorylation of elongation factor EF-2 by an EF-2 kinase causes arrest of translation, and dephosphorylation of EF-2 restores this activity (28, 29). Phosphorylation of EF-1β also impairs its function, i.e. the stimulation of guanine nucleotide exchange on EF-1α (8). A casein kinase II is responsible for this phosphorylation at serine 89 of EF-1β, but it is not yet known whether *in vivo* the kinase affects directly the elongation stage of translation, as much as perhaps another function of EF-1β, hitherto unknown.

4. Purification of elongation factor 2

4.1 Assays for EF-2

The presence of EF-2 in column chromatographic fractions can be monitored by diphtheria toxin-mediated adenylribosylation of EF-2, or the stimulation of polyphenylalanine synthesis (30). The procedures for these assays are given in *Protocols* 7 and 8. SDS-PAGE can also be used; EF-2 from *Artemia* migrates at the same position as phosphorylase *a* (94 000 D). SDS-PAGE patterns of the proteins before and after hydroxylapatite are given in *Figure 2*.

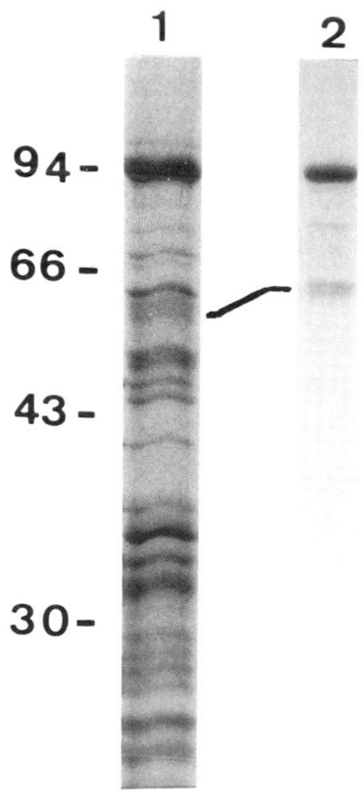

Figure 2. Analysis by SDS-PAGE of the purification of EF-2. Lane 1; 0.5 M KCl split fraction (*Protocol 9*, step 7) of crude 80S ribosomes: lane 2; followed by purification over hydroxylapatite (step 9). Numbers on the left are $M_r \times 10^{-3}$.

Protocol 7. Adenylribosylation of EF-2

1. Use crude diphtheria toxin (*poison!*), which is not inactivated. Dialyse against 50 mM potassium phosphate (pH 7.0). Store at −70 °C at a protein concentration of 1 mg/ml. Suitable diphtheria toxin (e.g., for vaccine production) may be obtained from the Rijksinstituut voor Volksgezondheid, Bilthoven, the Netherlands or purchased from commercial sources.
2. Incubate 100 µl 50 mM Tris–HCl (pH 7.6), 6 mM EDTA, 30 mM DTT,

containing 1 μg toxin, 400 pmol [^{14}C]NAD$^+$ (266 mCi/mmol; Amersham, UK) and a few μl column fractions for 1 h at 37 °C.

3. Add 1 ml 10% (w/v) TCA, incubate 15 min at 90 °C, filter the protein precipitate through a GF/C filter, wash the filter with 10% TCA and ethanol and, finally, determine its ^{14}C content by liquid scintillation counting.

Protocol 8. Polyphenylalanine synthesis

1. Incubate 1.6 pmol of salt-washed 80S ribosomes (protocol for purification of 80S ribosomes from *Artemia* step 11), 5 μg of poly(uridylic acid) (Boehringer), 5 nmol GTP, 100 pmol [^{3}H]phe–tRNA (4000 d.p.m./pmol) (protocol for preparation of [^{3}H]phe–tRNA), 3 μg (60 pmol) EF-1α (protocol for purification of EF-1α) and approximately 0.5 μg (5 pmol) EF-2 in 30 μl buffer A (see footnote *a* in protocol for the isolation of EF-2 from crude 80S ribosomes).
2. Incubate 20 min at 37 °C.
3. Add 1 ml of 10% TCA and incubate 15 min at 90 °C.
4. Cool to room temperature and collect precipitated protein by filtration through Whatman GF/C filters.
5. Wash the filter several times with 10% TCA followed by ethanol.
6. Determine radioactivity retained by the filter. Typically 6 pmol of phenylalanine is incorporated.

4.1.1 Diphtheria toxin-mediated adenylribosylation

Adenylribosylation of EF-2 by diphtheria toxin is a very simple assay for screening column fractions. It is based on the modification of EF-2 with the adenylribose moiety of [^{14}C]NAD$^+$. This reaction is catalysed by diphtheria toxin. It proceeds stoichiometrically and can be used to quantitate fractions for EF-2 content.

4.1.2 Polyphenylalanine synthesis

Though the adenylribosylation assay is a very simple one, it is not strictly specific for EF-2. A specific test for EF-2 is the stimulation of polyphenylalanine synthesis in a system containing poly(uridylic acid) as mRNA, ribosomes, EF-1, and [^{3}H]phe–tRNA. In the absence of EF-2, hardly any radioactivity is incorporated as monitored by precipitation with hot TCA. Hot TCA is used because precipitation with cold TCA also yields [^{3}H]phe–tRNA in the precipitate. Addition of EF-2 to the incubation mixture gives approximately a 15-fold stimulation of [^{3}H]phenylalanine incorporation.

4.2 Purification protocol

A simple method, which yields EF-2 in reasonable quantities (about 1.5 mg in a single run starting from 425 g of cysts), is based on the observation that crude 80S ribosomal pellets from undeveloped *Artemia* contain EF-2 (31), which is released by a high-salt wash. Crude 80S ribosomes are dissolved in a high-salt buffer and the ribosomes are removed by centrifugation through a sucrose gradient while EF-2 is retained on top of the gradient. This material is subsequently purified to near homogeneity by chromatography over hydroxylapatite and DEAE-Sephadex A-50. The protocol is given below.

Protocol 9. Isolation of EF-2 from crude 80S ribosomes

1. All procedures are carried out at 4 °C.
2. Isolate crude 80S ribosomes as described in the protocol for the purification of 80S ribosomes from *Artemia*, step 6. Typically, 425 g of unhydrated cysts are used.
3. Dissolve the crude ribosomal pellet (about 500 mg of ribosomes) in buffer A[a] at a concentration of approximately 75 mg/ml. 1 mg of 80S ribosomes/ml yields an A_{260} value of 15.
4. Add 4 M KCl to a final concentration of 0.5 M.
5. Centrifuge through 15–30% (w/v) sucrose gradients (made up in buffer A) in a Beckman SW 27 rotor (16 h, 26000 r.p.m.).
6. Collect the upper part of the gradients, test for EF-2 by SDS-PAGE or adenylribosylation.
7. Pool the EF-2 containing fractions (about 21 mg of protein) and dialyse against HA[b] buffer. Then load onto a 10 ml hydroxylapatite column (Biorad, Bio-gel HTP), equilibrated with HA buffer.
8. Run the column with HA buffer, containing a linear phosphate gradient (2 × 800 ml, 20–500 mM potassium phosphate pH 7.0). EF-2 elutes at a conductivity of 8 mS, approximately 340 mM potassium phosphate.
9. Pool the EF-2 containing fractions (approximately 7 mg of protein) and concentrate in a dialysis bag by placing the bag in dry Sephadex G-200. Dialyse against D-25[c] buffer and load onto a DEAE-Sephadex A-50 column (50 ml bed volume), equilibrated with D-25.
10. Develop the column with a linear gradient of KCl (2 × 500 ml, 25–250 mM KCl). EF-2 elutes from the column at a conductivity of 8.5 mS, i.e. approximately 120 mM KCl.
11. Collect EF-2 containing fractions and concentrate the protein to a concentration of about 1 mg/ml. Dialyse the protein against D-25 buffer. Typically, 1.5 mg of EF-2 is obtained with a purity of approximately 90%

according to SDS-PAGE. EF-2 is stable for several years when stored at −70 °C in D-25 buffer.

[a] Buffer A: 50 mM Tris–HCl (pH 7.6), 70 mM KCl, 9 mM magnesium acetate, 0.1 mM EDTA, 10 mM 2-mercaptoethanol and 250 mM sucrose.
[b] HA buffer: 20 mM potassium phosphate (pH 7.0) 10% glycerol 12 mM 2-mercaptoethanol, 0.1 mM EDTA.
[c] D-25 buffer: 20 mM Tris–HCl (pH 7.6), 25 mM KCl, 0.1 mM EDTA, 12 mM 2-mercaptoethanol, 10% glycerol.

References

1. Maessen, G. D. F. (1989). Ph.D. dissertation, Laboratory for Medical Biochemistry, University of Leiden.
2. Fischer, I., Arfin, S. M., and Moldave, K. (1980). *Biochemistry*, **19**, 1417.
3. Walter, P., Gilmore, R., and Blobel, G. (1984). *Cell*, **38**, 5.
4. Hardesty, B. and Kramer, G. (ed.) (1986). *Structure, Function and Genetics of Ribosomes*. Springer Verlag, NY.
5. Slobin, L. I. and Möller, W. (1975). *Nature*, **258**, 452.
6. Janssen, G. M. C. and Möller, W. (1988). *Eur. J. Biochem.*, **171**, 119.
7. Möller, W. and Amons, R. (1989). In *The Guanine-Nucleotide Binding Proteins, Common Structural and Functional Properties*, (ed. L. Bosch, B. Kraal, and A. Parmeggiani) Plenum Press, NY.
8. Janssen, G. M. C., Maessen, G. D. F., Amons, R., and Möller, W. (1988). *J. Biol. Chem.*, **263**, 11063.
9. Slobin, L. I. and Möller, W. (1976). *Eur. J. Biochem.*, **69**, 351, 367.
10. Slobin, L. I. and Möller, W. (1978). *Eur. J. Biochem.*, **84**, 69.
11. Finamore, F. J. and Clegg, J. S. (1969). In *The Cell Cycle, Gene–Enzyme Interactions*, (ed. G. M. Padilla, G. L. Whitson, and I. L. Cameron), p. 249. Academic Press, NY.
12. Sorgeloos, P., Bengtson, D. A., Decleir, W., and Jaspers, E. (1987). *Artemia Research and its Applications*, Vol. 1. Universa Press, Wetteren.
13. Decleir, W., Moens, L., Slegers, H., Sorgeloos, P., and Jaspers, E. (1987). *Artemia Research and its Applications*, Vol. 2. Universa Press, Wetteren.
14. Moldave, K. (1985). *Ann. Rev. Biochem.*, **54**, 1109.
15. McLennan, A. G. and Miller, D. (1987). In *Artemia Research and its Applications*, Vol. 2, p. 253. Universa Press, Wetteren.
16. Moldave, K. and Grossman, L. (1979). *Methods in Enzymology*, Vol. 60, Section II. Academic Press, NY.
17. Miller, D. L. and Weissbach, H. (1977). In *Molecular Mechanisms of Protein Biosynthesis*, (ed. H. Weissbach and S. Pestka), p. 323. Academic Press, NY.
18. Brot, N. (1977). In *Molecular Mechanisms of Protein Biosynthesis*, (ed. H. Weissbach and S. Pestka), p. 375. Academic Press, NY.
19. Zasloff, M. and Ochoa, S. (1971). *Proc. Nat. Acad. Sci. USA*, **68**, 3059.
20. Nirenberg, M. W. and Matthaei, J. H. (1961). *Proc. Nat. Acad. Sci. USA*, **47**, 1588.
21. Nishizuka, Y., Lipmann, F., and Lucas-Lenard, J. (1968). In *Methods in Enzymology*, Vol. 12B, (ed. L. Grossman and K. Moldave), p. 708. Academic Press, NY.

22. Traub, P., Mizushima, S., Lowry, C. V., and Nomura, M. (1971). In *Methods in Enzymology*, Vol. 20, (ed. K. Moldave and L. Grossman), p. 391. Academic Press, NY.
23. Wagner, T. and Sprinzl, M. (1979). In *Methods in Enzymology*, Vol. 60 (ed. K. Moldave and L. Grossman), p. 615. Academic Press, NY.
24. Möller, W. and Maassen, J. A. (1987). In *Structure, Function and Genetics of Ribosomes*, p. 309. Springer Verlag, NY.
25. Moazed, D., Robertson, J. M., and Noller, H. F. (1988). *Nature*, **334**, 362.
26. Viel, A., Djé, M. K., Mazabraud, A., Denis, H., and le Maire, M. (1987). *FEBS Lett.*, **223**, 232.
27. Janssen, G. M. C. and Möller, W. (1988). *J. Biol. Chem.*, **263**, 1773.
28. Ryazanov, A. G., Shestakova, E. A., and Natapov, P. G. (1988). *Nature*, **334**, 170.
29. Nairn, A. C. and Palfrey, H. C. (1987). *J. Biol. Chem.*, **262**, 17299.
30. Raeburn, S., Collins, J. F., Moon, H. M., and Maxwell, E. S. (1971). *J. Biol. Chem.*, **246**, 1041.
31. Yablonka-Reuveni, Z. and Warner, A. H. (1979). In *Biochemistry of Artemia Development*, (ed. J. C. Bagshaw and A. H. Warner), p. 22. University Microfilms International, Ann Arbor.
32. Weissbach, A. and Weissbach, H. (ed.) (1986). *Methods in Enzymology*, Vol. 118, Sections III and IV. Academic Press Inc., Orlando, Florida.

4

Peptidyltransferase: the soluble protein EF-P restores the efficiency of 70S ribosome-catalysed peptide-bond synthesis

DAE-GYUN CHUNG, NASIR D. ZAHID, ROSS M. BAXTER, and M. CLELIA GANOZA

1. Introduction

The peptidyltransferase activity of ribosomes catalyses, among other reactions, the condensation of the aminoacyl ester of the nascent peptidyl–tRNA with the amino group of the incoming aminoacyl–tRNA. Two general hypotheses have been formulated regarding the mechanism underlying peptidyltransferase activity. Based on thermodynamic considerations, one of these proposes that correct steric alignment of the translation substrates is sufficient to drive protein chain elongation (1). The alternate hypothesis, supported by several lines of biochemical evidence, argues that a catalytic mechanism is required to ensure the rapid rate of peptide-bond synthesis *in vivo* (2).

The actual peptidyltransferase enzyme is considered to be an integral part of the 50S ribosomal subunit. Minimal reconstruction of this activity *in vitro* requires the ribosomal proteins L2, L3, L4, L15, L16, and L18 as well as 23S rRNA, whereas optimum reaction rates are achieved only when a host of other components including L6 and L11 are present (3). Although the absence of any one of these results in decreased peptide-bond synthesis, there are viable deletion mutants of L11 and L15, suggesting that they are in fact not essential for activity (4, 5). In addition, L6 and L11 are predominantly involved in transesterification whereas L2 exhibits esterase activity (6, 7). Although other components such as L3 and L4 have not been investigated fully as yet, ribosomal protein L16 has been consistently shown to occupy a central position within the peptidyltransferase complex. It also contains a unique histidine residue which has remained a potential candidate to furnish, at least in part, the catalytic activity of peptidyltransferase (6, 8–10). However, a histidine-carbonyl intermediate, analogous to that of other transpeptidation reactions, has never been observed. Also, the issue of whether L16 is required for assembly of the 50S subunit for active catalysis of peptide bonds or both is not clear (11, 12). There is also speculation that the 23S rRNA may participate directly in the catalytic process (13).

Recent studies involving antibiotics which inhibit protein synthesis, particularly those that mimic a peptide bond (e.g., chloramphenicol, anisomycin, and puromycin), suggest that domain V of 23S rRNA is essential (14). However, other regions of this rRNA bind some of the proteins that are essential for peptide-bond synthesis (15). It is thus possible that the 23S rRNA helps form a structural locus that accommodates the ribosomal proteins essential to the reaction.

Regardless of these mechanistic considerations, the 50S peptidyltransferase complex can be activated *in vitro* only by artificial conditions such as the addition of appropriate concentrations of methanol or ethanol (16). *In vivo*, the functional translation complex contains the 30S subunit as well. Therefore it is perhaps best to consider the overall environment of the native 70S ribosome as being optimal for peptide-bond synthesis. From this perspective, the issues of catalysis and substrate alignment may be viewed as integral aspects of a single problem. Thus the correct steric alignment of the aminoacyl-tRNA, for example, is crucial in positioning the $-NH_3^+$ moiety such that deprotonation prior to nucleophilic attack on the carbonyl carbon of the peptidyl-tRNA can be facilitated by residues like the unique histidine of L16. Unfortunately, progress in studying peptide-bond synthesis under more physiological conditions has been hampered to a large extent by our inability to detect reconstructed 70S ribosomal activity at low substrate concentrations and/or in the absence of methanol or ethanol.

Here we describe a large scale purification scheme for a previously described protein, EF-P, which functionally couples the 50S peptidyltransferase with the 30S subunit and restores the efficiency of the 70S ribosome complex. We also find that ribosomal protein L16 containing an intact histidine residue is required for the EF-P mediated 70S peptidyltransferase reaction.

2. Materials and methods

2.1 Materials

Puromycin dihydrochloride may be purchased from Sigma, [^{35}S]-Met (1040 Ci/mmol) from New England Nuclear, *E. coli* K12 or MRE600 cells (mid-log) from the Grain Processing Corp. (Muscatine, IA), *E. coli* B tRNA from General Biochemicals, DNase I (electrophoretically purified) from Worthington, AUG from Boehringer Mannheim, alumina (Alcoa 305) from Sigma, and calcium leucovorin from Lederle.

2.2 Methods

Perform all operations at 4 °C or on ice unless indicated otherwise.

Protocol 1. Preparation of ribosomes

1. To prepare 70S ribosomes, grind 50 g of E. coli MRE600 cells with 100 g of alumina (Alcoa 305) to an even consistency, add 50 μl of 1 mg/ml DNase I, mix, and then add 50 ml of 10 mM Tris–HCl (pH 7.4), 30 mM NH_4Cl, 10 mM $MgCl_2$, and 1 mM dithiothreitol (DTT).
2. Centrifuge at 12 000 r.p.m. for 15 min (discard pellet) and at 15 000 r.p.m. for 30 min (Sorvall SS34 rotor) to remove alumina and cell debris.
3. Centrifuge the supernatant at 55 000 r.p.m. for 2.5 h (Beckman Ti 70 rotor) and wash the resulting ribosomal pellet by resuspending in 10 mM Tris–HCl (pH 7.4), 1 M NH_4Cl, 10 mM $MgCl_2$, and 1 mM DTT (at an A_{260}nm of 400) and centrifuging at 30 000 r.p.m. for 12 min.
4. Discard the pellet, centrifuge the supernatant at 58 000 r.p.m. for 2 h to recover ribosomes and wash the pellets as above.
5. Rinse the final pellets with 10 mM Tris–HCl (pH 7.4), 50 mM NH_4Cl, 10 mM $MgCl_2$, 10% glycerol, and 1 mM DTT, and resuspend the washed ribosomes at a concentration of 50 mg/ml (1 mg/ml $\triangleq$ 14.4 A_{260} units). After a final spin at 30 000 r.p.m. for 12 min aliquot the supernatant ribosome suspension and store at −70 °C (17).

Ribosomal subunits can be obtained by dissociation of 70S particles in a buffer containing 0.3 mM Mg^{2+} and sedimentation through sucrose gradients (17). Core particles and split proteins can be prepared by 2 M LiCl extraction of 50S ribosomal subunits. Split proteins and ribosomal protein L16 may be purified and characterized as described in (17–20). Further details are also provided in Chapter 8.

i. Ethoxyformylation of ribosomal components

30S and 50S subunits and ribosomal proteins such as L16 can be treated with ethoxyformic anhydride as described in (10). A variety of other chemical modification and affinity labelling protocols are available to study different components of the peptidyltransferase centre (9).

Protocol 2. Preparation of translation substrates

1. Assay for f[^{35}S]Met–tRNA synthesis by incubating (15 min at 35 °C) 3 μg calcium leucovorin, 120 μg *E. coli* B tRNA, 25 mCi of L-[^{35}S]methionine, and 10 μl (containing about 50 μg of protein) of an S100 fraction (i.e. post-ribosomal supernatant dialysed 1.5 h against 10 mM Tris–HCl (pH 7.4), and 1 mM DTT) in 8 mM Tris–HCl (pH 7.4), and 8 mM $MgCl_2$ (60 μl final volume).

2. Stop the reaction with excess 5% trichloroacetic acid (TCA), collect the [^{35}S]Met–tRNA on glass-fibre filters, and wash with 5% TCA to remove unreacted label.
3. Place filters into a suitable scintillation cocktail and count. The initial rates of the EF-P dependent reaction may be estimated by kinetic analysis using the same amount of the 70S·AUG·fMet-tRNA complex and different amounts of EF-P.

The [^{35}S]met–tRNA is prepared in a 6 ml scaled-up version of the reaction described above and extracted with H_2O saturated phenol. Precipitate the charged tRNA with two volumes of cold ethanol for several hours at -20 °C. Centrifuge the precipitate and redissolve at 50 000–100 000 c.p.m./μl in distilled H_2O. Dialyse for 1.5 h against 1% potassium acetate (pH 5.0) and then for 2 h against distilled H_2O. To score for the amount of f[^{35}S]Met–tRNA, incubate an aliquot of the preparation for 15 min at 37 °C in 0.33 M KOH, neutralize by addition of an equal volume of 0.33 M HCl and add 1.5 ml ethyl acetate. Centrifuge briefly to separate the phases and withdraw 1 ml of the organic phase for liquid scintillation counting and for electrophoretic analysis as described in (21). Generally 50–60% of the total Met–tRNA should be formylated.

Prepare hydroxypuromycin by nitrous acid deamination of puromycin and determine its concentration spectrophotometrically by using an absorption coefficient of 2.0×10^4 cm^{-1} M^{-1} at 267.5 nm (22). Estimate protein concentrations by the Lowry method (23) or by protein-dye binding (24). Perform SDS-PAGE according to Laemmli (25) or equivalent.

Protocol 3. Reconstitution of 50S complexes and reassociation with 30S subunits

1. For reconstitution purposes, mix 7.5 A_{260} units of 50S core particles with L16 (32 μg) and/or other ribosomal proteins in 20 mM Tris–HCl (pH 7.4), 250 mM NH_4Cl, 20 mM $MgCl_2$, and 2 mM 2-mercaptoethanol (1.0 ml final volume) and incubate for 1.5 h at 50 °C.
2. Cool and then precipitate the complexes with an equal volume of cold ethanol.
3. Centrifuge and resuspend the particles in 10 mM Tris–HCl (pH 7.4), 50 mM NH_4Cl, 10 mM $MgCl_2$, and 2 mM 2-mercaptoethanol at a concentration of 50 A_{260}/ml.
4. Reassociate the reconstituted particles with 3.8 A_{260} units of 30S subunits to form 70S complexes prior to assaying for peptidyltransferase activity (10).

Peptide-bond formation and EF-P activity are assayed by monitoring f[^{35}S]Met–puromycin synthesis from ribosome·AUG·f[^{35}S]met-tRNA complexes and added puromycin (16, 26).

Protocol 4. Peptidyltransferase assays

1. Prepare the following stock solutions:
   (a) 50 μg/ml AUG in H_2O,
   (b) 200 mM Tris–HCl (pH 7.4), 500 mM NH_4Cl, and 100 mM $MgCl_2$,
   (c) *E. coli* MRE 600 70S ribosomes at approximately 80 mg/ml,
   (d) 12 μM puromycin, and
   (e) 10 mM Tris–HCl (pH 7.4) and 1 mM DTT.
2. For a typical 10 tube assay prepare a pre-mix solution by combining 360 μl H_2O, 60 μl (a), 60 μl (b), 10–20 μl (c), and 20 μl of f[^{35}S]Met–tRNA (*Protocol 2*). Incubate this pre-mix for 15 min at 20 °C.
3. In each assay tube combine 5 μl (d), 10 μl (e), and 5 μl of the EF-P fraction (typically at 0.5 mg/ml) to be tested. Incubate these for 30 sec at 37 °C, quickly add 40 μl of the pre-mix solution to each tube, and incubate for 10 min at 37 °C.
4. Add 0.5 ml of 1 M potassium phosphate (pH 7.2), vortex, add 1 ml ethyl acetate, vortex, spin briefly to separate the phases, and finally remove 0.75 ml of the top phase for liquid scintillation counting.

For individual assay tubes in the case of reassociated 70S ribosomes, mix one A_{260} unit of 50S complex with 0.5 A_{260} unit of 30S subunits in 100 μl of 15 mM Tris–HCl (pH 7.4), 75 mM NH_4Cl, 15 mM $MgCl_2$, and 2 mM 2-mercaptoethanol. Add 5 μl of AUG (50 μg/ml, solution (a) and 1 pmol of f[^{35}S]Met–tRNA (approximately 100 000 d.p.m.) and incubate the mixture for 0.5 h at 30 °C. This yields a ribosome·AUG·f[^{35}S]Met–tRNA complex which can then be incubated with EF-P and puromycin as above.

The following represents a substantially modified procedure for the purification of EF-P based on the protocol described in (26).

Protocol 5. Purification of EF-P

1. For a large scale preparation, suspend 700 g of frozen *E. coli* K12 cells in 700 ml of 10 mM Tris–HCl (pH 7.4), 10 mM $MgCl_2$, 50 mM NH_4Cl, and 1 mM DTT.
2. Disrupt the suspension in a French pressure cell (Aminco) at 12 000 p.s.i.
3. Add 700 μg DNase I to the lysed suspension and centrifuge at 10 000 × *g* for

20 min (Sorvall GSA rotor), at 17 000 × *g* for 15 min, and at 27 000 × *g* for 15 min (Sorvall SS-34 rotor) to remove cell debris.

4. Centrifuge the supernatant at 100 000 × *g* for 20 min (Beckman 30 rotor) to remove the ribosomes.
5. Dilute the resulting S100 extract to 10 mg/ml of protein with 10 mM Tris–HCl (pH 7.4) and 10 mM $MgCl_2$.
6. Add $(NH_4)_2SO_4$ to 80% saturation, stir for 30 min, and then centrifuge at 10 000 × *g* for 20 min. Discard the supernatant. Dissolve the pellet in 250 ml of 10 mM Tris–HCl (pH 7.4) and 10 mM KCl (buffer A), and then dialyse the suspension against at least two changes of 4 litre each of buffer A.
7. Load the dialysed S100 fraction on to a column (4.5 × 25 cm) of DEAE-cellulose (Whatman DE-23) equilibrated with buffer A. Wash the column with 800 ml of buffer A then apply a linear gradient of KCl [10–350 mM in 10 mM Tris–HCl (pH 7.4), 2 litre total volume], adjust the flow rate to 30 ml/h, and collect 12 ml fractions.
8. Assay 5 μl aliquots for EF-P, pool the active fractions, and precipitate with $(NH_4)_2SO_4$ as in step 6 above.
9. Dissolve the pellet in 40 ml buffer A and dialyse this EF-$P_{DEAE\text{-}1}$ suspension against at least two changes of 1 litre each of buffer A.
10. Apply the EF-$P_{DEAE\text{-}1}$ fraction to a second column of DEAE-cellulose (2.5 × 50 cm) also equilibrated with buffer A. Wash the column with 500 ml of buffer A then apply a linear gradient of KCl [10–350 mM in 10 mM Tris–HCl (pH 7.4), 1 litre total volume], adjust the flow rate to 20 ml/h and collect 12 ml fractions.
11. Assay for EF-P activity, pool appropriate fractions, and precipitate with $(NH_4)_2SO_4$ as in step 6 above. Dissolve the pellet in 6 ml of 50 mM Tris–HCl (pH 7.4) and 50 mM NH_4Cl (buffer B) and dialyse the resulting EF-$P_{DEAE\text{-}2}$ fraction against at least two changes of 0.5 litre each of buffer B.
12. Prepare six 5–20% linear sucrose gradients (12 ml each) in buffer B, remove 1 ml from the top of each gradient and carefully layer on 1 ml of the EF-$P_{DEAE\text{-}2}$ fraction. Centrifuge the gradients at 39 000 r.p.m. for 22 h at 8 °C in a Beckman SW40 rotor.
13. Remove the tubes, puncture the bottom with an 18 gauge needle, and slowly collect 1 ml fractions. The gradient elution may be carried out at room temperature. Assay for EF-P activity, pool active fractions and dialyse the resulting EF-$P_{sucrose}$ against two changes of 0.5 litre each of buffer B.
14. Concentrate the EF-$P_{sucrose}$ fraction by ultrafiltration (e.g., using an Amicon PM10 filter) to 5 ml and load the sample on a BioGel P60 column (2.5 × 85 cm) equilibrated with buffer B. Elute the column at 10 ml/h and collect 5 ml fractions (e.g. 75 fractions with 350–400 ml elution volume).
15. Assay 5 μl aliquots for the presence of EF-P, pool the activity, and

concentrate the resulting EF-P_{P60} fractions by ultrafiltration as in step 14 above to about 10 ml.

16. Separate the EF-P_{P60} fraction into two equal batches and apply each to a hydroxylapatite column (1.5 × 2.5 cm) equilibrated with buffer B. Elute the column with 5-ml aliquots of phosphate buffer step gradient [20–120 mM phosphate (pH 7.4), step increments of 20 mM]. Collect 1 ml fractions, monitor the absorbance at 280 nm, and assay 2 μl aliquots for EF-P activity.
17. Visualize the individual fractions of the EF-P activity peak by SDS-PAGE analysis and pool appropriate fractions to yield an electrophoretically homogeneous EF-P_{HAP} fraction.
18. Concentrate the EF-P protein by ultrafiltration to 250 μg/ml, dialyse against 10 mM Tris–HCl (pH 7.4), and store aliquots at −20°C.

3. Comments

3.1 The role of puromycin in studying peptidyltransferase

The antibiotic puromycin closely resembles an amino acid attached to the terminal adenosine of a tRNA. However, puromycin presents the ribosome with an amide linkage which is not susceptible to hydrolysis, in contrast to the ester linkage of an aminoacyl–tRNA. Thus chain elongation is terminated prematurely and the nascent polypeptide is released in the form of polypeptidyl–puromycin. In the classical model assay for peptidyltransferase activity, puromycin is allowed to react with radiolabelled fmet–tRNA to form fmet–puromycin, which can then be extracted and quantitated by liquid scintillation counting. Other assays for peptidyltransferase use fragments of tRNA such as CACCA–phe or CCA–phe. These substrates are bound to the 50S subunit in the presence of methanol (1–3, 9, 27).

3.2 Purification of EF-P

We take advantage of the above model assay to purify the translation factor EF-P based on its ability to stimulate f[^{35}S]Met–puromycin synthesis from ribosome·AUG·f[^{35}S]Met–tRNA complexes. *Figure 1* shows an SDS-PAGE analysis of the EF-P fractions obtained at various stages during the purification. A 100 000 × g or S100 ribosome-free supernatant was chosen in this case as a starting point since 90% of the EF-P activity is found in this fraction. Only about 10% of the activity is associated with the ribosome fraction. Thus EF-P is a soluble factor and typical parameters associated with its large-scale purification are given in *Table 1*. Electrophoretically homogenous EF-P exhibits an apparent molecular weight of 27 000 and previous physical studies using gel filtration and sedimentation analyses have shown that EF-P behaves as an asymmetric

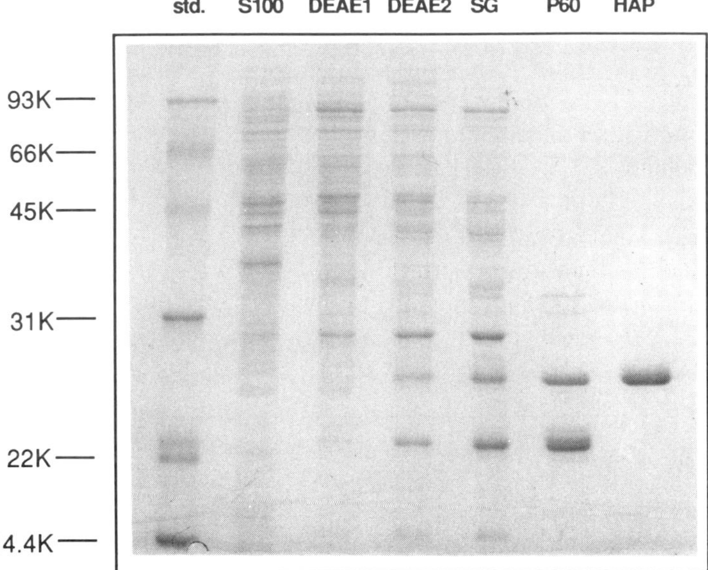

Figure 1. SDS-PAGE analysis of the various fractions containing EF-P activity. Lanes 2 to 7 represent EF-P fractions at the S100, DEAE-1, DEAE-2, sucrose gradient, BioGel P60, and hydroxylapaptite stages of purification, respectively (see *Protocol 5* and Section 3.2 for details concerning the purification). The molecular weight markers in lane 1 (from top to bottom) were phosphorylase B (92.5 kD), bovine serum albumin (66.2 kD), ovalbumin (45.0 kD), carbonic anhydrase (31.0 kD), soybean trypsin inhibitor (21.5 kD), and lysozyme (14.4 kD).

Table 1. Purification of protein EF-P. This table shows the results of a purification of EF-P from 700 g of *E. coli* K12 cells (wet weight). Each activity unit refers to 10 000 d.p.m. of f [^{35}S]methionyl–puromycin formed in the EF-P dependent peptidyl-transferase assay (see *Protocols 4* and *5* for details). N.B. S100 or S65 post-ribosomal supernatants may be used as starting material.

EF-P fraction	Protein conc. (mg/ml)	Total protein (mg)	Total activity units $\times 10^{-3}$	Specific activity units/mg
S100	65.0	15 600	1 140	73
DEAE-1	50.0	1 750	963	550
DEAE-2	23.0	356	504	1 420
Sucrose gradient	6.5	49	263	5 370
BioGel P60	1.0	7.0	261	37 300
Hydroxylapatite	0.53	1.6	114	71 300

molecule (26). The final preparations, aliquoted and stored at -20 °C, are stable for several months.

3.3 Effect of L16 and EF-P on f[^{35}S]Met–puromycin formation

Although peptide-bond synthesis as monitored by the puromycin reaction can occur in the absence of soluble factors or GTP (16, 27), 70S ribosomes are unable to stimulate peptide-bond synthesis at low concentrations of puromycin (e.g. <1.0 μM). EF-P is required for the peptidyltransferase to function optimally on 70S ribosomes. *Figure 2* shows a time course for the EF-P stimulated formation of f[^{35}S]Met–puromycin by 70S ribosomes at such low concentrations of acceptor. By 10 min approximately 80% of the bound f[^{35}S]Met–tRNA has reacted with puromycin. In contrast, less than 20% of the available substrate is utilized in the absence of EF-P.

To learn which proteins of the peptidyltransferase 50S domain affect the 70S coupled EF-P mediated reaction, ribosomes were reconstituted from L16-depleted core particles and 30S subunits. These show virtually no peptidyltransferase activity (*Figure 3*, panel A). With the incorporation of exogenous ribosomal protein L16, both the peptide-bond formation and transesterification reactions increase significantly. 30S subunits, 50S subunits, or EF-P alone show no detectable peptide-bond formation activity. In addition, EF-P is unable to substitute for protein L16. This underscores the importance and central role of ribosomal protein L16 in the peptidyltransferase centre. However, *Figure 3* (panel A) also shows that the activity of 70S ribosomes, at puromycin concentrations below 4 μM, is greatly increased by the addition of EF-P. Under appropriate conditions, a 10-fold enhancement in peptidyltransferase activity can be achieved by the presence of this protein (10). Thus the intrinsic activity of

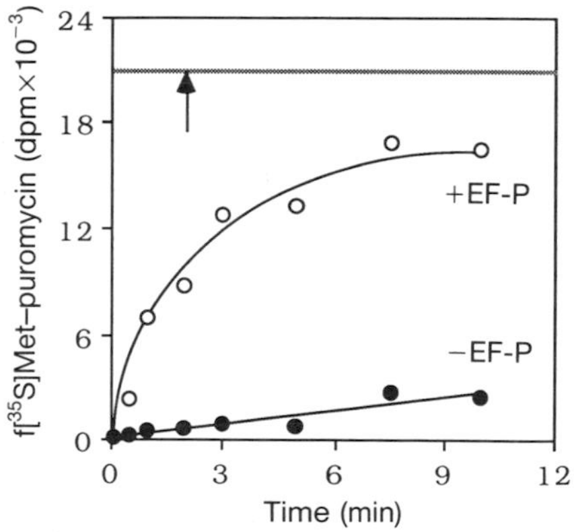

Figure 2. Effect of EF-P on f[^{35}S]Met–puromycin formation by 70S ribosomes. The assay was performed as described in *Protocol 4* except that the time of the incubation was varied. The amount of f[^{35}S]Met-tRNA bound to ribosomes available for reaction with puromycin is indicated by the arrow pointing to the horizontal line.

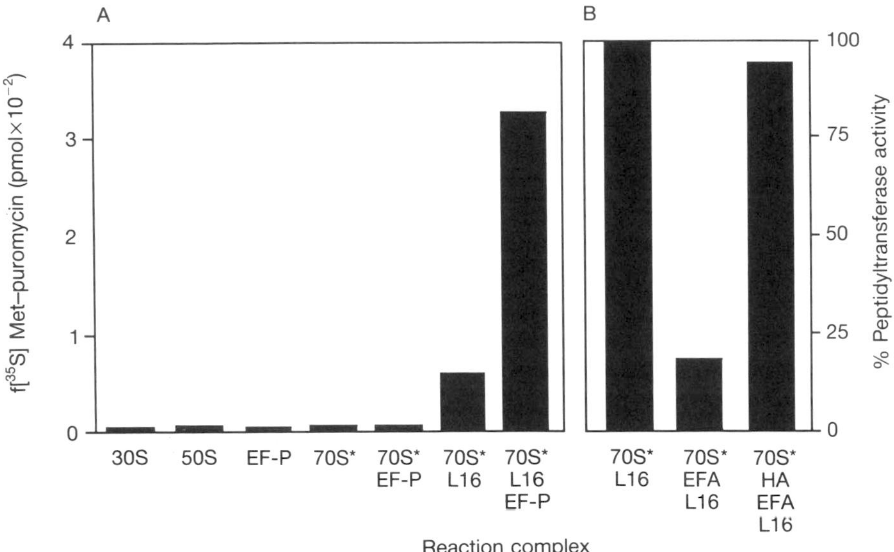

Figure 3. Stimulation of L16-dependent 70S peptidyltransferase activity by EF-P and chemical modification of L16. A: Core particles derived from 50S ribosomal subunits were reconstituted with ribosomal protein L16 or protein EF-P, reassociated with 30S subunits, and further complexed with AUG and f [^{35}S]Met–tRNA. The resulting 70S* complexes were monitored for peptide-bond formation using puromycin at 4 μM. 30S subunits, 50S subunits, and EF-P alone were also monitored for peptide-bond formation as controls. B: Core ribosomal particles were reconstituted with L16, ethoxyformylated (EFA) L16, and hydroxylamine (HA) treated EFA-L16. The resulting 50S subunits were reassociated with 30S subunits, and monitored for peptide-bond formation as in panel A. The results are expressed as a percentage of the activity observed with 70S ribosomes reconstituted with native L16.

70S ribosomes stripped of the soluble factor EF-P is quite low under these conditions.

3.4 Modification of the peptidyltransferase centre components

Chemical modification of proteins with ethoxyformic anhydride is relatively specific for histidine residues and can be reversed by the action of hydroxylamine (10, 12). This technique can be used to complement the reconstitution approach in demonstrating the importance of certain chemical groups in peptide-bond synthesis. *Figure 3* (panel B) shows that the activity of 70S complexes reconstituted with ethoxyformylated protein L16 is reduced to about 20% of that observed with native L16. Hydroxylamine treatment of ethoxyformylated L16 prior to assembly into 70S ribosomes regenerates the histidine residue and results in a concomitant recovery of peptidyltransferase activity.

In addition to modification of specific residues, a variety of other approaches exist for the study of the peptidyltransferase complex. Examples of such techniques which have also demonstrated the importance of L16 are chloramphenicol binding, affinity labelling with modified tRNAs, and antibody inhibition studies [9].

3.5 Possible mechanism of action of EF-P

It is clear that the established number of minimum proteins including L16 within a 70S ribosome complex are not sufficient for optimal catalysis of peptide-bond synthesis. In the absence of methanol or ethanol and at low puromycin concentrations, the soluble factor EF-P is strongly required for this reaction. EF-P enhances peptide-bond synthesis by both 70S ribosomes and reassociated 30S and 50S subunits containing the bound f Met-tRNA substrate. This makes it unlikely that this protein is involved in the direct coupling of ribosomal subunits. Previous evidence has shown that EF-P is not directly involved in the transport and binding of f[^{35}S]Met–tRNA or puromycin to the ribosome (28). Instead, EF-P alters the affinity of the ribosome for puromycin and increases the reactivity of this tRNA analogue as an acceptor for peptidyltransferase (29). In addition, 70S ribosomes are unable to condense certain model substrates, such as CA–gly, and are less efficient in synthesizing f Met–leu, for example, when f Met–tRNA and similar CA– amino acids are used as substrates. Interestingly, however, EF-P promotes the synthesis of dipeptides f Met–gly and f Met–leu, but not f Met–phe (29). Thus EF-P could be involved in fine-tuning the overall conformation of the peptidyltransferase centre, or in a more subtle realignment of the bound substrates, and the physical studies showing that EF-P behaves as an asymmetric protein long enough to span both ribosomal subunits (26) are consistent with such a notion.

4. Prospects

The study of the mechanism of peptide-bond formation, which is of fundamental importance in biochemistry, has been greatly facilitated by development of the classical model reaction using puromycin as a tRNA analogue to assay for peptidyltransferase activity. Exceptional progress has been made using biochemical approaches in general, and reconstitution experiments in particular, to define the minimum number of components necessary for this activity. Regardless of whether ribosomes possess catalytic residues or function merely as macromolecular templates, a serious limitation in the study of its function has been the inability to demonstrate significant activity of the 70S peptidyltransferase complex under physiologically more relevant conditions. This limitation can be overcome to a large extent by the addition of exogenous soluble factor EF-P. Although the molecular events underlying EF-P-mediated peptide-bond synthesis are still unclear, the availability of large amounts of this purified protein will allow the exploitation of EF-P as a probe to examine the mechanism of action of peptidyltransferase and its controlling elements.

Acknowledgements

The authors wish to thank K. Fallavollita and N. Barraclough for expert technical assistance. This work was supported by a grant and fellowship from the Medical Research Council of Canada to MCG and DGC, respectively.

References

1. Nierhaus, K. H., Schulze, H., and Cooperman, B. S. (1980). *Biochem. Int.*, **1**, 185.
2. Rychlik, I. and Cerna, J. (1980). *Biochem. Int.*, **1**, 193.
3. Hampl, H., Schulze, H., and Nierhaus, K. H. (1981). *J. Biol. Chem.*, **256**, 2284.
4. Dabbs, E. R. (1979). *J. Bacteriol.*, **140**, 734.
5. Lotti, H., Dabbs, E. R., Hasenbank, R., Stoffler-Meilicke, H., and Stoffler, G. (1983). *Mol. Gen. Genet.*, **192**, 295.
6. Baxter, R. M. and Zahid, N. D. (1986). *Eur. J. Biochem.*, **155**, 273.
7. Remme, J., Metspalu, E., Maimets, T., and Villems, R. (1985). *FEBS Lett.*, **190**, 275.
8. Schulze, H. and Nierhaus, K. H. (1982). *EMBO J.*, **5**, 609.
9. Wittmann, H. G. (1983). *Ann. Rev. Biochem.*, **52**, 35.
10. Ganoza, M. C., Zahid, N. D., and Baxter, R. M. (1985). *Eur. J. Biochem.*, **146**, 287.
11. Tate, W. P., Sumpter, V. G., Trotman, C. N. A., Herold, M., and Nierhaus, K. H. (1987). *Eur. J. Biochem.*, **165**, 403.
12. Baxter, R. M. and Zahid, N. D. (1978). *Eur. J. Biochem.*, **91**, 49.
13. Garrett, R. A. and Woolley, P. (1982). *Trends Biochem. Sci.*, **7**, 385.
14. Vester, B. and Garrett, R. A. (1988). *EMBO J.*, **7**, 3577.
15. Beauclerk, A. A. D. and Cundliffe, E. (1988). *EMBO J.*, **7**, 3589.
16. Monro, R. E. and Marcker, K. E. (1967). *J. Mol. Biol.* **25**, 347.
17. Traub, P. and Nomura, M. (1968). *J. Mol. Biol.*, **34**, 575.
18. Wystup, G., Teraoka, H., Schulze, H., Hampl, H., and Nierhaus, K. H. (1979). *Eur. J. Biochem.*, **100**, 101.
19. Howard, G. A. and Traut, R. R. (1974). In *Methods in Enzymology*, Vol. 30 (ed. K. Moldave and L. Grossman), p. 526. Academic Press, NY.
20. Baxter, R. M., Ganoza, M. C., Zahid, N. D., and Chung, D. G. (1987). *Eur. J. Biochem.*, **163**, 473.
21. Ganoza, M. C., Barraclough, N., and Wong, J. T. (1976). *Eur. J. Biochem.*, **65**, 613.
22. Fahnestock, S., Neumann, H., Shashoua, V., and Rich, A. (1970). *Biochemistry*, **9**, 2477.
23. Lowry, O. H., Roseborough, N. J., Farr, A. L., and Randall, R. J. (1951). *J. Biol. Chem.*, **193**, 265.
24. Bradford, M. M. (1976). *Anal. Biochem.*, **72**, 248.
25. Laemmli, U.K. (1970). *Nature*, **227**, 680.
26. Glick, B. R., Green, R. M., and Ganoza, M. C. (1979). *Can. J. Biochem.*, **57**, 749.
27. Monro, R. E., Staehelin, T., Celma, M. L., and Vasquez, D. (1969). *Cold Spring Harbor Symp. Quant. Biol.*, **34**, 357.
28. Glick, B. R. and Ganoza, M. C. (1975). *Proc. Nat. Acad. Sci. USA*, **72**, 4257.
29. Glick, B. R., Chladek, S., and Ganoza, M. C. (1979). *Eur. J. Biochem.*, **97**, 23.

5

Termination of protein synthesis

WARREN P. TATE and C. THOMAS CASKEY

1. Introduction

Peptide chain termination involves the recognition of signals in the mRNA specifying that the programme for the protein being synthesized on the ribosome has come to an end. These signals are one of the codons UAA, UAG, or UGA in most genetic systems, although mitochondria and unicellular eukaryotes like tetrahymena have different genetic codes. The recognition of the stop codons is mediated through protein factors, called release factors, unlike the recognition of codons in the other phases of protein synthesis where tRNAs act as the code adapters. Following codon recognition, involving the interaction of the factors with the ribosome, the ester bond between the completed polypeptide and the last tRNA is hydrolysed in a reaction at the peptidyltransferase centre of the ribosome, the same centre which mediates peptide bond formation. This centre shifts from a bond-forming mode, where water is excluded, to a bond-breaking mode, where water is acting as an attacking agent. A change in the conformation or microenvironment after the release factor has bound to the ribosome apparently allows water access to the site. Whether the factors themselves participate in the hydrolytic reaction, and thereby the release of the completed polypeptide, remains an open question.

In this chapter we have outlined how release factors from different organisms can be assayed and purified, and the approaches which have yielded information on how they function. The utilization of the cloned genes to elucidate the structure and function of the proteins is also described.

2. Release factors—how to detect them

The release factors mediating the termination event are proteins and therefore several approaches can be taken to detect their presence. The availability of a functional assay where the release factors are acting catalytically allows for the detection of relatively small amounts. The development of sophisticated immunological methods in recent times has meant that the factors can be monitored independently of the more complex functional assay once specific antibodies against the proteins have been isolated. Genetic assays, which will detect specifically the release factors expressed from the cloned genes, are also potentially very useful.

2.1 Functional assay

2.1.1 Overall termination event

The development of an assay by Caskey *et al.* (1) which can be used routinely to analyse large numbers of samples has enabled the termination event to be studied in detail. This assay measures the release of a model peptide (representing the completed polypeptide) from a ribosomal complex. The complex simulates the state of the ribosome when the stop signal in the mRNA enters the ribosomal A site, instead of another sense codon for an additional amino acid for the growing polypeptide chain. In place of a complete mRNA only the initiation codon, AUG, and the termination codons UAA, or UAG, or UGA, as individual triplets are necessary in this assay. This overcomes a difficulty of trying to measure the termination event following the initiation and elongation of protein synthesis. It is a very complicated task to construct a system from natural components to achieve this, and it is not appropriate for a routine assay. No effective inhibitor is available to arrest protein synthesis when the stop codon of a natural mRNA enters the ribosomal A site.

The substrate for the assay aims at simulating the state of the ribosome as it would be when the termination of synthesis of the protein is occurring *in vivo*. It contains the bacterial 70S or eukaryotic 80S ribosome with f[^{3}H or ^{35}S]Met–tRNA in the P site together with the codon for methionine, AUG. The radiolabel provides the tracer to follow the reaction and, since the termination event is independent of the length of the polypeptide, the single amino acid can be used as the model for the completed polypeptide. Either of the triplets, UAA, UAG, or UGA, can be used to signal stop in the prokaryotic assay with the bacterial ribosome, but tetramers containing these stop codons are the minimum length oligonucleotides that can be used with the eukaryotic ribosomes. A release factor will recognize the stop codon, and the ester bond between the amino acid and tRNA is cleaved to release the radiolabelled amino acid. This can be separated from the substrate and other components after acidification by extraction into ethyl acetate. The quantitative release of the amino acid is then measured from the radioactivity in the ethyl acetate phase.

The components for this assay are prepared as follows:

Protocol 1. Ribosomes

1. Take a colony or a streak of *E. coli* MRE600 (Public Health Laboratory Service, CAMR, Porton Down, UK) and seed it into 50 ml of sterile media, for example, Luria broth.
2. Grow at 37 °C overnight with shaking.

3. Seed 5 ml of this culture into 500 ml of the sterile media in a 1 litre flask (four of these per preparation is a convenient amount).
4. Grow the cultures until the late log phase of growth, monitoring the cultures at 650 nm measuring turbidity. Usually this takes 4 h.
5. Pellet the bacteria at $10\,000 \times g$ for 20 min at 4 °C and resuspend in a minimum volume of 100 mM Tris–HCl (pH 7.8), 100 mM $MgCl_2$, 300 mM NH_4Cl, 60 μM 2-mercaptoethanol, 40 μg/ml DNase 1.
6. Repellet at $10\,000 \times g$ for 20 min at 4 °C. Typical yields are 2–4 g.
7. Grind the bacteria in twice the amount of alumina (e.g., 1 g of bacteria, 2 g of alumina) for about 15 min and add the buffer described in step 5–2 ml per 1 g of bacteria ground.
8. Remove the alumina and cell debris by centrifuging the sample at $15\,000 \times g$ for 20 min at 4 °C.
9. Pellet the ribosomes at $100\,000 \times g$ in a Beckman ultracentrifuge for 3 h using a type Ti60 rotor at 4 °C and resuspend in 1–2 ml of 20 mM Tris–HCl (pH 7.8), 60 mM NH_4Cl, 10 mM $MgCl_2$, 10 mM 2-mercaptoethanol by gently stirring overnight at 4 °C.
10. Remove material not in solution by centrifuging at $12\,000 \times g$ for 20 min and pellet the ribosomes from the supernatant by centrifuging through a 1.1 M sucrose cushion made up in 50 mM Tris–HCl (pH 7.5), 10 mM Mg acetate, 500 mM NH_4Cl.
11. Resuspend the pellet in 0.5–1.0 ml of the buffer as in step 9. The yield should be in the order of 10 nm (400–500 A_{260} units).

To prepare eukaryotic ribosomes a crude extract is made by breaking the tissue or cells and after typical low-speed centrifugation steps the ribosomes are pelleted as in step 9 and the subsequent steps carried out as for the prokaryotic ribosomes. See also Chapter 1.

Protocol 2. f[^{3}H]Met–tRNA

1. In a 2 ml reaction add 100 mM sodium cacodylate (pH 6.7) (280 μl), 400 μM of each of the aminoacids (without methionine) (100 μl), [^{3}H]-methionine, 15 Ci/mmol (Amersham) (320 μl), 20 mM ATP (120 μl), 0.3 mg $tRNA_f^{met}$ (Sigma), leucovorin (formyl donor) (pH 6.8), (560 μl—see below), aminoacyl-tRNA ligase/formylating crude enzyme mixture (2, 3) (volume to be determined by titration—see below), 100 mM dithiothreitol (20 μl), 200 mM $MgCl_2$ (100 μl), and deionized H_2O to make up the volume.
2. Incubate at 37 °C for 30 min and remove three 5 μl samples for analysis of the efficiency of formylation, the efficiency of aminoacylation and the extent of

non-specific hydrolysis during the preparation and subsequent purification (see below).

3. Add 1/20th volume of 2M Na acetate (pH 5.4) to the remaining volume and then 2 ml of water-saturated phenol and extract with continual mixing for 15 min.
4. Separate the phases by centrifuging at 5000 × *g* for 15 min at 4 °C and remove the aqueous phase.
5. Re-extract the phenol phase with 2 ml of 200 mM Na acetate (pH 5.4) and combine this aqueous phase with that from the first extraction.
6. Precipitate the tRNA from the combined aqueous phases overnight at −20 °C with two volumes of ethanol.
7. Pellet the f[^{3}H]Met–tRNA by centrifuging at 5000 × *g* for 10 min, wash the pellet with 5 ml of ethanol at −20 °C several times until the absorbance of the washings from residual leucovorin at 260 nm is negligible.
8. Briefly lyophilize the pellet to remove traces of ethanol and take up in 1–1.5 ml deionized water.

Protocol 3. Formyl donor (leucovorin)

1. Dissolve 50 mg folinic acid (Sigma) in 4 ml 50 mM 2-mercaptoethanol and add 0.44 ml of 1 M HCl with slight warming.
2. Leave at room temperature for 3 h and measure the absorbance at 355 nm ($25A_{355}$ units = 1 mM).
3. Adjust the concentration to 15 mM with 0.1 M HCl. Store aliquots at −20 °C.
4. Before use in f[^{3}H]Met–tRNA preparation neutralize with KOH, 10 M initially until the pH is 5.0, then 1 M in 1 μl aliquots until the pH is about 7.0.

Protocol 4. Oligonucleotides (UAA, UGA, UAG, AUG)

UAA or UGA

1. Preincubate 400 μl polynucleotide phosphorylase (5 mg/ml) (Boehringer-Mannheim) with 400 μl trypsin (20 μg/ml) (Sigma) and 240 μl 5 M NaCl in a volume of 2 ml.
2. After 30 min at 30 °C add 120 μl of trypsin inhibitor (2 mg/ml) (Sigma) and then 360 μl 1 M Tris-HCl (pH 9.5), 24 μl 1 M $MgCl_2$, 24 μl 40 mM EDTA, 56 μl 5 M NaCl, 20 mg UA or UG in 200 μl, 320 μl 100 mM ADP or GDP and incubate for a further 72 h at 37 °C.

3. Heat the samples for 10 min at 90 °C, then add 2 μl alkaline phosphatase (28 U/μl) (Boehringer-Mannheim) and complete a further 4 h incubation at 37 °C.
4. Dilute the reaction to 500 ml with deionized water to reduce the salt concentration and apply to a DE-32 column (2 × 12 cm). Elute the homologous series of oligonucleotides (UA, UAA, UAAA, etc.) with a gradient of 0–500 mM triethyl ammonium bicarbonate using 1 litre.
5. Remove the volatile salt from the appropriate fractions by several cycles of rotary evaporation and take up the residue in sterile MilliQ water (MilliQ is the process for water purification developed by the Millipore corp.).

UAG, AUG

1. Use polynucleotide phosphorylase (267 μl) without trypsin pretreatment together with 300 μl 1M Tris–HCl (pH 9.4), 20 μl 1M Mg acetate, 20 mg of UA or AU in 200 μl, 320 μl 100 mM GDP, 16.5 μl RNase T1 (200 000 U/ml–Sigma) and MilliQ water to 2 ml.
2. Proceed as for UAA and UGA except that the DE-32 column is eluted with 500 ml of 0–250 mM triethyl ammonium bicarbonate since the RNase T1 treatment prevents the formation of members of the series greater than UAG/AUG in length.

The substrate: f[^{3}H]Met–tRNA–AUG–ribosome complexes

A release factor can catalyse the release of a product only from the ribosomal P site, and therefore the substrate must have the f[^{3}H]Met–tRNA to be hydrolysed in this site rather than the A site. This is achieved by binding the aminoacylated tRNA at relatively high magnesium ion concentration in an initiation–factor independent reaction. The quality of the substrate can be checked as described in steps 2–4 below. A good substrate has a high ratio (>20) of the bound aminoacylated-tRNA in the P site to free fMet. Fortunately the site specificity can be determined using the aminoacyl-tRNA analogue, puromycin, which will form a peptide bond if the fMet is in the P site.

Protocol 5. f[^{3}H]Met–tRNA–AUG–ribosome complex formation

1. Incubate 50 pmol of ribosomes, 100 000 c.p.m. f[^{3}H]Met–tRNA (about 25 pmol), 2.5 nmol AUG in binding buffer, 20 mM Tris–HCl (pH 7.4), 10 mM $MgCl_2$, 150 mM NH_4Cl within a volume of 50 μl at 30 °C for 30 min.
2. Check the substrate for free f[^{3}H]Met, produced by non-specific hydrolysis during the preparation. Add 250 μl 0.1 M HCl to a 5 μl sample and then 1 ml of ethyl acetate, vortex briefly, and determine the radioactivity in the ethyl acetate layer (4).

3. Check the substrate for ribosome-bound f[^{3}H]Met–tRNA by plating a sample (5 μl) in 1 ml of the above binding buffer on a glass fibre filter (GF/C), and determining the radioactivity bound to the filter. The ratio of radioactive aminoacyl-tRNA bound to free formylated amino acid (step 2) should exceed 20.
4. Check the substrate for the site specificity of the ribosome bound f[^{3}H]Met–tRNA by incubating a sample (5 μl) with 3 mM puromycin dihydrochloride in 50 μl of the binding buffer, incubating for 30 min at 30 °C, adding 250 μl of 100 mM KH_2PO_4 (pH 6), and then extracting the f[^{3}H]Met–puromycin product into ethyl acetate.

Assay for release factor

The release factor will recognize a termination codon of appropriate specificity and catalyse the release of f[^{3}H]Met from the ribosomal substrate to be quantitated by its extraction into ethyl acetate. A typical response to a sample containing release factor is shown in *Figure 1A*. It is important to determine that any response is codon dependent (using codon-free controls) since there are many activities in crude extracts which can non specifically hydrolyse the substrate.

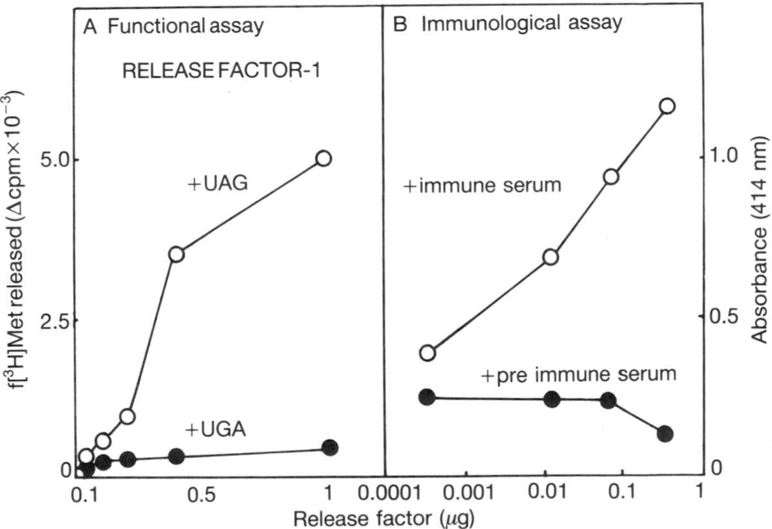

Figure 1. Assays for bacterial release factors. A: The functional assay is demonstrated with RF-1 where there is release of f [^{3}H] Met from the substrate in response to added factor and to the specific codon UAG, but not UGA. B: The ELISA immunological assay demonstrates the binding of antiRF to increasing amounts of the release factor, detected as described in Section 2.2.3. The pre-immune serum shows no reactivity.

Protocol 6. Release factor assay

1. Add the sample containing release factor within a 50 μl reaction containing 2 pmol of the f[^{3}H]Met–tRNA–AUG–ribosome substrate, 4 nmol of termination codon in a buffer of 50 mM Tris–HCl (pH 7.5), 75 mM NH_4Cl, 30 mM $MgCl_2$, and incubate at 20 °C for up to 30 min.
2. Stop the reaction by adding 250 μl of 0.1 M HCl, add 1 ml of ethyl acetate, vortex for 15 sec, microfuge for 30 sec, and remove 750 μl of the top organic phase for determination of radioactive f[^{3}H]Met released from the substrate.

2.1.2 Partial steps of the termination event

Although the exact mechanism of how the release factor recognizes the termination codon and mediates the release of the polypeptide is still to be elucidated, two useful partial reactions have been developed (6, 7). The first measures the codon-specific binding of the release factor to the ribosome, and the second circumvents the codon recognition requirement by using ethanol to promote both factor interaction with the ribosome's functional centre and peptidyl–tRNA hydrolysis. This provides a way of differentiating between the termination codon recognition event and the release of the completed polypeptide. An additional step *in vivo* may involve a stimulatory factor, RF-3 in bacteria (7), which *in vitro* influences the interaction of the two factors recognising the termination codons, that is RF-1 and RF-2.

In the eukaryotic assay f[^{3}H]Met–tRNA–AUG–ribosome (80S) substrate is used, 5 nmol of tetranucleotide, for example UAAA, in place of the triplet, and 1 mM GTP is added. The buffer is 20 mM Tris–HCl (pH 7.4), 60 mM KCl and 11 mM $MgCl_2$.

Codon recognition

This is a stoichiometric assay and it requires a higher amount of release factor than the assay measuring the overall event where the factor is acting catalytically. Moreover, the stability of the complex is low, particularly with the bacterial factor RF-1, although ethanol is generally included to stabilize the complex. It is possible to measure binding, however, in the absence of ethanol.

Protocol 7. Codon recognition–stoichiometric assay

1. Add the sample containing release factor within a 50 μl reaction containing 50 pmol ribosomes, 400 pmol UA[^{3}H]A of known specific activity, a buffer of 20 mM Tris–HCl (pH 7.5), 100 mM NH_4Cl, 20 mM $MgCl_2$, and then add ethanol to give a final concentration of 10% (v/v). Incubate for 15 min at 4 °C. If a eukaryotic factor is to be assayed the codon is replaced by UA[^{3}H]AA

and 0.1 mM GMP–PCP is added and the ethanol increased to 20% final concentration.

2. Add 1 ml of the buffer containing 10% ethanol to the reaction and collect the radioactive complex on a GF/C glass fibre filter, washing 2–3 times with the buffer (e.g., 3–5 ml).
3. Dry the filters and then count to determine the amount of complex formed.

The radioactive codon (100 c.p.m./pmol) used in this assay is prepared as for the oligonucleotides described above, but in a reaction scaled down 20-fold. The products are separated on Whatman 3MM paper developed in 1 M ammonium acetate:ethanol (1 : 1) for 16 h instead of on the column. The appropriate spot, when cut out of the chromatogram, is rinsed in 3 × 25 ml of ethanol at −20 °C. After drying the radioactive oligonucleotide can be eluted in 2 ml MilliQ water.

Peptidyl–tRNA hydrolysis

This assay is identical to that described for the overall event with the exception that the termination codon is replaced by ethanol [10% (v/v)—prokaryotic assay, 20% (v/v)—eukaryotic assay] which is added last to the reaction.

Enhanced release factor activity

The hydrolysis of peptidyl–tRNA is assayed at concentrations of the release factor supporting a very slow catalysis and an enhancement of the reaction is measured in the presence of the stimulatory factor. The factor is unstable to freeze–thawing, a problem which has received little attention. A typical assay contains in 50 μl 50 mM Tris acetate (pH 7.2), 30 mM Mg acetate, 50 mM KCl, 2 pmol of f[^{3}H]Met–tRNA–AUG–ribosome substrate, 2 nmol termination codon, 0–1 μg bacterial RF-1 or RF-2, and the stimulatory factor, RF-3.

2.2 Immunological assay

The release factors can be detected immunologically in assays of differing complexity which yield different information. For example, the simplest, double immunodiffusion (7), indicates the presence of the factor, whereas radial immunodiffusion (8) can provide information on the amount of factor present, while the most sophisticated, an ELISA or competitive ELISA (9), is considerably more sensitive. These assays are particularly suitable for following the factor through a purification and the ELISA is more sensitive than the functional assay.

2.2.1 Double immunodiffusion

A positive signal in this assay is a precipitin line resulting from aggregates of a complex between antigen and antibody precipitating, and some indication is given of the amount of antigen by where the precipitin line forms. If close to the centre well, the antigen concentration is relatively strong; if close to the outer well, it is relatively weak.

Protocol 8. Double immunodiffusion assay

1. Prepare 2% (w/v) agarose (Sigma), mix with an equal volume of 0.05% diethylbarbituric acid (pH 8.2), containing 0.85% (w/v) NaCl, heat to 90 °C with stirring, and use 13 ml aliquots to pour each glass plate.
2. When the gel has set, punch rosettes with a centre well and six surrounding wells.
3. Place the antibody against the release factor in the centre well and place fractions to be analysed for release factor in the outer wells.
4. Develop the plates for 24 h to allow diffusion and precipitin lines to develop, indicating where antigen (release factor) and antibody have reached a critical concentration.

2.2.2 Radial immunodiffusion

The difference in this procedure is that the antibody is placed within the gel, and single wells are punched with a precipitin ring forming where there is a reaction between antigen and antibody. The diameter of the ring can be related to the amount of the factor present. If a sample of pure factor of known concentration is available then this can be used to create a standard curve to determine the exact amount of factor in the unknown samples.

Protocol 9. Radial immunodiffusion assay

1. Cool the dissolved agarose (see above) to 60 °C, add an amount of antibody as appropriate (e.g., 50 μl of serum), and then pour the plate.
2. Punch 2.5 mm holes for the antigen samples, and allow to diffuse for 2 days.
3. Place filters on the gel and press under a heavy object for a few min (this reduces background on staining).
4. Soak the gel in NaCl (11.6 g/l) for 2 days, press again, and then stain and destain as appropriate for protein (e.g. Coomassie).

2.2.3 ELISA

This assay can detect as little as 0.5 ng of release factor which is almost 1000-fold more sensitive than the functional assay. A typical response curve is shown in *Figure 1B*.

Protocol 10. ELISA

1. Place the release factor sample within 100 μl of 0.2 M $NaHCO_3$ (pH 9.5) in each well of a microtitre plate (Falcon) in duplicate or triplicate and incubate overnight at 4 °C.

2. Wash wells three times in 40 mM Tris–HCl (pH 7.4), 150 mM NaCl, 0.05% Tween 20, 0.01% Thiomersal, 1% Non-fat Milk Powder (MT^3N buffer), and then incubate for 1 h in the same buffer before washing a further time.
3. Add to each well 100 μl of a suitable dilution of antibody against release factor (e.g. 1 : 500–1 : 20 000) in MT^3N buffer and incubate for 2 h at 35 °C.
4. Wash three times again with buffer and then incubate for 1 h at room temperature with horse radish peroxidase (HRP) linked to an antibody, (raised against the antibody fraction from the animal species in which the test antibody was raised—e.g., HRP-goat anti rabbit).
5. Wash the well three times with buffer without the milk powder and incubate with a substrate [e.g., 1.0 mM 2, 2-Azido-bis (3-ethylbenzthiazoline-6-sulphonic acid), 0.1% Tween 20, 0.1 M Na acetate, 0.05 M NaH_2PO_4 (pH 4.2)—adding 2.5 mM peroxide just before use] at room temperature in the dark until a suitable colour develops.
6. Read the absorbance at 414 nm in an ELISA plate reader (LKB).

2.3 Genetic assay

Suppressor tRNAs which can compete with the release factors for reading stop codons provide the basis for a genetic assay to detect the release factors. The placing of a nonsense mutation within a gene means that the product is not made because the mutant codon is read as stop unlcss a suppressor tRNA is present. An abundance of release factor, however, can compete out the suppressor and then no gene product is formed. This assay has been particularly useful for the cloning of the *E. coli* gene for release factor-1 (10). Here DNA fragments of *E. coli* were cloned into a plasmid which was introduced into a strain carrying both a suppressor tRNA and having an amber stop codon within the β-galactosidase gene. The production of β-galactosidase can be monitored with a colour reaction on a bacterial plate and so one looks for colonies of bacteria which do not develop colour as potential overproducers of release factor. Although this has some general application the specific example for RF-1 is given in *Protocol 11*.

Protocol 11. Genetic assay to detect RF-1

1. Transform competent cells of *E. coli* Cp79 118 UAG SupE with plasmids containing a cloned fragment of DNA, suspected of coding for RF-1, and plate on McConkeys media.
2. Select transformants from an antibiotic resistance marker on the plasmid and examine the colour of the colonies, red colonies expressing β-galactosidase and white colonies not.
3. Check white colonies for release factor-1 using an immunological or activity assay on crude extracts of 25 ml cultures of the bacteria.

---

3. Purification of release factors

3.1 Prokaryotic release factors

Since there are two bacterial factors which recognize termination codons, RF-1 (UAA, UAG) and RF-2 (UAA, UGA), and a third stimulatory factor (RF-3) which amplifies their activities, the aim of a purification scheme is firstly to separate the three species and, secondly, where appropriate, isolate each as a pure protein. Traditionally the factors have been purified on soft-gel systems, but while successfully employed there is always some batch-wise variation with the gel materials, and the procedures are quite long so that several days are required to complete a purification. The advent of FPLC (Fast Performance Liquid Chromatography) with solid supports for column material has allowed rapid purification, sharper peaks, and very consistent results in the elution profiles for proteins. This system has recently been applied to the purification of release factors with excellent results. It is now possible to predict which fraction the factor will be in, thus reducing the time and materials required for assay.

3.1.1 Soft gel chromatography

The procedure is as essentially described in Caskey *et al.* (1). There is a critical anion exchange step which allows all three factors to be separated. The factor RF-2 which elutes last of the three can be almost pure at this stage, whereas RF-1 and RF-3 are still quite heavily contaminated with other proteins. RF-1 is retained on a cation exchange column whereas almost all of the contaminants are

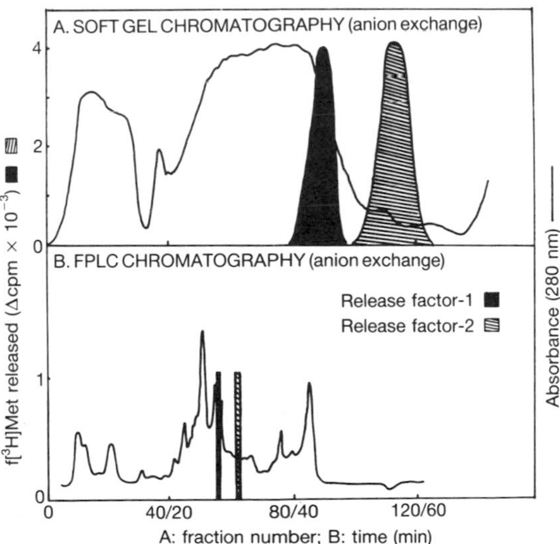

Figure 2. Anion exchange chromatography separation of RF-1 and RF-2. A: Fractionation of RF-1 and RF-2 on DEAE Sephadex A50. B: Fractionation of RF-1 and RF-2 on Protein Pak DEAE-5PW with a 60 min elution. Protein profile (line) the RF-1 activity (solid peak) and RF-2 activity (hatched peak).

excluded. The purification scheme is outlined in *Protocol 12* and a profile of the anion exchange column shown in *Figure 2A*.

Protocol 12. Purification of the prokaryotic release factors by soft gel chromatography

1. Grind 50–500 g *E. coli* cells in 30 g portions with 60 g alumina for 20 min and add 60 ml of 10 mM Tris–HCl (pH 7.8), 10 mM $MgCl_2$, 30 mM NH_4Cl, 0.6 mM 2-mercaptoethanol, 40 μg/litre DNAse 1.
2. Centrifuge at 10 000 × *g* for 20 min to remove debris and then at 100 000 × *g* for 4 h to remove ribosomes.
3. Collect the precipitate from a 0–55% ammonium sulphate fractionation (350 g/litre) of the supernatant by centrifuging at 10 000 × *g* for 1 h.
4. Suspend the pellet in a minimum volume of 20 mM Tris–HCl (pH 8), 200 mM KCl, 2 mM dithioerythritol, and dialyse against several changes of the same buffer.
5. Load the sample on a 3 × 95 cm column of Sephadex G100 and elute with the same buffer collecting 10 ml fractions. The combined release factor activity elutes in a peak towards the end of the main protein peak.
6. Pool the active fractions, concentrate by ultrafiltration, and apply to a DEAE Sephadex A 50 column (2 × 20 cm), washing with the same buffer until the absorbance at 280 nm is low.
7. Elute the column with a gradient of 200–600 mM KCl in 1 litre of the buffer, assaying each fraction for RF-1 (with UAG), for RF-2 (with UGA), and RF-3, as described in the text.
8. Concentrate the appropriate fractions with Aquacide II (Calbiochem) and dialyse into 50 mM imidazole (pH 6.0), 0.1 M KCl, 2 mM dithioerythritol for RF-1, and 10 mM potassium phosphate (pH 7.2), 2 mM dithioerythritol for RF-2. RF-3 is used without further purification.
9. Load RF-1 to a column (1.5 × 16 cm) of CM Sephadex equilibrated in the imidazole buffer and elute with a gradient of 500 ml 0.1–0.8 M KCl, and load RF-2 to a column (1.5 × 10 cm) of hydroxylapatite equilibrated in the phosphate buffer, eluting successively with 10 mM, then 30 mM, and then 80 mM potassium phosphate buffer (pH 7.2).
10. Concentrate the fractions containing the release factor activities with Aquacide II and dialyse against 50 mM Tris–HCl (pH 8.0), 50 mM KCl, 3 mM dithioerythritol. Store aliquots at −80 °C.

3.1.2 Fast performance liquid chromatography

The soft-gel chromatography steps can be carried out with equivalent FPLC column matrices. For a large preparation of factors (starting with greater than

10 g of *E. coli*) it is necessary to do the preliminary fractionation on a gel permeation column using Sephadex G100 as outlined in the traditional scheme. This reduces the protein amount to an acceptable level whereby the FPLC columns are not overloaded. As illustrated in *Figure 2B*, the protein peaks are much sharper on the anion exchange column (Waters) than on the equivalent soft-gel column, and the factors elute in a small volume. The modified scheme is detailed in *Protocol 13*.

Protocol 13. Purification of the prokaryotic release factors by FPLC chromatography

1. Proceed with the first five steps of the purification as in *Protocol 12*.
2. Load up to 500 mg of protein containing the release factors to a Waters Protein Pak DEAE-5PW column (0.8 × 7.5 cm) in 50 mM Tris–HCl (pH 7.5), 50 mM KCl, 3 mM dithioerythritol, wash with 5 ml of the buffer, and then elute with 35 ml of a gradient of 50–500 mM KCl in the same buffer. Assay fractions as step 7 of *Protocol 12*.
3. Dialyse the RF-1 fractions into 50 mM 2-[N morpholino] ethane sulphonic acid (pH 6.0), 50 mM KCl, 3 mM dithioerythritol, 0.1 M EDTA, and load on to a Pharmacia MonoS 10/10HR column (0.1 × 10 cm), wash the column with 5 ml of the buffer and elute with a gradient of 35 ml of 50–500 mM KCl in the same buffer.
4. Dialyse the RF-2 fractions into the buffer as in step 2 and repeat this step.
5. Dialyse the preparations of RF-1 and RF-2 as in step 10, *Protocol 12*.

3.2 Eukaryotic release factor

The eukaryotic equivalent of the bacterial RF-1 and RF-2 is a single factor with quite different biophysical properties. The single polypeptides of about 40 000 M_r typical of prokaryotes are replaced in eukaryotes with a dimer with a subunit relative mass of about 55 000. This structure is found in species ranging from the invertebrate *Artemia* (12) to mammalian species (13). The eukaryotic factor is asymmetric and has an apparent relative mass of 200 000 on gel permeation columns, whereas on sedimentation it has an apparent value of 50 000 (14). This unusual behaviour is very useful in the purification of the protein. A purification scheme is given in *Protocol 14*. A stimulatory factor, the apparent equivalent of the bacterial RF-3, has been isolated, but like RF-3 it is highly unstable and little work has been carried out on its purification and characterization.

Protocol 14. Purification of the eukaryotic release factor

1. Prepare a tissue or cell extract according to the protocol relevant to the source of the starting material and remove cell debris by centrifuging at 30 000 *g* for 20 min.

2. Remove the polysomes and free ribosomes by centrifuging at 100 000 g for 3 h and dissolve a 0–55% ammonium sulphate fraction obtained from the supernatant in DEAE buffer: 20 mM Tris–HCl (pH 7.8), 100 mM KCl, 2 mM dithioerythritol, 0.1 mM EDTA, dialysing extensively against the same buffer.
3. Load on to a DEAE Sephadex A 50 column (2.5 × 50 cm) equilibrated in the buffer (step 2), then elute with 2 litres of a gradient of 100–700 mM KCl in the same buffer, assaying the fractions as described in the text for a eukaryotic factor.
4. Concentrate the fractions containing the release factor by ultrafiltration and dialyse into the DEAE buffer containing 100 mM KCl before loading on to a phosphocellulose column (1.5 × 30 cm) and elute with a gradient of 400 ml of 100–400 mM KCl in the same buffer.
5. Concentrate the active fractions by ultrafiltration and load on to a Sephadex G-200 column (2.5 × 100 cm), then elute with the DEAE buffer containing 50 mM KCl.
6. Concentrate the active fractions and layer on to a sucrose gradient (5–20% w/v) sedimenting at 40 000 r.p.m. for 48 h (e.g. using a Beckman SW 40 Ti rotor). Fractionate the gradient, monitoring the fractions for activity.
7. Store the active fractions at −80 °C directly; the 10–15% sucrose does not adversely affect the stability of the release factor.

3.3 Mitochondrial release factor

A factor has been isolated from rat mitochondria, the only example of an organelle release factor reported to date (15). It has physical characteristics more closely related to the bacterial factors and is quite distinct from the rat cytosolic factor. The factor recognizes UAA and UAG, but not UGA, and therefore it has been designated mtRF-1. A factor of the bacterial RF-2 type was not detected and this is consistent with the changed genetic code in mitochondria for termination. In this mammalian species only UAA has been found as the stop codon for mitochondrial genes whereas UGA is used for tryptophan. Although UAG seems not to be used to code for a stop signal or an amino acid the factor has retained an ability to recognize the codon. The factor has been partially purified as described in *Protocol 15*.

Protocol 15. Purification of the mitochondrial release factor

1. Isolate mitochondria from 100 g of rat liver by standard techniques except combine the 750 × g and 12 000 × g supernatants to maximize yield (see example in Chapter 1).
2. Purify the mitochondria by sedimenting through a 30% sucrose cushion in

15 mM Tris–HCl (pH 7.5), 150 mM KCl at 27 000 × g for 30 min, resuspend the pellet in 5 mM Tris–HCl (pH 7.5), 20 mM KCl, 30 mM $MgCl_2$, 3 mM 2-mercaptoethanol, before lysing with 0.75% Nonidet P-40.

3. Centrifuge the already partially clarified solution at 27 000 × g for 30 min and then at 100 000 × g for 4 h to remove the ribosomes and obtain the post ribosomal supernatant.
4. Dialyse the supernatant into 50 mM Tris–HCl, 50 mM KCl, 10 mM 2-mercaptoethanol (pH 7.5) and obtain the 0–70% ammonium sulphate cut precipitate (which often stays at the top of the container on centrifugation).
5. Dissolve the precipitate in the same buffer and after extensive dialysis load on to a DEAE Sephadex A 50 column (2 × 18 cm) eluting with a stepwise gradient containing increasing concentrations of KCl (50–400 mM) in the buffer of step 4. Assay the fractions as for the bacterial factors but include both codon UAA and 10% ethanol in the assay.
6. Concentrate the mtRF-1 fractions by ultrafiltration (Centricon) and load on to a Sephadex G 100 column (1 × 100 cm) eluting with the same buffer but containing 50 mM KCl.
7. Purify the concentrated mtRF-1 further on a 5–20% (w/v) sucrose gradient made up in the buffer of step 4 centrifuging at 36 000 r.p.m. for 48 h in an SW 40 Ti rotor.

---

4. Ribosomal domains for release factors

4.1 Mutants

A number of *E. coli* mutants have yielded ribosomes with altered activity in release-factor function. Often the mutants have arisen as a result of selection with antibiotics, and an examination of the various partial steps of protein synthesis have revealed in which phase of the event the altered ribosomal component is important. Examples of this type are the mutants, isolated by Dabbs (16), whose ribosomes lack the large ribosomal subunit protein L11 or the relC mutant which carries an alteration in L11 (17).

4.1.1 L11-lacking ribosomes

These ribosomes have been valuable in defining the 50S ribosomal binding domain for the release factors (18). Release factor RF-2 is more active with the mutant ribosomes than on wild-type ribosomes, whereas RF-1 has very low activity at normal concentrations (*Figure 3A*). The affinity of RF-2 for the mutant ribosome is increased and the RF-2–U[^{3}H]AA–ribosome complex can be isolated in the absence of ethanol (see Section 2.1.2). Moreover, since ribosomal protein L11 can be reconstituted back into the altered ribosomes, specific effects of the defect can be monitored closely (*Figure 3B*).

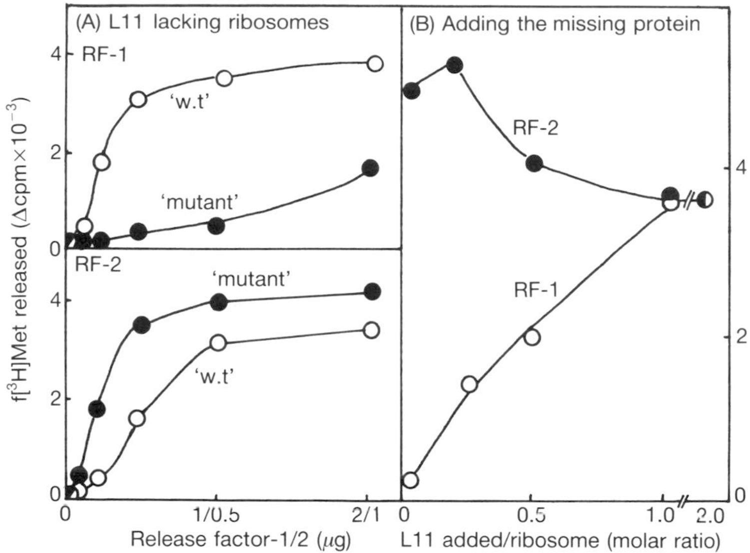

Figure 3. L11-lacking ribosomes as a tool to probe the release factor binding site. (A [upper]) The activity of RF-1 on wild-type (open circles) and L11-lacking ribosomes (closed circles) can be contrasted to that of RF-2 (A[lower]). B) The wild-type phenotype for release factor activity is restored by titrating back the missing L11 to the L11-lacking particles.

The L11-lacking ribosomes have been powerful tools in the investigation of the role of specific proteins at the peptidyltransferase centre in the termination event. It appeared initially that one protein, L16, was involved in peptidyl–tRNA hydrolysis, but on using the L11-lacking ribosomes, where the affinity of RF-2 is so much greater, the dependence on L16 dropped dramatically, suggesting that this protein was not critical for catalysis but is probably involved for promoting the correct conformation of the centre (19).

Once the use of these mutants established that L11 was an important determinant of the release factor binding domain and that the protein L11 could be reconstituted back, a strategy of modifying individual residues of protein L11, and then putting the modified protein back on the particles, was possible. In this way specific residues of L11 could be examined against an unmodified background, that is, the rest of the proteins and the RNA of the ribosome. This approach (20), for example, established that a tyrosine of the L11 was important for RF-1 binding.

4.2 Antibiotics/antibodies/depleted ribosomal particles

Antibiotics whose site of interaction with the ribosome is known; antibodies against specific ribosomal components, and the factors themselves, together with ribosomal particles depleted of a specific protein, have all been powerful tools to investigate how the release factors function. Examples include the use of the peptide antibiotic thiostrepton, and spectinomycin. The binding of thiostrepton is profoundly altered by changing nucleotide 1067 of the 23SrRNA of the bacterial ribosome and this antibiotic is a potent inhibitor of release factor binding. Spectinomycin, whose binding has been localized to 1192 of the 16SrRNA of the small ribosomal subunit, stimulates 2-fold the activity of RF-2, but does not affect RF-1 (Tate and Brown, unpublished data). Once it was established that large ribosomal subunit protein L11 was important in termination, antibodies against fragments of this protein were used to show that it was the N terminal domain of L11 that was important in release factor binding (18). A combination of antibodies against ribosomal proteins and those against the release factor have been instrumental in mapping the RF-2 binding site to the 70S ribosome (21). The first realization that L11 greatly influenced release factor function came using depleted ribosomal particles reconstituted without L11. These particles were up to six times more active with RF-2 than those that contained L11 in the reconstitution, but were almost completely inactive with RF-1(22) (see *Figure 3*).

4.3 Future studies with purified release factors

Detailed information is still lacking on the nature of the functionally important domains of the release factors and the positions of these domains within the molecules. The development of specific structural probes against regions of the factor will assist in this analysis. Specific antibodies raised against regions of the molecule, either large fragments or small subregions of primary sequence, can be used to identify which parts of the structure of the release factors are functionally important. Preliminary studies have already revealed one important region with this approach (23). Ultimately, however, the availability of significant quantities of the purified factors (see section 5.2) will allow attempts to obtain crystals of the proteins as a prerequisite to a detailed X-ray analysis of their three dimensional structure.

5. Release factor genes

5.1 Isolation, characterization, and mapping

The isolation of the genes for the bacterial release factors, RF-1 and RF-2, has provided new impetus for understanding how codon recognition occurs and how RF interacts with the peptidyltransferase centre of the ribosome. The two genes were isolated by quite different approaches, RF-1 with a genetic screen for overproduction of the factor (as described in section 2.3) where there was potential for competition between a suppressor tRNA in the host strain and RF-1 expressed from one of a library of random DNA fragments introduced on a high-copy number plasmid. The *E. coli* host strain carried an amber mutation in the *LacZ* gene which was the site of the competition (11). The RF-2 gene was isolated by labelling bacterial cultures containing a similar library of random fragments with [^{35}S] methionine and immunoprecipitating the labelled extract with a specific antibody recognizing RF-2 (24). In this case the colony containing the introduced RF-2 gene was scored from the radiolabelled factor on an autoradiograph.

After subcloning, the coding regions of the two genes were isolated and the sequences obtained by standard techniques. Southern blotting showed that the genes were present in a single copy with a low level of cross hybridization, reflecting sequence homology. The amino acid sequences of the two factors were derived from the DNA sequence (25), and revealed areas of strong homology between the two factors and other areas where the homology is rather weak, with an overall homology of about 30%. Presumably the former represent structural and functional domains reflecting the common functions shared by the two factors.

A dramatic finding revealed from a comparison of the N terminal protein sequence of RF-2 and the gene sequence was the presence of an in-frame stop codon, UGA, in the gene at codon position 26 necessitating a +1 frameshift during translation to synthesize the protein (25). This provides a potential novel mechanism for gene regulation whereby the factor competes with the frameshifting event at the UGA in codon position 26 of its own mRNA, the extent of competition reflected in the concentration of the factor in the cell. Consistent with this hypothesis Craigen and Caskey (26) have shown that RF-2, but not RF-1, can decrease the synthesis of RF-2 *in vitro*. This frameshift site provides a new tool to unravel the mechanism of how the release factor can recognize the stop codon.

The chromosomal location of the RF-1 and RF-2 genes have recently been determined (27). The RF-1 gene is located at 26.7 min on the map of the *E. coli* chromosome while the RF-2 gene is located quite distant from this at 62.3 min. This wide spacing on the genome would indicate that it is unlikely the two genes are regulated as part of the same operon.

5.2 Utilization of the release factor genes

Subcloning the release factor genes into multicopy high-level expression vectors has provided the potential to produce large amounts of protein as a source of the factors. Those used to date have employed *E. coli* as their host, the natural domain of the factors and unfortunately high expression has lead to the selection of cells which produce inactive factor after a few growth cycles.

Fusion proteins containing part of the release factor molecule expressed from a gene fusion such as TrpE, under the control of a strong and inducible promoter, have been isolated in large amounts (23), and these have been useful as antigens for the production of antibodies against specific regions of the protein. Once more information on the structurally and functionally important domains of the release factors are obtained, sensible strategies can be devised for altering specific residues of the protein, by site-directed mutagenesis of the appropriate gene. A combination of protein chemistry utilizing immunological techniques and recombinant DNA technology should provide the necessary tools to resolve at least some of the important unresolved questions of how the release factors mediate the termination of protein biosynthesis.

Acknowledgements

Support is gratefully acknowledged to W.P.T. from the Medical Research Council of New Zealand, Alexander von Humboldt Foundation, Otago Medical Research Foundation, and University Grants Committee for the work discussed in this review. C.T.C. is an investigator of the Howard Hughes Medical Institute and recipient of PHS grant No. GM34438 from the NIH.

References

1. Caskey, C. T., Scolnick, E., Tompkins, R., Milman, G., and Goldstein, J. (1971). In *Methods in Enzymology*, Vol. 20 (ed. L. Grossman and K. Moldave), p. 367. Academic Press Inc., London.
2. Kelmers, A. D., Novelli, G. D., and Stulberg, M. P. (1965). *J. Biol. Chem.*, **240**, 3979.
3. Muench, K. H. and Berg, P. (1966). In *Procedures in Nucleic Acid Research* (ed. G. L. Cantoni and D. R. Davies), p. 375. Harper and Row, New York.
4. Leder, P. and Bursztyn, H. (1966). *Biochem. Biophys. Res. Commun.*, **25**, 233.
5. Scolnick, E. M. and Caskey, C. T. (1969). *Proc. Nat. Acad. Sci. USA*, **64**, 1235.
6. Caskey, C. T., Beaudet, A. L., Scolnick, E. M., and Rosman, M. (1971). *Proc. Nat. Acad. Sci. USA*, **68**, 3163.
7. Goldstein, J. L. and Caskcy, C. T. (1970). *Proc. Nat. Acad. Sci. USA*, **67**, 430.
8. Ouchterlony, O. (1958). *Prog. Allergy*, **5**, 1.
9. Mancini, G., Carbonara, A. O., and Heremans, J. F. (1965). *Immunochemistry*, **2**, 235.
10. Ziltener, H. J., Clark-Lewis, I., Hood, L. E., Kent, S. B. H., and Schrader, J. W. (1987). *J. Immunol.*, **138**, 1099.
11. Weiss, R. B., Murphy, J. P., and Gallant, J. A. (1984). *J. Bacteriol.*, **158**, 362.
12. Reddington, M. A. and Tate, W. P. (1979). *FEBS Lett.*, **97**, 335.
13. Konecki, D. S., Aune, K. C., Tate, W. P., and Caskey, C. T. (1977). *J. Biol. Chem.*, **252**, 4514.
14. Caskey, C. T., Beaudet, A. L., and Tate, W. P. (1974). In *Methods in Enzymology*, Vol. 30 (ed. L. Grossman and K. Moldave), p. 293. Academic Press Inc., London.
15. Lee, C. C., Timms, K. M., Trotman, C. N. A., and Tate, W. P. (1987). *J. Biol. Chem.*, **262**, 3548.
16. Dabbs, E. R. (1977). *Mol. Gen. Genet.*, **151**, 261.
17. Parker, J., Watson, R. J., Friesen, J. D., and Fiil, N. P. (1976). *Mol. Gen. Genet.*, **144**, 111.
18. Tate, W. P., Dognin, M., Noah, M., Stoffler-Meilicke, M., and Stoffler, G. (1984). *J. Biol. Chem.*, **259**, 7317.
19. Tate, W. P., Sumpter, V. G., Trotman, C. N. A., Herold, M., and Nierhaus, K. H. (1987). *Eur. J. Biochem.*, **165**, 403.

20. Tate, W. P., McCaughan, K. K., Ward, C. D., Sumpter, V. G., Trotman, C. N. A., Stoffler-Meilicke, M., Maly, P., and Brimacombe, R. (1986). *J. Biol. Chem.*, **261**, 2289.
21. Kastner, B., Trotman, C. N. A., and Tate, W. P. (1986). *Proc. Otago Med. Univ. Sch.*, **64**, 42.
22. Tate, W. P., Schulze, H., and Nierhaus, K. H. (1983). *J. Biol. Chem.*, **258**, 12816.
23. Moffat, J. G., Trotman, C. N. A., and Tate, W. P. (1988). *Proc. Otago Med. Univ. Sch.*, **66**, 11.
24. Caskey, C. T., Forrester, W. C., Tate, W. P., and Ward, C. (1984). *J. Bacteriol.*, **158**, 362.
25. Craigen, W. J., Cook, R. G., Tate, W. P., and Caskey, C. T. (1985). *Proc. Nat. Acad. Sci. USA*, **82**, 3616.
26. Craigen, W. J. and Caskey, C. T. (1986). *Nature*, **227**, 38.
27. Lee, C. C., Kohara, Y., Akiyama, K., Smith, C. L., Craigen, W. J., and Caskey, C. T. (1988). *J. Bacteriol.*, **170**, 453.

6

Design and use of a fast and accurate *in vitro* translation system

MÅNS EHRENBERG, NEŞE BILGIN, and CHARLES G. KURLAND

1. Introduction: a general approach to *in vitro* translation

Our present concern is to describe methods suitable for *in vitro* studies of translation, in particular, the elongation mode. All examples are for *E. coli* translation components but the general approach is relevant to other translation systems. The performance of the *in vitro* system described here is quantitatively comparable to that of the bacteria: the elongation rates measured *in vitro* are near ten amino acids per sec per ribosome as for *in vivo* polypeptide synthesis (1) and the missense error levels are below 10^{-3} (2).

1.1 Optimization of *in vitro* translation

The demand that ribosomes perform *in vitro* as they do *in vivo* forces us to confront two types of problems. The first is to learn how to tune an *in vitro* translation system so that it works 'properly'. The other is to learn how to measure kinetic parameters when the ribosomes are working at cycle times in the millisecond range.

We begin by assuming that the ribosomes of *E. coli* have evolved under the particular constraints of pH, temperature, and ionic conditions typical of that organism. Accordingly, the development of our *in vitro* translation system was guided as far as possible by knowledge of these parameters. For *E. coli* this means a neutral pH, a temperature around 37 °C, and a complex ionic mixture (3). Early experiments *in vitro* showed that it is possible to combine a high translation rate, high accuracy, and high ribosomal activity in a carefully balanced buffer system (polymix), with ions mimicing the interior of *E. coli* (4, 5). The same procedure can be implemented for the translation systems of other organisms.

1.2 Measurements in a fast and precise translation system

In this methods review we describe how to measure fast elongation as well as missense errors *in vitro* (Section 2), and these techniques are applied to a comparison between poly(U) translation in polymix (4) and in a conventional

Tris–magnesium buffer (Section 3). Next we characterize the kinetic properties of elongation factors EF-Tu (6, 7), EF-G (8), and EF-Ts (9) in translation (Section 4). The study of elongation factor function requires knowledge of their concentrations as well as stoichiometries of the complexes that participate in protein synthesis. Methods to determine the concentrations of active EF-Tu (6, 7), EF-G (8), as well as EF-Ts are described (Section 5). We show how to measure the number of GTPs hydrolysed per peptide bond in EF-Tu function for correct (6, 7) as well as incorrect (Diaz, in preparation) tRNAs, and how to partition the accuracy of the ribosome in an initial and a proof-reading selection step after GTP-hydrolysis on EF-Tu (Diaz, in preparation) (Section 6). We analyse stoichiometries and binding constants for the formation of a complex between EF-Tu and aminoacyl-tRNA (6, 7) (Section 7). Finally, we summarize some important results obtained with the *in vitro* system described here (Section 8).

2. Fast elongation burst

Wagner *et al.* (10), following Lucas-Lenard and Lipmann (11), showed that ribosomes incubated with ^{3}H-labelled N-acetyl-Phe-tRNA^{Phe} ([^{3}H]-NAc) and polyuridylic acid [poly(U)] during 10 min form an initiation complex. When all other factors, necessary for complete translation, are added, between 10% and 35% of the ribosomes participate in a fast elongation burst with rates in the range of 10 amino acids per sec per ribosome.

2.1 Components of *in vitro* translation

(a) Enzymes: Elongation factors (EF-Tu, EF-Ts, EF-G) and tRNA aminoacylating enzymes [Phe-tRNA synthetase (PheS) and Leu-tRNA synthetase (LeuS)] are prepared from *E. coli* MRE 600 cells. The purification procedures of these enzymes including the original references are given in Section 9. Pyruvate kinase (EC 2.7.1.40) (PK) may be purchased from Boehringer and myokinase (EC 2.7.4.3) (MK) from Sigma. Ribosomes are purified from *E. coli* 017 cells according to Jelenc (12) by passing them through a Sephacryl S-300 column (Pharmacia). The ribosomes are dialysed against polymix buffer (see below) and kept at −80 °C in aliquots.

(b) tRNAs: tRNA^{Phe} and tRNA^{Leu} isoacceptors are purified from tRNA (bulk) from *E. coli* MRE 600 cells using standard procedures with benzoylated DEAE cellulose (Boehringer) (13) and sepharose 4B (14). The tRNAs are dialysed against polymix buffer and kept in aliquots at −80 °C. N-acetyl-Phe-tRNA^{Phe} (NAc) is prepared from tRNA^{Phe} (15) with either [^{3}H]Phe (150 c.p.m./pmol) or [^{14}C]Phe (1 c.p.m./pmol) (10).

(c) Chemicals: Poly(U) (of highest available molecular weight) can be purchased from Pharmacia. It is dissolved in and dialysed against polymix buffer at a concentration of 20 mg/ml and kept in aliquots at −80 °C. GTP (e.g., from Pharmacia) is prepared as a 100 mM solution in water at pH 7 and is stored at −20 °C. A/P solution contains 10 mM ATP (Pharmacia) and 100 mM phosphoenolpyruvate (PEP) (Sigma) in water at pH 7, and is stored in aliquots at −80 °C.

(d) Polymix buffer (polymix): The buffer, used for our experiments, has final ion concentrations as follows: 5 mM magnesium acetate, 0.5 mM $CaCl_2$, 95 mM KCl, 5 mM NH_4Cl, 8 mM putrescine, 1 mM spermidine, 5 mM potassium phosphate (KP) pH 7.3, and 1 mM DTE. Ten times concentrated polymix (10 × polymix) is prepared without KP (to avoid precipitation) and DTE, the pH is adjusted to 7.3. KP is kept at a concentration of 100 mM (20 × KP) and DTE at 50 mM (50 × DTE). The correct working strength of polymix is obtained in the incubation mixture. To make 1 ml of polymix buffer: mix 830 μl water, 100 μl 10 × polymix, 50 μl 20 × KP, and 20 μl 50 × DTE. To avoid precipitation do not add KP directly to 10 × polymix. Radioactive amino acids are purchased from Amersham. 3 mM [^{14}C] Phe is prepared with a specific activity of 5 c.p.m./pmol. All other chemicals are of analytical grade (e.g., from Merck).

Protocol 1. Fast elongation burst

1. Make two separate mixes on ice; an initiation mix (70S-mix) and a factor mix (see Step 2). To prepare the 70S-mix, add the appropriate volumes of water, 10 × polymix, 20 × KP and 50 × DTE to balance the final polymix. Pipette in addition (per 50 μl) 10 pmoles of active ribosomes, 20 μg of poly(U) and [^{3}H]NAc (150 c.p.m./pmol), 30% in excess over the total number of 70S particles. This choice of NAc concentration leads to near maximal activation of the ribosomes (10).
2. Pipette appropriate volumes of polymix compounds into the factor mix tubes and add (per 50 μl) 10 μl A/P, PK (5 μg), MK (0.3 μg), [^{14}C]Phe (30 nmoles, 5 c.p.m./pmol), PheS (100 units), $tRNA^{Phe}$ (250 pmol), EF-Tu (600 pmol), EF-G (250 pmol) and EF-Ts (150 pmol).

   One unit of PheS can charge one pmol of $tRNA^{Phe}$ per sec. 100 units of PheS is sufficient to ensure a high level of $tRNA^{Phe}$ charging during the burst of poly(Phe) synthesis. Unfractionated tRNA ($tRNA_{bulk}$) containing the same amount of $tRNA^{Phe}$ can be used instead of purified $tRNA^{Phe}$. The elongation factor concentrations used here lead to wild-type ribosomal elongation rates close to the maximum rate.
3. Pre-incubate the two mixes separately for 10 min. Add at time zero after the pre-incubation 50 μl of the 70S-mix to the same volume of factor mix and start poly(Phe) synthesis.
4. Stop the reaction with 5 ml of 5% TCA, containing 15% casaminoacids (Difco), at different incubation times from 5 sec to 2 min.
5. Heat the reaction tubes at 95 °C for 10 min and filter the samples through glass fiber filters (GF/A, Whatman).

6. Rinse the filters first with 5% TCA, then with 2-propanol, put them in vials and dry them for 15 min at 105 °C.
7. Add a scintillation cocktail containing 0.5% PPO, 0.0125% bisMSB (p-bis-(o-methylstyrl) benzene; Beckman), 10% NCS tissue solubilizer (Amersham) in toluene, shake the vials on a rotary shaker for 5 min before counting. Use 10 μl of 3 mM [^{14}C]Phe and 10 μl of that [^{3}H]Phe (150 c.p.m./pmol), used to make [^{3}H]NAc, as standards for specific activities.

At elongation rates near 10 amino acids per sec per ribosome, the length of poly(U) messengers will influence the results (*Figure 1*). The extent of poly(Phe) synthesis increases linearly with time only at the beginning of the incubation. The rate of poly(Phe) synthesis per ribosome (v_{el}, in Equation 1 below), calculated from [^{14}C]-labelled poly(Phe) and [^{3}H]NAc in hot TCA precipitable polypeptide chains, is therefore apparently decreasing during the time of the experimental incubation as the ribosomes reach the end of poly(U) (10). Extrapolation of the curve in *Figure 1* to zero time gives an approximately correct elongation rate. Correction for finite poly(U) lengths is discussed elsewhere (6).

2.2 Measurement of error frequencies

The missense error frequency is studied by measuring the misreading of UUU-codons by near-cognate tRNAs such as $tRNA^{Leu}_{2}$ and $tRNA^{Leu}_{4}$ competing with $tRNA^{Phe}$ (10). The experiment is a burst as described in *Protocol 1* with a fixed

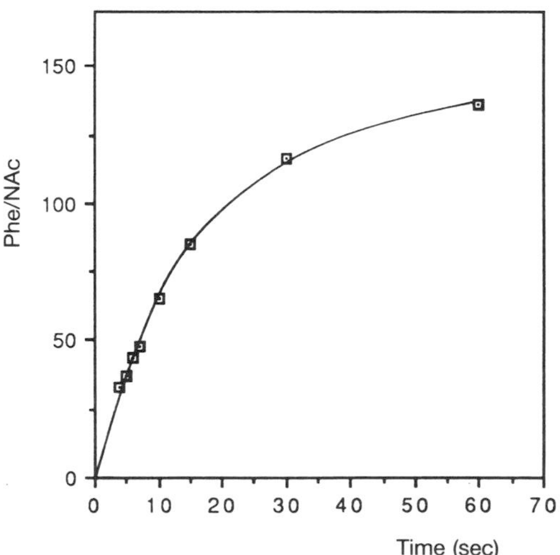

Figure 1. Poly(Phe)-synthesis is measured by the amount of [^{14}C]-Phe retained on the filter. The number of elongating chains is determined by the amount of [^{3}H]-NAc-Phe on the filter and this also provides data on the number of ribosomes that are active in elongation in the experiment (see Section 2 and *Protocol 1*).

incubation time (30 sec). The factor mix contains the competing amino acid, in addition to the Phe components, synthetase, and varying amounts of the chosen non-cognate tRNA. The ratio between correct (J_c) and incorrect (J_w) peptide bond formation rates varies linearly with the ratio between correct (ch_c) and incorrect (ch_w) tRNA charging levels:

$$J_c/J_w = A \cdot ch_c/ch_w$$

The slope of the line (A) is the normalized accuracy:

$$A = (J_c/J_w) \cdot (ch_w/ch_c).$$

The normalized error frequency (P_E) is the incorrect rate of peptide bond formation divided by the total (correct + incorrect) measured when $ch_w = ch_c$. P_E can be calculated from the normalized accuracy as:

$$P_E = 1/(1 + A).$$

Protocol 2. Missense errors

1. Prepare a 70S-mix as in *Protocol 1* (step 1) with [^{3}H]NAc replaced by [^{14}C]NAc (1 c.p.m./pmol). Note that here we are not concerned with the number of elongating ribosomes.
2. Prepare a factor mix which contains (per 50 μl of complete mix) polymix, [^{14}C]Phe (0.6 mM, 5 c.p.m./pmol), [^{3}H]Leu (0.04 mM, 1500 c.p.m./pmol), $tRNA^{Phe}$ (250 pmol), EF-Tu (1000 pmol), EF-Ts (150 pmol), EF-G (250 pmol), PheS (100 units), LeuS (10 units), $tRNA_2^{Leu}$ or $tRNA_4^{Leu}$ (10 μl containing between 0 and 800 pmol, in triplicate).
3. Pre-incubate the mixes for 10 min.
4. Add, to the first set of factor mix tubes, 50 μl of 70S-mix without poly(U) for backgrounds. Add to the second set 50 μl of 70S-mix with poly(U). Stop all incubations after 30 sec with 5 ml 5% TCA. The third set of factor tubes is quenched directly after the pre-incubation with ice cold TCA for charging analysis. Keep the charging samples on ice prior to filtering.
5. Filter the third set directly (without boiling), rinse with ice cold TCA and then with 2-propanol.
6. Dry and count as described in *Protocol 1* from step 6.
7. Process the polypeptide synthesis samples as in *Protocol 1* from step 5.

---

3. Magnesium optimizations in polymix and tris buffers

The composition of polymix is given in Section 2.1. The description of how to obtain a physiological buffer is presented in reference (4). Here we compare the physiological polymix buffer with a Tris–Mg^{2+} buffer (TMK) containing 50 mM Tris–HCl (pH 7.3), 50 mM KCl, 10 mM $MgCl_2$ and 1 mM DTE (16, 17).

We have determined the optimal Mg^{2+} concentration in the burst assay for polymix and TMK (see *Figure 2*). Several advantages are associated with the use of polymix. These are:

- A higher percentage of ribosomes that participate in poly(Phe) synthesis (*Figure 2b*).
- A higher ribosomal elongation rate (Phe incorporation per second per ribosome (v_{el})) (*Figure 2c*).
- A higher total poly(Phe) synthesis (*Figure 2a*).
- When the poly(Phe) synthesis rate is maximized (*Figure 2c*) at a Mg^{2+} concentration between 4 and 5 mM in polymix, and between 8 and 9 mM in TMK, the accuracy of poly(Phe) synthesis is higher in polymix than in TMK (*Figure 2d*: P_E is 1.7×10^{-4} in polymix and 4.7×10^{-4} in TMK) (*Figure* 2).

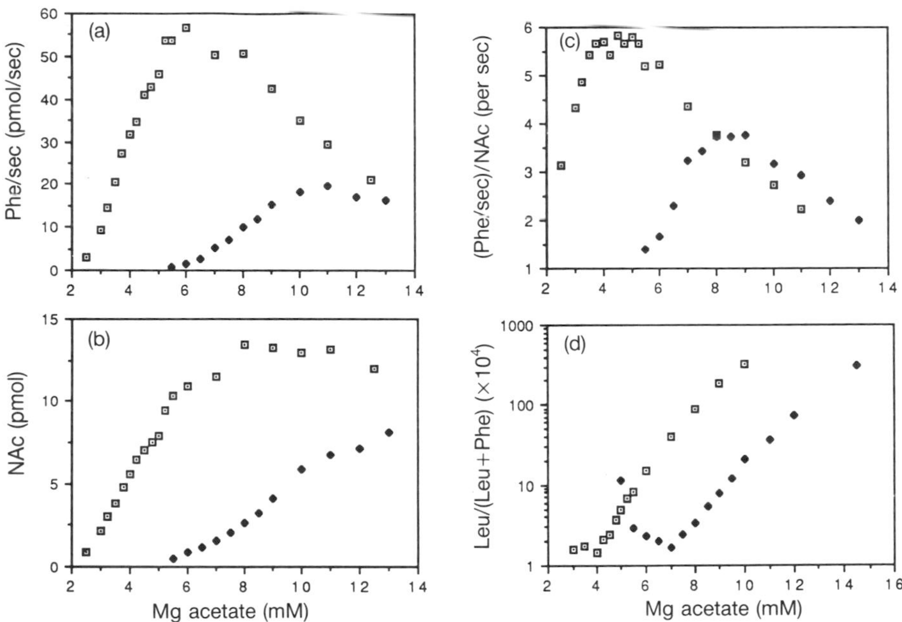

Figure 2. In (a) the total rate of poly(Phe)-synthesis, in (b) the number of active ribosomes, in (c) the ribosomal elongation rate, and in (d) the normalized error for $tRNA_4^{Leu}$ are plotted at different Mg^{2+}-concentrations, for polymix (□–□) and TMK-buffer (◆–◆). The standard burst (Section 2.1) and error (Section 2.2) assays, are modified to make the factor and 70S-mixes balanced to 2 mM Mg^{2+} acetate in polymix or TMK (using 10 × polymix or 10 × TMK buffers without Mg^{2+}). Appropriate dilutions of Mg^{2+} acetate (in water) were added in 10 μl to tubes with 40 μl factor mix, or in 15 μl to tubes with 60 μl 70S-mix to give the final Mg^{2+} concentrations in this figure. After 10 min pre-incubation, translation was started by pipetting 50 μl of 70S-mix to 50 μl of factor mix (with the same Mg^{2+} concentration). The incubation time was 7 sec for burst and 30 sec for error measurements.

4. Elongation factor characterizations

4.1 How factors influence ribosomal elongation

Efficient translation (18) requires that elongation factors EF-Tu and EF-G associate rapidly with the ribosome and that they spend a short time in complex with the 70S particle. The association rate of an elongation factor to the ribosome is determined by the concentration of its relevant complex ($[EF_1]$ for EF-Tu and $[EF_2]$ for EF-G) and by the value of the parameters k_{cat1}/K_{M1} and k_{cat2}/K_{M2} for EF-Tu and EF-G, respectively. To a first approximation *in vitro* translation follows ordinary enzyme kinetic behaviour. The elongation rate (v_{el}) can be written accordingly:

$$v_{el} = \frac{1}{(1/k_{cat}) + (K_{M1}/k_{cat1}[EF_1]) + (K_{M2}/k_{cat2}[EF_2])} \qquad (1)$$

The first term in the denominator ($1/k_{cat}$) is the elongation time at very high factor complex concentrations ($[EF_1]$, $[EF_2]$). In $1/k_{cat}$ the times that EF-Tu and EF-G spend in complex with the ribosome are included. The other two terms describe the time that a ribosome spends with an open A-site waiting for an aminoacyl-tRNA ($K_{M1}/(k_{cat1}[EF_1])$) and the time it spends in a pre-translocational state waiting for the EF-G complex ($K_{M2}/(k_{cat2}[EF_2])$).

4.2 Elongation factor titrations

When the ribosomal elongation rate is varied as a function of the concentration of an elongation-factor complex ($[EF_i]$), the effective association rate of that factor (k_{cati}/K_{Mi}) can be determined (Equation 1). If the value of k_{cat}/K_M and the concentration of the other factor are known, such a titration will also give the maximal elongation rate (k_{cat} in Equation 1) (*Figure 3*).

Protocol 3. Elongation factor titrations

1. Make two mixes where the 70S mix is exactly as in *Protocol 1*. The factor mix is modified as follows: For EF-Tu titrations put (per 50 μl of complete factor mix) 700 pmol of $tRNA^{Phe}$, 500 pmol of EF-G, and dilutions of EF-Tu (0–500 pmol) in 10 μl polymix. For EF-G titrations put (per 50 μl of complete factor mix) 700 pmol of $tRNA^{Phe}$, 500 pmol of EF-Tu, and dilutions of EF-G (0–500 pmol) in 10 μl polymix.
2. Pre-incubate the 70S and the different factor mixes for 10 min.
3. Start translation by pipetting 50 μl of 70S-mix into one of the factor mix tubes.
4. Stop the reaction after T_i sec with 5 ml 5% TCA and treat the samples as in *Protocol 1*.

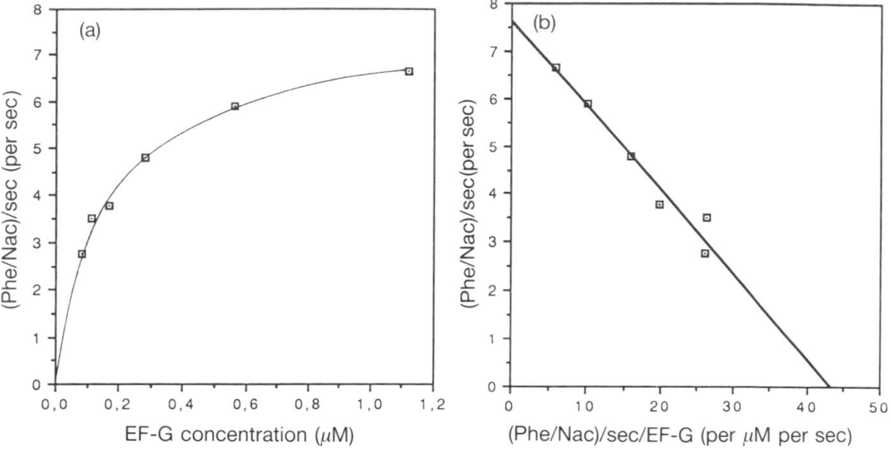

Figure 3. In (a) the ribosomal elongation rate (Phe/sec)/active 70S particle (NAc) is plotted as a function of the EF-G concentration (see Sections 4.1–2), and in (b) we show an Eadie-Hofstee plot of the same data. The *y*-axis intercept gives the maximal rate of 7.6 sec^{-1} and the *x*-axis intercept gives $k_{cat2}/K_{M2} = 4.3 \times 10^7\ M^{-1}sec^{-1}$.

T_i is chosen to make the poly(Phe)-chain length approximately equal for every $[EF_i]$, and short enough to make the extent of poly(Phe)-synthesis approximately linear in time. For a poly(Phe) chain length of 15 choose T_i according to (cf. Equation 1):

$$T_i = 15\{(1/k_{cat}) + (K_{M1}/k_{cat1}[EF_1]) + (K_{M2}/k_{cat2}[EF_2])\} \tag{2}$$

4.3 Elongation factor cycle times

Further information about EF-Tu and EF-G comes from probing how fast they go through their full cycles. One way is to investigate how the rate of poly(Phe)-synthesis per active elongation factor complex (v_1 for EF-Tu and v_2 for EF-G) varies with the concentration of active ribosomes ([70S]). At low [70S]-values the rate-limiting step in a factor cycle is the rate of association to ribosomes; this is the k_{cat}/K_M range. At very high [70S]-values the time the factor spends in complex with a ribosome, plus the time to regenerate a translationally active factor-GTP complex from the factor-GDP complex, dominates v_i. The elongation factor cycling rates v_1 and v_2 for EF-Tu and EF-G can be written as:

$$v_i = \frac{1}{(1/k_{cati}) + (K_{Mi}/k_{cati}[70S])} \tag{3}$$

The k_{cati}/K_{Mi} parameters in Equation 3 are the same as in Equation 1. $1/k_{cat1}$ is the time EF-Tu spends in complex with a ribosome plus the duration of the EF-Ts catalysed exchange of GDP for GTP, and subsequent binding of

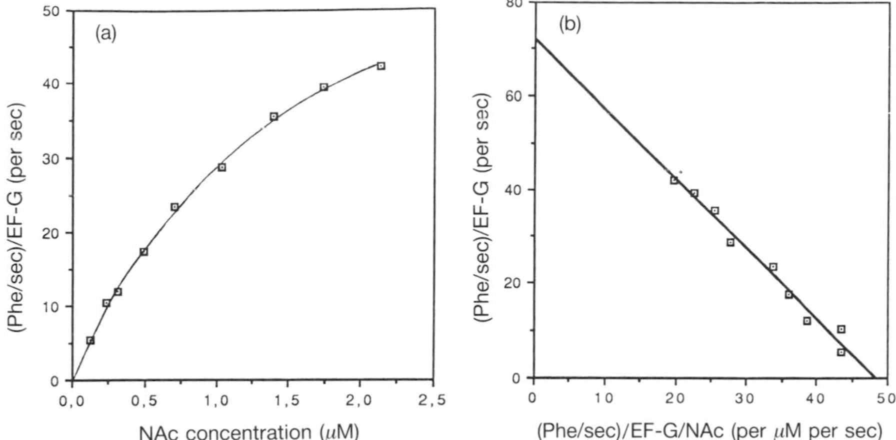

Figure 4. The rate of poly(Phe)-synthesis per EF-G as a function of the concentration of active 70S-particles (NAc) (a). In (b) an Eadie-Hofstee plot of the data in (a) shows that the maximal EF-G cycling rate is 70 sec^{-1} (y-intercept) and that $kcat_2/K_{M2}=4.7\times10^7\ M^{-1}s^{-1}$ (x-intercept).

aminoacyl-tRNA to EF-Tu-GTP. For EF-G $1/k_{cat2}$ is simply the time required to exchange GDP for GTP plus the duration of existence of the factor-ribosome complex (see *Figure 4* for the EF-G cycle).

Protocol 4. Elongation factor cycle times in 70S titrations

1. Prepare a 70S-mix with (per 50 μl) 800 pmol (total) ribosomes, 1000 pmol [^{3}H]NAc and 200 μg poly(U) in balanced polymix. Make dilutions from this 70S-mix with polymix to give 70S-mixes with between 40 and 800 pmol of ribosomes per 50 μl. (By diluting in this way, the ratios of poly(U) and NAc to ribosomes will stay constant.)
2. Prepare a factor mix with components as in *Protocol 1*, modified as follows: the studied elongation factor is kept at a concentration so low that the rate of poly(Phe) synthesis in a burst varies linearly with the factor concentration (for wild-type ribosomes 10 pmol of EF-Tu or 5 pmol of EF-G can be used). Add 1200 pmol $tRNA^{Phe}$, a concentration in excess over the highest total concentration of 70S particles, 200 units of PheS, and 600 pmol of EF-Tu together with 150 pmol EF-Ts (in EF-G limited) or 250 pmol of EF-G together with 1000 pmol EF-Ts (in EF-Tu limited experiments).
3. Pre-incubate the factor mix and 70S-mixes for 10 min as in *Protocol 1*.
4. Start translation by pipetting 50 μl of the factor mix into one of the 70S-mixes.
5. Quench with 5% TCA after an assay time chosen for about 15 amino acids per poly(Phe)-chain at all 70S concentrations (cf., Equation 2). Process the samples as described in *Protocol 1* from step 5.

4.4 Translation under EF-Tu and EF-Ts limitation

At low concentrations of EF-Tu the rate limiting step in protein synthesis is attachment of an aminoacyl-tRNA to the A-site. If, in addition, the EF-Ts concentration is small. most EF-Tu is in its GDP complex (9). Introducing explicitly the role of EF-Ts in the cycling rate of EF-Tu we get from Equation 3:

$$v_3 = \frac{1}{(1/k_1) + (K_{M1}/k_{cat1}[70S]) + (K_{M3}/k_{cat3}[\text{EF-Ts}])} \tag{4}$$

At low concentrations of EF-Ts the cycling rate of EF-Tu is determined by the effective association rate between EF-Tu and EF-Ts (k_{cat3}/K_{M3}) and [EF-Ts]. We have analysed the EF-Ts activity, using Equation 4 to interpret experiments where EF-Tu and EF-Ts are rate limiting. To obtain the absolute EF-Ts concentration from this assay k_{cat3}/K_{M3} must be known (see Section 5.3 on how to estimate [EF-Ts]).

Protocol 5. EF-Ts titration

1. Make a 70S-mix as in *Protocol 1*.
2. Make a factor mix as in *Protocol 1* with two modifications. Add (per 50 μl) 5 pmol of EF-Tu and 10 μl of EF-Ts (0–300 pmol) diluted in polymix.
3. Pre-incubate the 70S- and factor mixes for 10 min.
4. Start translation by pipetting 50 μl of 70S mix into the factor mixes.
5. Quench the reaction with TCA after incubation times chosen for a poly(Phe) chain length of about 15 amino acids (cf., Equation 2).
6. Process the samples as described in *Protocol 1*. Calculate v_3 as a function of [EF-Ts] by normalizing the extent of poly(Phe)-synthesis (poly(Phe) pmol) at each point to the constant amount of EF-Tu ([EF-Tu]$_0$ pmol—written as Tu_0 henceforward) as well as to the incubation time (T_i sec): $v_1 = poly(Phe)/(Tu_0 \cdot T_i)$. Plot v_3 versus v_3/[EF-Ts] according to Eadie–Hofstee. The intercept at the x-axis of the straight line gives k_{cat3}/K_{M3}.

---

5. Determination of active elongation factor concentrations

The characterization of elongation factors in the previous section requires precise knowledge of their active concentrations, which in turn requires special treatment.

5.1 Concentration of EF-Tu by a nucleotide exchange method

Nitrocellulose binding of EF-Tu, and then counting radiolabelled GDP on the filters, is a rapid method to determine the concentration of EF-Tu (19). However, this technique is sometimes unreliable. Therefore, we have developed a method which takes advantage of the slow rate of exchange of GDP on EF-Tu in the absence of EF-Ts. At time zero of the incubation we add to EF-Tu in complex with [^{3}H]GDP, components for fast phosphorylation of all free GDP-molecules to GTP. The rate-limiting step in the conversion of GDP, originally on EF-Tu, to GTP, is the release rate (k_d) of GDP from EF-Tu. The disappearance of ^{3}H-labelled GDP is registered as a decrease in time of the ratio: $r_0(t)=$ [[^{3}H]GDP]/([[^{3}H]GDP]+[[^{3}H]GTP]) which can be obtained from thin-layer plates, as described in the next section. At times longer than 10 min $r_0(t)$ takes a constant value r_b. The difference $\Delta r_0(t)$ between $r_0(t)$ and r_b is an exponential:

$$r_0(t)-r_b=\Delta r_0(t)=(Tu_0/G_T)e^{-k_d t}. \tag{5}$$

Tu_0 is the amount of EF-Tu and G_T is the total amount of GDP—free as well as in complex with EF-Tu. G_T can be estimated by adding a known amount of unlabelled GDP (G_0) to *EF-Tu*. Here the guanine nucleotide difference ratio $r_1(t)-r_b$, corresponding to $\Delta r_0(t)$ in Equation (5) is an exponential with the same exponent and with a smaller amplitude.

$$r_1(t)-r_b=\Delta r_1(t)=[Tu_0/(G_T+G_0)]e^{-k_d t}. \tag{6}$$

G_T can be obtained from $\Delta r_0(t)/\Delta r_1(t)$:

$$G_T=\frac{G_0}{(\Delta r_0(t)/\Delta r_1(t))-1} \tag{7}$$

The natural logarithm of $\Delta r_0(t)$ in Equation 5 plotted versus time is a straight line:

$$\ln\Delta r_0(t)=\ln(Tu_0/G_T)-k_d t. \tag{8}$$

The intercept at the *y*-axis at zero time can be used to estimate Tu_0/G_T and with G_T known from Equation 7 Tu_0 can be determined.

Protocol 6. Nucleotide exchange method

1. Make two EF-Tu mixes. Balance both with respect to polymix as in *Protocol 1*. To the first add (per 50 μl) EF-Tu (about 500 pmol) and [^{3}H]GDP (about 80 pmol, 1800 c.p.m./pmol). To the second add (per 50 μl) in addition to EF-Tu and [^{3}H]GDP, 1500 pmol of unlabelled GDP.
2. Prepare also an energy mix, balanced with respect to polymix which contains (per 50 μl) 10 μl A/P, GTP (2 mM), PK (5 μg), MK (0.3 μg).

3. Pre-incubate the three mixes for 10 min at 37 °C.
4. Start the nucleotide conversion by pipetting a suitable volume from the energy mix to an equal volume of one of the EF-Tu mixes and vortex immediately.
5. Withdraw 20 μl aliquots from the reaction after 10, 20, 30, 45, 60, 90, 120, 180, and 600 sec, and quench in 20 μl 20% formic acid in Eppendorf tubes placed on ice.
6. Spin the samples for 15 min in an Eppendorf centrifuge and apply about 10 μl from the supernatant to thin layer plates (Polygram Cel^{300} PI, Macherey-Nagel). Apply also 1 μl of a marker nucleotide mixture containing 10 mM GDP and 10 mM GTP to each spot.
7. Run the plates in 0.5 M KH_2PO_4 at pH 3.5 until the buffer front is near the top of the plate.
8. Dry the plates and identify the GDP and GTP spots under UV light.
9. Cut them out and put them into separate vials. Add scintillation mix without tissue solubilizer and count.
10. Calculate the ratios $r_0(t)$ and $r_1(t)$ from [[^{3}H]GDP]/([[^{3}H]GDP]+[[^{3}H]GTP]) at each time point t_i. Use the 10 min ratio (r_b) where all GDP, originally on EF-Tu, has been completely exchanged to GTP as a 'background' and form the ratio in Equation (5) as $\Delta r_0(t_i) = r_0(t_i) - r_b$ and in Equation 6 as $\Delta r_1(t_i) = r_1(t_i) - r_b$.

5.2 Concentration of active EF-G

The concentration of active EF-G is determined from an active site titration (cf., 20) with ribosomes in the presence of 2×10^{-4} M fusidic acid (FA) (8). The absolute calibration of the method depends on the assumption that there is one EF-G·FA complex per active ribosome (measured from [^{3}H]NAc). At 2×10^{-4} M FA, EF-G spends a long time on the ribosome and the K_M-value for the ribosome EF-G interaction is therefore small. The total concentration of EF-G [EF-G_0] is kept fixed and the concentration of 70S-particles is varied from zero to high values. The total rate of poly(Phe)-synthesis (v) is then given by:

$$v = v_{max} \cdot ([70S]/([70S] + K_M). \tag{9}$$

[70S] is the concentration of free active ribosomes and v/v_{max} measures the fraction of EF-G in complex with 70S-particles. If the input concentration of EF-G is much larger than the K_M-value, then v is, to a good approximation, given by:

$$\begin{aligned} v &= v_{max}[70S_0]/[\text{EF-G}_0] \text{ when } [70S_0] < [\text{EF-G}_0] \\ v &= v_{max} \qquad\qquad\qquad\quad \text{ when } [70S_0] > [\text{EF-G}_0] \end{aligned} \tag{10}$$

where [$70S_0$] is the total active ribosome concentration estimated from

[^{3}H]NAc. The concentration of EF-G can be found from that [$70S_0$]-value where v changes from linearly increasing with [$70S_0$] to a constant (cf., *Figure 5a*). If $v_{max}/(v_{max}-v)$ is plotted versus ($v_{max}/v \cdot [70S_0]$ (cf., *Figure 5b*) a straight line is always obtained and the intercept at the x-axis gives [EF-G_0]. When [EF-G_0] is much larger than K_M the line is almost vertical and the intercept at the x-axis can be estimated with high precision.

Protocol 7. Determination of active EF-G concentration

1. Prepare and dilute the 70S-mix exactly as described in *Protocol 4* with 20, 30, 40, 50, 80, 100, 120, 200, 300, 400, 500, and 800 pmol of ribosomes per 50 μl.
2. Prepare a factor mix as described in Section 2.1 with the following modifications: per 50 μl, add 600 pmol EF-Tu, 150 pmol EF-Ts, 1000 pmol $tRNA^{Phe}$, 100 units of PheS, 400 μM fusidic acid and 20–50 pmol of EF-G (as estimated from the protein content).
3. Pre-incubate the 70S-mixes and the factor mix for 10 min.
4. Start translation by pipetting 50 μl of the factor mix into a 70S tube and choose incubation times to incorporate about 15 amino acids per ribosome at all 70S concentrations.
5. Quench with TCA and process the samples as described earlier in *Protocol 1*.
6. Plot as in *Figure 5a* to determine v_{max} from the plateau. Plot as in *Figure 5b* to determine [EF-G_0].

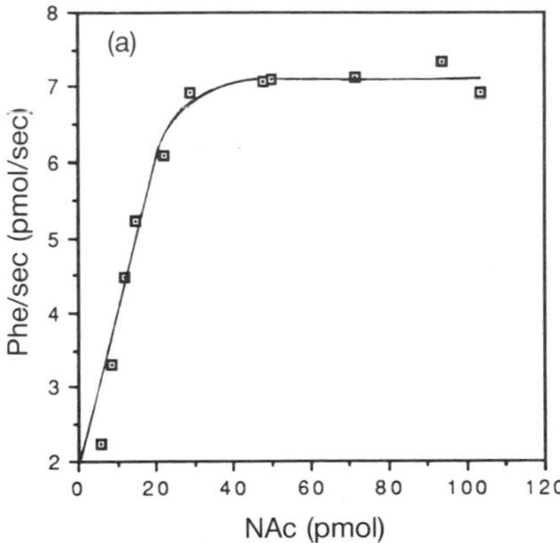

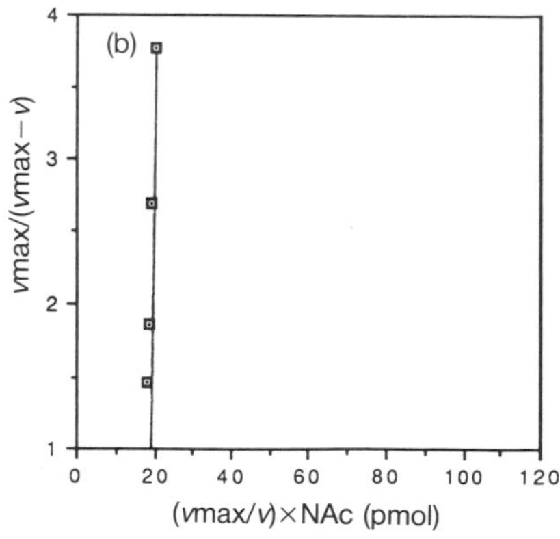

Figure 5. Determination of active EF-G. In (a) the total rate of poly(Phe)-synthesis (v=Phe/sec) is plotted as a function of the number of active ribosomes ($70S_0$) measured by the amount of NAc retained on the filter (see text and the *Figure 1* legend). There are 25 pmol of EF-G (from protein content) and 0.2 mM fusidic acid and $v = v_{max}$ at the plateau. The data are replotted as a Scatchard in (b) showing that there are about 20 pmol of active EF-G (x-axis intercept).

5.3 Concentration of active EF-Ts

We take advantage of the accurate estimates of active EF-Tu that can be obtained from nucleotide exchange (Section 5.1) together with the known 1:1 stoichiometry between EF-Tu and EF-Ts (21) to determine active EF-Ts in our preparations by equilibrium dialysis. We use the observations that EF-Ts has a much higher affinity for EF-Tu than GTP (22) and that the binary EF-Tu · EF-Ts complex has a much smaller affinity for GTP than EF-Tu (22).

Dialysis chamber A contains EF-Tu, GTP, and varying amounts of EF-Ts, and chamber B only GTP. In the absence of EF-Ts, almost all EF-Tu in chamber A is bound to GTP. When EF-Ts is titrated to higher concentrations, the GTP-molecules on EF-Tu are successively displaced by EF-Ts, and when EF-Ts is in excess no GTP is bound to EF-Tu. The amount of ^{3}H-labelled GTP in chamber A (GTP_A) decreases, and the amount of [^{3}H]GTP in chamber B (GTP_B) increases with increasing concentrations of EF-Ts (*Figure 6*). The ratio $GTP_B/(GTP_A+GTP_B)$ increases linearly with EF-Ts when [EF-Ts]<[EF-Tu] and is constant when [EF-Ts]>[EF-Tu] (see *Figure 6*). From the breaking point (*Figure 6a*), or from a Scatchard analysis (*Figure 6b*) the EF-Ts concentration can be obtained.

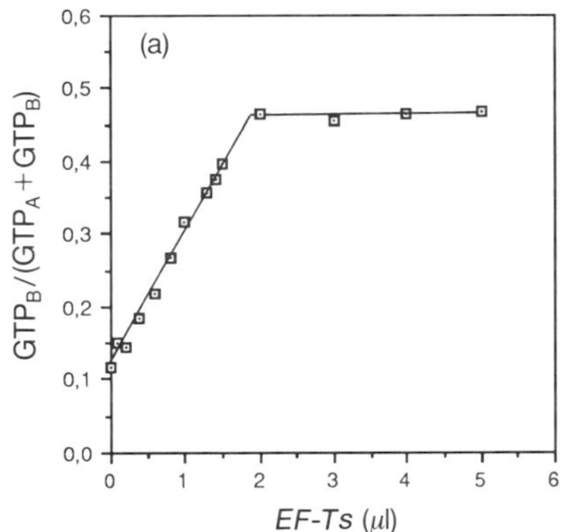

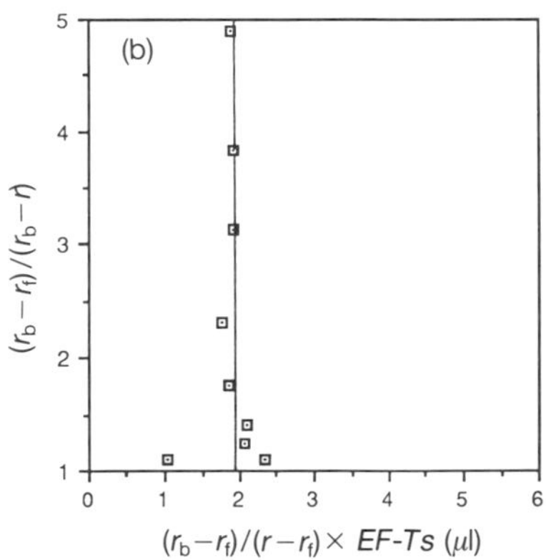

Figure 6. Determination of active EF-Ts. The ratio *(r)* between [^{3}H]-GTP in chamber A (containing 1000 pmol of EF-Tu) divided by [^{3}H]-GTP in both chambers is plotted as a function of added EF-Ts in μl (a). With no EF-Ts $r=r_f$ (*y*-axis intercept) and with EF-Ts in excess $r=r_b$ (plateau). In (b) the same data are presented in a Scatchard plot. The ratio $(r_b-r_f)/(r_b-r)$ corresponds here to total EF-Tu divided by free EF-Tu. The intercept at the *x*-axis gives the amount in μl of EF-Ts sample that corresponds to the amount of EF-Tu present. We obtain 526 pmol of EF-Ts/μl. (Note that the analysis is valid only under the experimental conditions as specified in Section 5.3).

Protocol 8. Concentration determination of EF-Ts

1. Prepare equilibrium dialysis cells with dialysis membranes.
2. Prepare three mixes in balanced polymix: mix 1 contains (in 50 μl) 10 μl A/P and 2.5 μg pyruvate kinase. Mix 2 contains (in 40 μl) 10 μl A/P, 2.5 μg pyruvate kinase, and 85 pmol [^{3}H]GDP (55 Ci/mmol). Mix 3 contains (in 40 μl) 10 μl A/P, 2.5 μg pyruvate kinase, 85 pmol [^{3}H]GDP (55 Ci/mmol) and 1000 pmol EF-Tu.
3. Incubate all mixes for 10 min at 37 °C.
4. Make EF-Ts dilutions with polymix in 10 μl containing 30, 60, 120, 180, 240, 300, 600, 900, 1200, 1500, and 2000 pmol EF-Ts estimated from its protein content.
5. Pipette 40 μl of mix 2 and either 10 μl of polymix (−Tu−Ts) or 10 μl of Ts dilutions (−Tu+Ts) into the tubes.
6. Pipette 40 μl of mix 3 and 10 μl of polymix (+Tu−Ts) or 10 μl of each of the EF-Ts dilutions (+Tu+Ts) into tubes.
7. Incubate all tubes for another 10 min at 37 °C.
8. Pipette 50 μl of mix 1 (−Tu, −GDP) to one side of each dialysis cell.
9. Pipette 50 μl of mix 2 or mix 3 into the other side of the cells.
10. Place the cells on a rotary shaker at 4 °C. Shake for 24 h.
11. Take 20 μl duplicate samples from both sides of each cell, directly on to GF/A filters, dry, and count in the scintillation cocktail containing tissue solubilizer as described in *Protocol 1*.
12. Determine $GTP_B/(GTP_A+GTP_B)$ for each dialysis cell. Plot as shown in *Figure 6* and determine the EF-Ts concentration as described in the legend. Use the (−Tu−Ts) points to detect whether equilibrium has been achieved and the (−Tu+Ts) points as controls for possible Tu contamination in the Ts preparation.

6. Stoichiometry of GTP-hydrolysis in EF-Tu function and poly-(Phe)-elongation

In this assay we relate the number of GTPs hydrolysed on EF-Tu to the extent of poly(Phe)-synthesis in translation.The number of GTPs hydrolysed per peptide bond (f_c) is one of the parameters that control the accuracy of translation (18, 23, 24). The recent discovery that there is never less than two GTP-molecules hydrolysed per peptide bond, even when the accuracy of the ribosome is low, raises a number of new and intriguing questions (6, 7).

6.1 The number of GTPs hydrolysed in EF-Tu function per correctly formed peptide bond (f_c)

When EF-Tu–GDP is incubated with an energy regenerating system and with aminoacyl-tRNA in excess, practically all EF-Tu · GDP will be converted to EF-Tu · GTP and form a complex with aminoacyl-tRNA (6, 7). As this complex is mixed with activated ribosomes and other translation components (except EF-Ts), there will be a rapid burst of GTP-hydrolysis and poly(Phe)-synthesis. Subsequently, when all EF-Tu · GTP–aminoacyl-tRNA complexes have been consumed, translation will enter a steady state. Here, most EF-Tu will exist as an EF-Tu · GDP binary complex, and the rate limiting step in translation will be the release rate (k_d) of GDP from EF-Tu (26, 27). The extent of poly(Phe)-synthesis as a function of time ($Phe(t)$) can be written (*Figure 7a*):

$$Phe(t) = Phe_0 + Ct. \tag{11}$$

The first term Phe_0 is the extent of poly(Phe)-synthesis in the burst-phase of the experiment which is over when the sampling begins (typically when $t \geq 10$ sec). Phe_0 is the number of peptide bonds catalysed by a single round of the original amount of EF-Tu · GTP in complex with aminoacyl-tRNA. The constant C is the rate of poly(Phe)-synthesis during the steady-state phase. The amount of GTP that is hydrolysed in EF-Tu function during the burst (ΔTu_0) is obtained using ^{3}H-labelled guanine nucleotides as in Section 5.1. Just before translation is started the ratio (r_{f0}) between [^{3}H]GDP and total guanine nucleotide in the factor mix is sampled. During the burst the corresponding ratio $r_0(t)$ rapidly increases to its highest value ($r_0(0)$) and then it decreases exponentially, as [^{3}H]GDP on EF-Tu is diluted out by non-labelled GDP, down to its

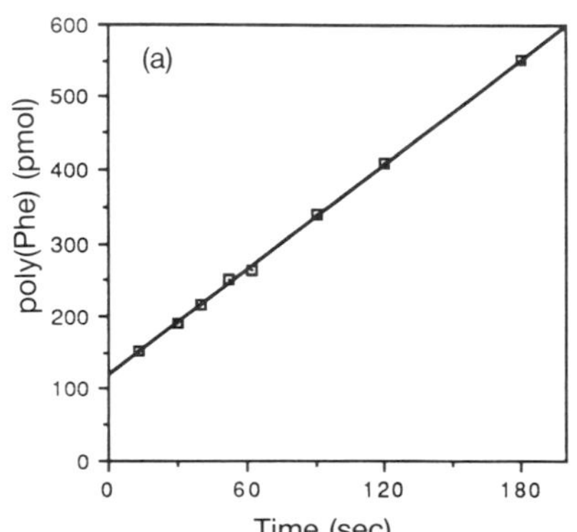

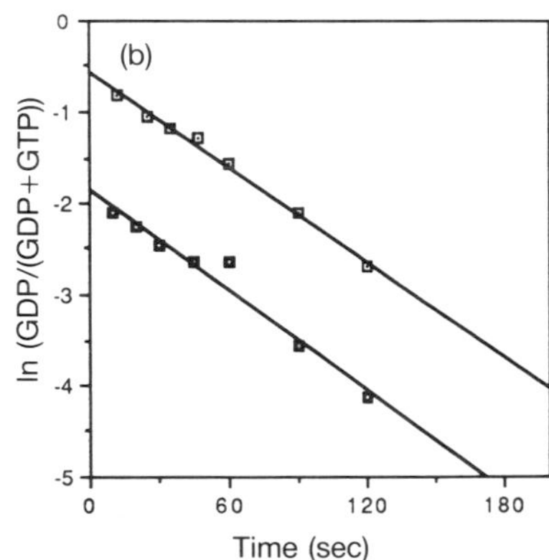

Figure 7. Determination of the number of GTPs hydrolysed per cognate peptide bond in EF-Tu function. In (a) the extent of poly(Phe)-synthesis is plotted as a function of time. The intercept at the *y*-axis is Phe_0 and the slope is C in Equation 11. In (b) a ln-plot of the fraction of GDP on EF-Tu as a function of time with (lower curve) and without (upper curve) cold GDP added to the factor mix. The calculations are described in Section 6.1.

background value (r_b) for times longer than 10 min. With $\Delta r_0(t)=r_0(t)-r_b$ we have when $t \geq 10$ *sec*.

$$\Delta r_0(t)=\Delta r_0(0)e^{-k_d t} \tag{12}$$

where $\Delta r_0(0)=r_0(0)-r_b$. A log plot of $\Delta r_0(t)$ gives a straight line which can be used to obtain $\Delta r_0(0)$ by extrapolation (cf. Equation 8 and *Figure 7b*).

The amount of GTP that has been hydrolysed on EF-Tu (ΔTu_0) during the burst is given by:

$$\Delta Tu_0=[r_0(0)-r_{f0}]G_T \tag{13}$$

where G_T is the total amount of guanine nucleotide in the factor mix containing EF-Tu. G_T can be determined by adding a known amount of GDP (G_0) to an identical factor mix and to obtain from a parallel experiment:

$$\Delta Tu_0=[r_1(0)-r_{f1}]\,(G_T+G_0). \tag{14}$$

Here, $r_1(t)$ is the background uncorrected nucleotide ratio and r_{f1} is the nucleotide ratio in the factor mix. The unknown guanine nucleotide content in the factor mix can now be obtained from:

$$G_T=G_0/\{[r_0(0)-r_{f0}]/[r_1(0)-r_{f1}]-1\} \tag{15}$$

and can be used to calculate ΔTu_0 from Equation 13. The validity of Equation 15 requires that the amount of GTP-hydrolysis is the same in Equations 13 and 14. The criterion we use for this to be true is that the extent of poly(Phe)-synthesis (Phe_0 in Equation 11) is the same with and without the extra guanine nucleotide.

For wild-type EF-Tu, estimates for ΔTu_0 correlate well with estimates for the total concentration of EF-Tu (Tu_0) (Section 5.1). This means that the amount of EF-Tu that can bind GDP can also form EF-Tu·GTP and hydrolyse GTP in translation. We have shown, elsewhere, that those EF-Tu molecules that can hydrolyse GTP are capable of binding to Phe-tRNA^{Phe} (6, 7).

There are two ways to determine the number of GTPs hydrolysed per peptide bond (f_c) from this assay:

(a) From the burst

$$f_c=\Delta Tu_0/Phe_0. \tag{16}$$

(b) From the steady-state translation, provided that $\Delta Tu_0=Tu_0$,

$$f_c=\Delta Tu_0 k_d/C. \tag{17}$$

Protocol 9. f_c measurement

1. Prepare a 70S-mix as in *Protocol 1* which, in addition, contains 2 mM unlabelled GTP.
2. Make two different EF-Tu-mixes (A and B), which are balanced with respect to polymix, and which contain (per 20 μl) EF-Tu (500 pmol), and [^{3}H]GDP (about 100 pmol, 1800 c.p.m./pmol).

3. Add 1500 pmol of unlabelled GDP to mix B.
4. Incubate A and B for 10 min at 37 °C to allow complete exchange between free and bound GDP-molecules. Put A and B on ice after the incubation.
5. Prepare two factor mixes (F_A and F_B) which contain (per 50 μl of final volume) polymix, 10 μl A/P, PK (5 μg), MK (0.3 μg), [^{14}C]Phe (0.6 mM, 5 c.p.m./pmol), PheS (100 units), $tRNA^{Phe}$ (700 pmol), EF-G (250 pmol).
6. Pipette to F_A (per 50 μl of final volume) 20 μl of pre-incubated EF-Tu mix A and to F_B 20 μl of mix B.
7. Split the 70S-mix into four parts (R_{A1}, R_{A2}, R_{B1}, R_{B2}). Split F_A and F_B into two parts (F_{A1}, F_{A2}, and F_{B1}, F_{B2} respectively).
8. Pre-incubate each mix exactly 10 min before translation start.
9. Remove two 20 μl aliquots from F_{A1} and F_{B1} and quench them in 20 μl 20% HCOOH for thin layer chromatographic analysis to obtain r_{f0} and r_{f1} as described in Section 6.1.
10. Start translation for guanine nucleotide analysis by pipetting a suitable volume from F_{A1} (or F_{B1}) into an equal volume of R_{A1} (or R_{B1}) and vortex immediately. Withdraw 20 μl aliquots at times as in *Protocol 6* to determine $r_0(t)$ and $r_1(t)$ and proceed as described (steps 5–10) there.
11. Start translation for poly(Phe)-analysis by pipetting a suitable volume of F_{A2} (or F_{B2}) into an equal volume of R_{A2} (or R_{B2}) and vortex immediately.
12. Withdraw 50 μl aliquots, at the same times as used for the guanine nucleotide analysis. To determine *Phe*(t) quench the samples in 5% TCA. Proceed with sample handling and counting as described in *Protocol 1*.

Calculate r_b from $r_0(t)$ or $r_1(t)$ at 10 min. If r_0(10 min) and r_1(10 min) differ significantly this indicates a faulty assay. Calculate the differences $\Delta r_0(t) = r_0(t) - r_b$ and $\Delta r_1(t) = r_1(t) - r_b$ at all time points. Plot the natural logarithms of $\Delta r_0(t)$ and $\Delta r_1(t)$ versus time. For a homogeneous population of EF-Tu these plots should give two straight lines with identical slopes. Determine the GDP exchange rate (k_d) from the slope of the upper (and more accurate) line according to Equation 8. Extrapolate the upper line to zero time to obtain $\Delta r_0(0)$ and the lower one to obtain $\Delta r_1(0)$. Calculate with the help of these parameters and the guanine nucleotide ratios r_{f0} and r_{f1} in the respective factor mixes G_T from Equation 15 and, subsequently, ΔTu_0 from Equation 13. Plot the extent of poly(Phe)-synthesis as a function of time for the A2 as well as the B2 case. If there is a significant difference between the two lines the experiment is probably faulty. Extrapolate the straight line to zero time to obtain Phe_0 according to Equation 11 and calculate f_c from Equation 16. Alternatively, determine the slope C of the line and calculate f_c from Equation 17.

6.2 The number of GTPs hydrolysed in EF-Tu function per incorrectly formed peptide bond (f_w)

Experiments to characterize the role of kinetic proof-reading (23, 24) in *E. coli* translation suggest a second, GTP-driven, tRNA selection step on the ribosome (25, 26). The number of GTPs hydrolysed per incorrectly formed peptide bond (f_w) normalized to the number of GTPs per correct peptide bond (f_c, Section 6.1) is a measure of how the accuracy (A, Section 2.2) is partitioned in initial selection (I) and proof-reading ($F=f_w/f_c$):

$$A=IF=If_w/f_c. \tag{18}$$

Steady-state methods to determine F have been described previously (26, 27). The present experiment measures the number of GTP-molecules that are hydrolysed in the formation of an incorrect N-acetyl–Phe–aa dipeptide in the absence of EF-Ts and EF-G (9). ^{3}H-labelled EF-Tu·GTP in complex with aa–tRNA (EF_1) is consumed with a rate determined by the concentration of active ribosomes $[R_0]$ and k_{cat}/K_M for GTP-hydrolysis (R_{GTP}):

$$\frac{dEF_1}{dt}=-R_{GTP}[R_0]EF_1=-qEF_1. \tag{19}$$

We have defined $q=R_{GTP}[R_0]$.

The amount of ribosomes (R_0) is chosen to be much larger than the amount of dipeptides that can be formed from a single round of EF-Tu so that q is approximately constant. The time development of the concentration of [^{3}H]GDP on EF-Tu ($TuGDP$) is described by:

$$\frac{dTuGDP}{dt}=2qEF_1-k_dTuGDP. \tag{20}$$

Here, k_d is the dissociation rate of GDP from EF-Tu in the absence of EF-Ts as before. The $2q$ that appears with EF_1 to the right is because there are two molecules of EF-Tu·GTP in EF_1 (6, 7). The solutions to Equations 19 and 20 are:

$$EF_1(t)=EF_1(0)e^{-qt}=\tfrac{1}{2}\Delta Tu_0e^{-qt}. \tag{21}$$

$$TuGDP(t)=(Tu_0-\Delta Tu_0)e^{-k_dt}+\Delta Tu_0q(e^{-qt}-e^{-k_dt})/(k_d-q). \tag{22}$$

When q is much larger than k_d it is a good approximation to write for short times:

$$TuGDP(t)=Tu_0-\Delta Tu_0+\Delta Tu_0(1-e^{-qt}). \tag{23}$$

For dipeptide formation:

$$dip(t)=\Delta Tu_0/f_w(1-e^{-qt}) \tag{24}$$

where f_w is the number of GTPs hydrolysed per incorrect peptide bond. The parameters in Equations 22 and 23 are all related to guanine nucleotide ratios discussed in Sections 5.1 and 6.1 according to:

$$TuGDP(t) = [r_0(t) - r_b]G_T$$
$$Tu_0 - \Delta Tu_0 = [r_{f0} - r_b]G_T$$
$$\Delta Tu_0 = [r_0(0) - r_{f0}]G_T. \tag{25}$$

G_T can be obtained by isotope dilution as described in Section 6.1. The proof-reading parameter (f_w) is obtained by comparing the extent of GTP-hydrolysis ($\Delta Tu_0(1-e^{-qt})$) in Equation (23) with the amount of dipeptides formed (*dip*(t) in Equation 24). The rate constant q reveals how well the ribosome can initially discriminate against an incorrect tRNA provided that R_0 is known.

Protocol 10. f_w-determination

1. Prepare a 70S-mix with balanced polymix and (per 50 μl) ribosomes (30 pmol or more of active ones), [^{3}H]NAc (30% in excess over total ribosomes, 500 c.p.m./pmol), poly(U) (660 μg or more), GTP (2 mM).
2. Split the 70S-mix into two mixes (R_A, R_B).
3. Prepare factor mixes A and B as in Section 6.1 with $tRNA^{Phe}$ replaced by the isoacceptor under investigation.
4. Incubate the mixes exactly 10 min.
5. Before translation start remove from A and B two 20 μl samples to determine r_{f0} and r_{f1}.
6. Pipette a volume of A (or B) into R_A (or R_B), vortex, and remove 100 μl aliquots for quenching in 100 μl 20% formic acid.
7. Spin the samples 15 min in an Eppendorf centrifuge.
8. Remove the supernatant for thin-layer analysis of guanine nucleotide ratios (as in *Protocol 6*).
9. Add to each pellet 200 μl 0.5 M KOH, incubate for 15 min at 37 °C to hydrolyse the aminoacyl- and peptidyl-tRNA's.
10. Add 5 μl concentrated formic acid, vortex, and spin each sample for 15 min in the Eppendorf centrifuge.
11. Remove 150 μl of the supernatant for HPLC analysis with the column Li Chrosorb (RP-18, Merck) equilibrated with 43% methanol and 0.1% TFA in water.
12. Quantitate the amount of dipeptides by comparing the ^{3}H-activity in the dipeptide peak with the ^{3}H-activity in dipeptide plus [^{3}H]NAc peaks, which sum up to the total (known) amount of [^{3}H]NAc originally present in the 70S-mix. Calculate $r_0(t)$, r_b, and r_{f0} to obtain $TuGDP(t)$ and $Tu_0 - \Delta Tu_0$ according to Equation 25. Relate $TuGDP(t) - Tu_0 + \Delta Tu_0$ in Equation 23

to $dip(t)$ in Equation 24, and obtain f_w from their ratio as a suitable average over the measured time points.

---

7. Binding of EF-Tu · GTP to aminoacyl-tRNA

The purpose of this experiment (6, 7) is to provide a fast and simple method to determine binding constants and stoichiometries for the binding between EF-Tu · GTP and aminoacyl-tRNAs under conditions which are optimal for translation (cf., Section 3).

7.1 How the binding assay works

The assay is based on the observation that the spontaneous rate of deacylation of free tRNA isoacceptors (k_f) is much faster than the deacylation rate (k_b) of tRNAs in complex with EF-Tu · GTP (28, 29). The fractions of free aminoacyl–tRNA (T_f/T_0) and tRNA bound to EF-Tu · GTP (T_b/T_0) can be deduced from the current deacylation rate (k) as follows (6, 7, 29).

$$T_f/T_0 = (k - k_b)/(k_f - k_b), \qquad T_b/T_0 = (k_f - k)/(k_f - k_b) \tag{26}$$

To measure k a charging system is used to aminoacylate all tRNA with ^{14}C-labelled amino acid during a pre-incubation period. The aminoacyl–tRNA synthetase concentration is kept high enough to charge all tRNA, and low enough so that the spontaneous deacylation rate dominates over the enzyme catalysed one. This is important to ensure that the deacylation rate of free aminoacyl–tRNA (k_f) is the same at all values of $[T_f]$ when the concentration of EF-Tu · GTP is varied. At time zero a small amount of high specific activity ^{3}H-labelled amino acid is added. The conditions are chosen such that the dilution of the ^{14}C-label by the addition of [^{3}H]-amino acid is negligible. This ensures that the specific activity of the ^{14}C-amino acid is unchanged during the incubation. The ^{3}H-label is, at short times, present only in the pool of free amino acids. As time goes on the ^{14}C-labelled amino acid originally on tRNA is replaced by ^{14}C- (of the same specific activity as before) and ^{3}H-labelled amino acids from the free pool. The rate at which ^{3}H-labelled aminoacyl–tRNA (a_h) increases from zero to a constant value (a_c) is determined by the deacylation rate (k) (24).

$$a_h = a_c(1 - e^{-kt}) \tag{27}$$

a_c is the (constant) amount of ^{14}C-labelled aminoacyl–tRNA. The rate k can be obtained from a single observation time T, since both a_h and a_c are known at each time point (*Protocol 11*):

$$k = -(1/T)\ln(1 - a_h/a_c). \tag{28}$$

From a Scatchard plot (6, 7), where $(T_b/Tu_0)(1/T_f)$ is plotted versus T_b/Tu_0, it follows that the binding constant for the EF-Tu · GTP binding to Phe-tRNA^{Phe} is about 10^{-7} M (slope of the line) and that there are two molecules of EF-Tu · GTP bound to one molecule of Phe–tRNA^{Phe} (inverse of the intercept at the x-axis).

Protocol 11. EF-Tu · GTP binding to aminoacyl-tRNA

1. Prepare a charge-mix (balanced in polymix) which, after the addition of [^{3}H]-amino acid, contains (per 100 μl) tRNA (normally between 10^{-7} M and 10^{-6} M), ^{14}C-labelled amino acid (0.3 mM, specific activity between 5 and 30 c.p.m./pmol), 10 μl A/P, GTP (1 mM), PK (5 μg), MK (0.3 μg), aminoacyl-tRNA synthetase (typically 1 unit).
2. Add to one part (Ch_A) of the mix (per 100 μl final volume), 10 μl polymix, and 10 μl ^{3}H-labelled amino acid (typically 200 pmol, 1500 c.p.m./pmol).
3. Make EF-Tu dilutions in 10 μl polymix and add 80 μl from the other part of the charge mix (Ch_B) to each EF-Tu dilution.
4. Pre-incubate Ch_A and all Ch_B-tubes for 20 min. Then quench aliquots (100 μl) from Ch_A in 5% ice cold TCA for subsequent analysis of charging as in *Protocol 2*.
5. Add after 20 min pre-incubation, to the different Ch_B-tubes, 10 μl of [^{3}H]-amino acid which has been pre-warmed to 37 °C.
6. Incubate the Ch_B-tubes before quenching with ice cold 5% TCA. A suitable incubation time is one which gives a precise estimate of k in Equation 28. It depends on the tRNA isoacceptor and on the amount of EF-Tu · GTP. The ratio a_h/a_c should not be too low (the determination of a_h becomes inaccurate) and not too near a value of one (a small error in a_h/a_c leads to a large error in k).
7. Treat the samples as in *Protocol 1* except that they should be kept in ice cold TCA until filtering.
8. Rinse the filters with ice cold TCA and subsequently with ice cold 2-propanol before drying them in a baking oven. Take 10 μl samples from the ^{14}C-labelled amino acid directly on to filters for ^{14}C-specific activity standards. Use the charging levels from the Ch_A-samples (where it is known that $a_c = a_h$ since both isotopes were added simultaneously) to relate the ^{3}H-specific activity to the ^{14}C-specific activity. Calculate the a_h/a_c ratio and determine k (Equation 28) for each EF-Tu · GTP concentration in the Ch_B-mixes.

---

8. Summary

The steady-state methods described above have shown that poly(U) can be translated with a rate (k_{cat}) near 10 Phe/sec (Sections 3, 4.1–2). Two molecules of EF-Tu · GTP form a complex (EF_1) with one molecule of aminoacyl–tRNA

(Section 7.1). EF_1 enters the ribosomal A-site with an effective rate (k_{cat1}/K_{M1}) near $3 \times 10^7\ M^{-1}sec^{-1}$ (Sections 4.2, 5.1). Two molecules of GTP are hydrolysed per peptide bond in EF-Tu function (Section 6.1). The EF-G·GTP binary complex (EF_2) enters the ribosome with $k_{cat2}/K_{M2} = 4.3 \times 10^7\ M^{-1}sec^{-1}$ (Sections 4.2–3, 5.2). The EF-G cycle rate is near 70 sec^{-1} (Sections 4.2–3, 5.2). EF-Ts associates slowly with EF-Tu·GDP with $k_{cat3}/K_{M3} = 5 \times 10^6\ M^{-1}sec^{-1}$ (Sections 4.4, 5.3). At the Mg^{2+} concentration which maximizes the rate of ribosomal elongation (Sections 2, 3) Leu–$tRNA_4^{Leu}$ gives, in competition with Phe–$tRNA^{Phe}$, a normalized error level near 2×10^{-4} (Sections 2.2–3). The accuracy is enhanced by a factor of 50 by a proof-reading step (Sections 6.1–2).

9. Purification methods for enzymes necessary for *in vitro* poly(Phe) synthesis

Here we review purification procedures for EF-G (30), EF-Tu (31, 10), EF-Ts (32), Phe–tRNA synthetase (PheS) and Leu–tRNA synthetase (LeuS) (10) and describe our own modifications. All purifications described below are made at 4 °C. It is convenient to separate EF-G, EF-Tu, PheS, and LeuS during the same preparation and to purify EF-Ts separately.

9.1 Buffers

(a) Buffer A: 64.4 mM Tris–HCl (pH 7.5), 10 mM $MgCl_2$, 0.5 mM DTE, 100 μM phenylmethylsulphonyl fluoride (PMSF), 3 mM NaN_3.

(b) Potassium-phosphate buffer: 50 mM potassium phosphate (pH 7.5), 10% glycerol, 100 μM PMSF, 30 μM GDP, 3 mM NaN_3, 1 mM DTE.

(c) TEA buffer: 10 mM triethanolamine (TEA), 100 mM KCl, 14 mM 2-mercaptoethanol, the pH is adjusted with HCl to 6.9.

(d) TMK buffer: 20 mM Tris–HCl (pH 7.5), 10 mM $MgCl_2$, 50 mM KCl, 1 mM DTE.

9.2 Purification of EF-Tu, EF-G, EF-Ts, PheS, and LeuS

Protocol 12. Preliminary separation for EF-Tu, EF-G, PheS, and LeuS

1. Suspend 500 g frozen *E. coli* MRE 600 cells in 300 ml of buffer A. Add 6000 units of DNase 1. Break the cells using the French-press (5000–7000 p.s.i.). Add three volumes of buffer A to the lysate and centrifuge 2 h at $17\,000 \times g$.
2. Pool all supernatants and apply them at a rate of 160 ml/h to a DEAE-sepharose CL-6B column (5×85 cm, Pharmacia) pre-equilibrated with buffer A.
3. Wash the column with 500 ml buffer A and elute with a 2×3 litre salt

gradient in buffer A from 0 to 0.4 M NaCl and test those fractions which have conductivities between 2 and 4 mS for PheS and LeuS activity. (See *Protocol 18* for charging assay.)

4. Pool the fractions with PheS activity (PheS elutes before LeuS) so that contamination by LeuS activity is minimized.
5. Fractionate the PheS pool with $(NH_4)_2SO_4$. (For all $(NH_4)_2SO_4$ fractionations, add solid $(NH_4)_2SO_4$ gradually into the solution during 30 min, stir for another 30 min, and centrifuge for 30 min at 10 000 × *g*.) Fractionate the PheS pool with 0.23 g/ml $(NH_4)_2SO_4$ by discarding the pellet and saving the supernatant. Then fractionate with 0.063 g/ml $(NH_4)_2SO_4$, discard the supernatant, and save the pellet containing PheS.
6. Dissolve the pellet in a minimal volume of potassium phosphate buffer containing 50% glycerol and store it at −20 °C for further purification (see *Protocol 15*).
7. Pool the LeuS activity from the DEAE-sepharose CL 6B column.
8. Fractionate with 0.3 g/ml $(NH_4)_2SO_4$, save the pellet as EF-G and EF-Tu.
9. Fractionate the supernatant with 0.2 g/ml $(NH_4)_2SO_4$. Save the pellet as LeuS and dissolve it in a minimal volume of potassium phosphate buffer containing 50% glycerol and store it at −20 °C for further purification (see *Protocol 16*).
10. To continue with the purification of EF-G and EF-Tu, dissolve their pellets in a minimal volume of potassium phosphate buffer, and apply to an Ultrogel AcA 44 (LKB-Beckman) column (5 × 90 cm) pre-equilibrated with the same buffer.
11. Elute EF-G and EF-Tu at a flow rate of 40 ml/h.
12. Assay the EF-G activity as GTP binding to the ribosome in complex with EF-G and fusidic acid on nitrocellulose filters (33) and assay the EF-Tu activity from GDP retention on nitrocellulose filters (19).
13. Pool both factors avoiding cross-contamination.
14. Precipitate the EF-Tu pool with 0.5 g/ml $(NH_4)_2SO_4$. Dissolve the pellet in a minimal volume of potassium phosphate buffer. For further purification, either rerun the EF-Tu pool on the AcA column or, alternatively, crystallize the EF-Tu.

Protocol 13. Crystallization of EF-Tu with $(NH_4)_2SO_4$

1. Dialyse the EF-Tu fraction against 25 mM Tris–HCl (pH 7.8), 5 mM $MgCl_2$, 1 mM DTE, 50 μM GDP and 25% $(NH_4)_2SO_4$.
2. Centrifuge at 30 000 × *g* for 30 min to remove precipitates.

3. Measure the volume and place it in a tube. Add 22.4 μl of 80% $(NH_4)_2SO_4$ per ml of EF-Tu solution per day to the above crystallization buffer. Continue for 10 days and collect the EF-Tu crystals by centrifugation.
4. Finally, dissolve in 3–5 ml and dialyse EF-Tu against polymix buffer containing 10 μM GDP and store in aliquots at −80 °C.

Protocol 14. Further purification of EF-G

1. Apply the EF-G pool from the AcA column (*Protocol 12*, step 13) directly at a rate of 60 ml/h to DEAE-cellulose (DE-52, Whatman) (2.5 × 30 cm), equilibrated with potassium phosphate buffer.
2. Wash the column with 100 ml of the same buffer. Apply a 500 ml × 500 ml gradient of potassium phosphate buffer from 50 mM potassium phosphate at pH 7.5 to 250 mM potassium phosphate at pH 6.9.
3. Pool the EF-G activity and precipitate with 0.5 g/ml $(NH_4)_2SO_4$.
4. Dissolve the EF-G pellet in a minimum volume of TEA buffer.
5. Dialyse against TEA and apply at a rate of 60 ml/h to a glass column (1.5 × 60 cm, CPG 007, Electro-Nucleonics, Inc.), equilibrated with the same buffer.
6. Apply a 400 ml × 400 ml gradient from pH 7.0 to pH 9.0 in TEA buffer.
7. Pool the EF-G activity and precipitate with 0.5 g/ml $(NH_4)_2SO_4$.
8. Dissolve in 2–3 ml of polymix buffer and dialyse EF-G against polymix buffer for 12 h and store aliquots at −80 °C.

Protocol 15. Further purification of PheS

1. Apply the PheS pool from the DEAE-sepharose CL 6B column (*Protocol 12* step 6) directly to a Sephacryl S-300 column (Pharmacia) (5 × 90 cm) equilibrated with potassium phosphate buffer.
2. Develop the column at a rate of 40 ml/h.
3. Assay the fractions for PheS and LeuS activity.
4. Pool the PheS avoiding LeuS contamination.
5. Precipitate PheS with 0.5 g/ml $(NH_4)_2SO_4$.
6. Dissolve in a minimal volume of potassium phosphate buffer and apply to an AcA column (5 × 90 cm) in the same buffer.
7. Pool the PheS monitored by its charging activity (see below).

8. Apply at a rate of 60 ml/h directly to a DEAE-cellulose (DE-52) column (2.5 × 20 cm) equilibrated in the same buffer.
9. Apply a 200 ml × 200 ml gradient of potassium phosphate buffer from 50 mM potassium phosphate at pH 7.5–250 mM potassium phosphate at pH 6.5.
10. Pool the PheS activity and precipitate with 0.5 g/ml $(NH_4)_2SO_4$.
11. Dissolve in polymix and dialyse first against polymix and then against polymix containing 50% glycerol and store at −20 °C.

Protocol 16. Further purification of LeuS

1. Apply the LeuS pool from a DEAE-sepharose CL 6B column (*Protocol 12* step 9) directly on to an AcA column equilibrated with potassium phosphate buffer.
2. Run the column at a rate of 30 ml/h.
3. Assay the fractions for Leu charging and pool the LeuS activity, avoiding EF-Tu.
4. Apply the LeuS pool directly on to a DEAE-cellulose column (DE-52, 2.5 × 30 cm), equilibrated in the same buffer. Elute with a 400 × 400 ml gradient of potassium phosphate buffer from 50 mM at pH 7.5–250 mM at pH 6.9.
5. Pool the LeuS activity and precipitate it with 0.5 g $(NH_4)_2SO_4$.
6. Dissolve and dialyse it in TEA buffer.
7. Apply to a porous glass column (CPG 007) (1.5 × 60 cm, in TEA buffer) at a rate of 60 ml/h. Elute with a 250 ml × 250 ml gradient of TEA buffer from pH 6.9 to pH 9.2.
8. Pool the LeuS activity avoiding EF-G, which elutes after LeuS.
9. Precipitate the LeuS with 0.5 g/ml $(NH_4)_2SO_4$, dissolve in polymix, dialyse first against polymix, and then against polymix containing 50% glycerol, and store at −20 °C.

Protocol 17. Purification of EF-Ts

1. Suspend 400 g frozen *E. coli* MRE 600 cells in 250 ml TMK buffer. Add 4000 units DNase I. French press at 5000–7000 p.s.i. Add 250 ml TMK to the lysate and centrifuge for 70 min at 17 000 × *g*.
2. Fractionate with 0.22 g/ml $(NH_4)_2SO_4$, save the supernatant, fractionate again with 0.18 g/ml $(NH_4)_2SO_4$.

3. Discard the supernatant, dissolve the pellet in about 200 ml TMK buffer, and dialyse against 3 × 5 litres of TMK buffer over about 36 h.
4. Centrifuge for 15 min at 17 000 × g to clarify the solution.
5. Apply at a flow rate of 150 ml/h to a DEAE-sepharose CL 6B column (5 × 30 cm, pre-equilibrated with TMK buffer).
6. Wash the column with 500 ml of the same buffer and apply a 1500 ml × 1500 ml gradient of TMK buffer from 50 mM KCl to 350 mM KCl. Collect 10 ml fractions and measure the conductivity.
7. Analyse the fractions with a conductivity between 3.5 and 4.5 mS on 10% SDS gels with EF-Tu and EF-Ts markers.
8. Pool the EF-Ts fractions.
9. Precipitate with 0.5 g/ml $(NH_4)_2SO_4$ then centrifuge for 40 min at 17 000 × g. Discard the supernatant.
10. Dissolve the pellet in a minimal volume of TMK buffer containing 30 μM GDP. Add an additional 5 mg of GDP into the EF-Ts solution.
11. Dialyse against TMK buffer containing 30 μM GDP (2 × 2 litre) for 12 h.
12. Centrifuge to clear the solution and apply it to another DEAE–sepharose CL 6B column (1.5 × 20 cm) equilibrated with TMK buffer containing 10 μM GDP.
13. Wash the column with 100 ml of the same buffer and elute the EF-Ts with a 100 ml × 100 ml gradient of TMK buffer containing 10 μM GDP from 50 mM to 350 mM KCl at a rate of 60 ml/h.
14. Analyse the fractions on 10% acrylamide gels.
15. Pool and collect EF-Ts by precipitating with 0.6 g/ml $(NH_4)_2SO_4$.
16. Dissolve in polymix (2–3 ml) and dialyse for 12 h against polymix. Store at −80 °C in aliquots.

Protocol 18. Assay for tRNA synthetase activity

1. Prepare a charge mix (balanced in polymix) containing (per 100 μl) 10 μl A/P, PK (5 μg), MK (0.3 μg), bulk tRNA (500 pmol $tRNA^{Phe}$, 1500 pmol $tRNA^{Leu}$ isoacceptors and 0.3 mM [^{14}C]-Phe and/or [^{3}H]-Leu (100 c.p.m./pmol).
2. Pipette 10 μl of fractions containing PheS or LeuS into tubes.
3. Pre-incubate the charge mix and synthetase samples at 37 °C for 2 min.
4. Pipette 90 μl charge mix into the synthetase fractions, vortex, and quench after 15 sec with 5 ml cold 5% TCA.
5. Proceed as in *Protocol 11*.

Acknowledgements

This work was supported by grants from the Swedish Natural Science Research Council and from the Swedish Cancer Society. We thank Karin Olsson for typing the manuscript.

References

1. Pedersen, S. (1984). *EMBO J.* **3**, 2895.
2. Kurland, C. G. and Gallant, J. A. (1986). In *Accuracy in Molecular Processes* (ed. T. B. L. Kirkwood, R. F. Rosenberger, and D. J. Galas). Chapman and Hall, London.
3. Luria, S. E. (1960). In *The Bacteria*, Vol. 1 (ed. I. C. Gunsalus and R. Y. Stanier), p. 1. Academic Press, New York.
4. Jelenc, P. C. and Kurland, C. G. (1979). *Proc. Nat. Acad. Sci. USA*, **76**, 3174.
5. Petterson, I. and Kurland, C. G. (1980). *Proc. Nat. Acad. Sci. USA*, **77**, 4007.
6. Ehrenberg, M. and Kurland, C. G. (1988). In *Methods in Enzymology*, Vol. 164 (ed. H. F. Noller and K. Moldave), p. 611. Academic Press, San Diego.
7. Tapio, S., Bilgin, N., and Ehrenberg, M. (1989). *Eur. J. Biochem.*, in press.
8. Bilgin, N., Kirsebom, L. A., Ehrenberg, M., and Kurland, C. G. (1988). *Biochimie*, **70**, 611.
9. Ruusala, T., Ehrenberg, M., and Kurland, C. G. (1982). *EMBO J.*, **1**, 75.
10. Wagner, E. G. H., Jelenc, P. C., Ehrenberg, M., and Kurland, C. G. (1982). *Eur. J. Biochem.*, **122**, 193.
11. Lucas-Lenard, J. and Lipmann, F. (1967). *Proc. Nat. Acad. Sci. USA*, **57**, 1050.
12. Jelenc, P. C. (1980). *Anal. Biochem.*, **105**, 369.
13. Gillam, J. C., Millward, S., Blew, D., Van Tigerstrom, M., Wimmer, E., and Tener, G. M. (1967). *Biochemistry*, **6**, 3043.
14. Holmes, N. M., Hurd, R. E., Reid, B. R., Rimerman, R. A., and Hatfield, G. W. (1975). *Proc. Nat. Acad. Sci. USA*, **72**, 1068.
15. Rappoport, S. and Lapidot, Y. (1974). In *Methods in Enzymology*, Vol. 29E (ed. L. Grossman and K. Moldave), p. 685. Academic Press, NY.
16. Thompson, R. C., Dix, D. B., Gerson, R. B., and Karim, A. M. (1981). *J. Biol. Chem.* **256**, 6676.
17. Dix, D. B. and Thompson, R. C. (1986). *J. Biol. Chem.*, **261**, 4868.
18. Ehrenberg, M. and Kurland, C. G. (1984). *Prog. Nucleic Acid Res. Mol. Biol.*, **31**, 191.
19. Miller, D. L., Cashel, M., and Weissbach, H. (1973). *Arch. Biochem. Biophys.*, **154**, 675.
20. Fersht, A. (1977). *Enzyme Structure and Mechanism.* W. H. Freeman and Company, San Francisco.
21. Arai, U., Kawakita, M., and Kaziro, Y. (1974). *J. Biochem., (Tokyo)*, **76**, 293.
22. Kaziro, Y. (1978). *Biochim. Biophys. Acta*, **505**, 95.
23. Hopfield, J. J. (1974). *Proc. Nat. Acad. Sci. USA*, **71**, 4135.
24. Ninio, J. (1975). *Biochimie*, **57**, 587.
25. Thompson, R. C. and Stone, P. J. (1977). *Proc. Nat. Acad. Sci. USA*, **74**, 198.
26. Ruusala, T., Ehrenberg, M., and Kurland, C. G. (1982). *EMBO J.*, **1**, 741.
27. Ehrenberg, M., Kurland, C. G., and Ruusala, T. (1986). *Biochimie*, **68**, 261.

28. Beres, L. and Lucas-Lenard, J. (1973). *Biochemistry*, **12**, 3998.
29. Pingoud, A., Urbanke, C., Krauss, G., Peters, F. K., and Maass, G. (1977). *Eur. J. Biochem.*, **78**, 403.
30. Wagner, E. G. H. and Kurland, C. G. (1980). *Mol. Gen. Genet.*, **180**, 139.
31. Leberman, R., Antonsson, B., Giovanelli, R., Guariguata, R., Schumann, R., and Wittinghofer, A. (1980). *Anal. Biochem.*, **104**, 29.
32. Arai, K. I., Kawakita, M., and Kaziro, Y. (1972). *J. Biol. Chem.*, **247**, 7029.
33. Bodley, J. W., Weissbach, H., and Brot, N. (1974). In *Methods in Enzymology*, Vol. 30 (ed. K. Moldave and L. Grossman), p. 235. Academic Press, NY.

7

New techniques for the analysis of intra-RNA and RNA-protein cross-linking data from ribosomes

RICHARD BRIMACOMBE, BARBARA GREUER, HEINZ GULLE, MICHAEL KOSACK, PHILIP MITCHELL, MONIKA OßWALD, KATRIN STADE, and WOLFGANG STIEGE

1. Introduction

Cross-linking methodology provides one of several approaches that can be applied to the investigation of the tertiary or quaternary structure of complex biological particles. In the case of the ribosome, cross-links can be introduced between the various ribosomal proteins (inter-protein cross-linking), between different regions of the RNA (intra-RNA cross-linking), or between individual proteins and RNA (RNA–protein cross-linking). Ligands such as tRNA, mRNA, or protein factors can of course also be implicated in cross-linking studies. This chapter is concerned specifically with cross-linking reactions involving the rRNA, i.e. with intra-RNA and RNA–protein cross-linking.

Cross-linking is a purely 'topographical' method, that is to say it defines neighbourhoods between the various components or domains of the system being studied, without necessarily implying any strong physical interaction between the components that have been linked together. The precision with which these neighbourhoods can be identified is obviously dependent on the level to which the analysis of the cross-linked products and cross-link sites can be pursued, and we will describe the techniques which have been developed in our laboratory for the determination of multiple cross-link sites on the rRNA at the nucleotide level. The result of these studies has been the derivation of a detailed model for the three-dimensional arrangement of the 16S rRNA of *E. coli in situ* in the 30S ribosomal subunit (1, 2). Our methodology has recently been reported in some detail (3), but the latter publication does not include a number of significant improvements to the techniques which have been incorporated during the last two years. We therefore concentrate in this chapter on these newer aspects of the methodology, and give detailed examples of their application in both intra-RNA and RNA–protein cross-linking experiments. However, for a 'Practical

Approach' book, it is appropriate to begin by discussing some general considerations that are important in deciding which experimental strategies can or cannot be applied to this type of cross-linking study, or indeed whether a cross-linking approach is a priori suitable for a particular research problem.

2. General considerations

2.1 Cross-linking versus 'higher-order structure analysis' or 'foot-printing' techniques

Higher-order structure analysis and foot-printing techniques (see e.g. references 4, 5, and Chapter 11 by Garrett *et al.* in this volume) offer an alternative approach to cross-linking for topographical studies on rRNA at the nucleotide level. However, the two types of methodology give somewhat different types of information, and each has its own advantages and disadvantages.

In the higher-order structure analysis approach, the substrate (either a ribosomal subunit or isolated RNA molecule) is subjected to a mild nuclease digestion or treatment with a nucleotide-specific reagent, and the sites on the RNA that have been attacked are then determined. If the analysis is carried out in the presence and absence of a ligand (such as tRNA, or an individual ribosomal protein or factor), then the difference between the two sets of data provides a 'foot-print' for the ligand concerned on the rRNA.

The question being asked in such an experiment is a simple one—namely, is a particular base susceptible to attack by a given nuclease or chemical reagent, or is it not?—and in consequence the data can be analysed in a relatively simple manner. Usually, this is now done by 'scanning' the RNA with reverse transcriptase, using a set of complementary deoxyoligonucleotide primers located at suitable intervals along the rRNA (5). The reverse transcriptase patterns are compared with those of suitable untreated control samples, and in this way a large amount of data can be rapidly collected. In the case of a higher-order structure analysis, for instance, sets of individual bases can be identified that are likely to be involved in tertiary interactions within the rRNA (cf. reference 6). The method, however, has the important disadvantage that it gives no hint a priori as to which base is partnered with which in these interactions, and the usefulness of the information clearly diminishes as the size of the RNA molecule under study becomes larger.

Similarly, the extension of the approach to a foot-printing investigation only identifies residues in the rRNA whose susceptibility to modification or attack is altered in the presence of the ligand. The altered susceptibilities may be a result of direct contacts with the ligand concerned, but they may equally well be allosteric effects caused by a ligand-dependent rearrangement of the RNA structure. This type of situation has been observed regularly both in tRNA (7) and ribosomal protein foot-printing studies (e.g. 8), and in the absence of corroborative information there is no way of distinguishing an allosteric effect from a 'genuine foot-print'.

In contrast, the cross-linking approach offers the dual advantage that only direct contacts or neighbourhoods between ligands and the rRNA are observed, uncomplicated by allosteric effects, and that both 'partners' involved in a particular contact or neighbourhood can in principle be positively identified. On the other hand, the analysis of the data from a cross-linking experiment is a lot more difficult than is the case from the approaches just described, at least in regard to the analysis of cross-link sites at the nucleotide level on the rRNA. This is due to the fact that many different cross-links are likely to be present in the same sample, and they cannot be located simply by 'scanning' the RNA for effects as in a higher-order structure analysis. Rather, each individual cross-linked complex must first be isolated in some manner if both components of the cross-link are to be identified and matched to one another. To this end a number of factors need to be considered, which are discussed in the following sections.

2.2 Choice of cross-linking reagent

The successful analysis of a cross-link site on the rRNA requires that the cross-link is kept intact up to the end of the experimental procedure. This is because, if the cross-link is prematurely reversed, then the cross-link site on the RNA fragment becomes analytically indistinguishable from other sites, where either a 'monovalent' reaction with the cross-linking agent or else some other irrelevant side-reaction has taken place. In general this has the consequence that the chance of localizing the cross-link site to an individual nucleotide is lost. We therefore see no advantage in applying reversible cross-linking agents to intra-RNA or RNA–protein cross-linking studies with the ribosome, and all our experimental procedures have been designed for use with irreversible cross-linking systems.

The simplest cross-linking method is direct irradiation with ultraviolet light (e.g. 9). This method has the advantages that the dose level can be controlled accurately, and the cross-linking reaction can if required be carried out *in vivo*, in growing cell cultures (10). Alternatively, bifunctional chemical reagents may be used, but although many such compounds have been reported in the literature, only a few have found a useful application. In our laboratory we use the simple symmetrical reagent *bis*-(2-chloroethyl)-methylamine ('nitrogen mustard' (11, 12)), and the two heterobifunctional reagents, 2-iminothiolane (13, 14) and methyl *p*-azidophenyl acetimidate ('APAI' (15, 16)). All three reagents are relatively short, with a maximum cross-linking span of *c*., 5–7 Å.

Nitrogen mustard reacts with both RNA and proteins, and therefore leads to simultaneous formation of protein–protein, protein–RNA and intra-RNA cross-links. The other two reagents both first react specifically via their imido functions with the ε-amino groups of lysine residues in the proteins, and a non-specific reaction with the RNA is then induced by a very brief exposure to ultraviolet irradiation. APAI and iminothiolane used in this way are thus exclusively RNA–protein cross-linking agents, and reagents based on this principle—namely, 'specific first step, non-specific second step'—have in our hands generated the largest amount of useful data (1, 14, 16). In contrast, bifunctional reagents that are too non-specific in both steps (e.g. various aldehydes) lead to many unwanted side-reactions, whereas reagents that are designed with too high a specificity in both steps may pre-suppose neighbourhoods between reactive groups which simply do not exist in the ribosome, and as a result no cross-linking will be observed. In this context it should also be noted that in the complex micro-environment of the ribosome, a cross-linking reagent may often exhibit a different reactivity than that expected from its 'standard' chemistry (see e.g. 12, 17). In the interpretation of cross-link site analyses, it is therefore dangerous to make any assumptions as to where the reagent 'should have' reacted (e.g., with a particular nucleoside or amino acid).

When working with bifunctional reagents, it must be remembered that a high level of 'monovalent' reaction will occur. For example, in the case of APAI it has been estimated that 20–50 molecules of the reagent react monovalently with the ribosomal proteins for every one molecule which leads to formation of a cross-link (18). If this ratio becomes too unfavourable, then severe aggregation problems are likely to be encountered, and in this connection some practical criteria for preliminary screening of potential cross-linking reagents have been described previously (3). It is in any event advisable to keep the level of reaction as low as possible, since this reduces both the level of unwanted side-reactions (which can interfere with the subsequent cross-link site analysis) as well as the danger of introducing conformational artefacts (which can lead to formation of non-physiological and irrelevant cross-links). As a corollary to this, it must also be remembered that there is a strong element of chance in any cross-linking approach. The experimenter is often interested in generating cross-links between two molecules which are of special interest to him, but there is, however, no guarantee that the desired cross-links will actually be formed or observed with a given cross-linking method. In such cases it is very inadvisable to try to 'force' the reaction by using drastic cross-linking conditions, as this will almost certainly increase the danger of side-reactions and artefacts.

2.3 Choice of cross-linking substrate

The question of possible artefacts leads to the question of the substrate to be used for the cross-linking reactions. Our objective has always been the study of the structure of the rRNA *in situ* in the ribosome, and we therefore use ribosomal subunits, 70S ribosomes, or (as mentioned above in Section 2.2) growing cell cultures as the substrate for cross-linking. In contrast, a number of other authors, for reasons of technical convenience, have used isolated rRNA as a substrate for intra-RNA cross-linking experiments (e.g. 17, 19). The data from such studies can only be interpreted in terms of the *in situ* structure of the RNA by assuming that the tertiary structure of the isolated RNA is identical to its structure in the ribosome; our reasons for rejecting this assumption have been discussed in detail elsewhere (e.g. 1, 10), and the ensuing Sections will only deal with cross-linking reactions in intact ribosomal particles.

2.4 Partial digestion and labelling of the rRNA

At some stage during the isolation procedure the cross-linked complexes must be subjected to a partial digestion step, so as to generate RNA fragments of a suitable size for sequence analysis. The fragments must be sufficiently long to enable the sequence region(s) they contain to be identified unambiguously, but not so long that the position of the cross-link site on the RNA becomes obscured. Finding suitable partial digestion conditions is not a trivial matter. The use of single-strand-preferring nucleases such as ribonuclease T1 (see e.g. 20) is not to be recommended, because the products of partial digestion with these enzymes are very heterogeneous, and, more important, many potentially interesting cross-links located in single-stranded regions of the RNA are lost due to preferential over-digestion. The double-strand specific cobra venom nuclease (21) can be usefully applied (see e.g. 10, 12), although here also the products of digestion are highly selected (again resulting in the loss of potentially important data), and it must be borne in mind that this enzyme only gives 'clean' digestion patterns when applied to intact ribosomal subunits (10, 22), as opposed to isolated rRNA. Unquestionably the best method for partial digestion of the RNA is to use ribonuclease H in combination with short oligodeoxynucleotide templates complementary to chosen sequences within the rRNA (cf. 23). With a modern oligonucleotide synthesizer these templates can be prepared in large numbers, and they allow 'tailor-made' RNA fragments to be produced in high yield.

For sequence analysis of the cross-linked RNA, the RNA fragments must obviously be labelled. However, in our experience, the use of conventional end-labelling techniques is not suitable for the analysis of cross-link sites on rRNA for several reasons. First and foremost, a cross-link site analysis at the nucleotide level requires that the cross-link must be left intact throughout the experiment, as already noted in Section 2.2 above. In the case of an intra-RNA cross-linking study, the product that is isolated will therefore either consist of two separate RNA fragments held together by the cross-link, or a single piece of RNA in which the cross-link forms a closed loop. End-labelling of the first type of complex would give rise to two labels, and thus to a product which cannot be sequenced by the usual methods, whereas in the second case sequencing within the closed loop is also excluded. (Some highly specialized techniques have been published which circumvent this problem (e.g. 24, 25), but they are unfortunately not generally applicable). In the case of an RNA–protein cross-linking experiment, the product that is isolated will consist of an RNA fragment linked to a protein, and such complexes are often very difficult to end-label satisfactorily, particularly if the cross-link site is near to the end of the RNA fragment. End-labelling procedures are in any case notoriously selective in this type of situation, so that the most highly-labelled product is not necessarily the most abundant, and this can lead to serious errors in the interpretation of the data. Last but not least, RNA fragments that have been generated by partial digestion (even in the case of the ribonuclease H method) usually have 'ragged ends', which means that the labelled products are heterogeneous and that laborious extra purification steps must be introduced before sequencing can be carried out.

All of these considerations argue strongly for the use of *in vivo* uniformly-labelled RNA for cross-linking studies, coupled with sequence analysis by classical 'fingerprinting' rather than end-labelling techniques. Provided that the rRNA sequence is known (and a large number of sequences are now available (26, 27)), the finger-printing method allows both the rapid identification of an RNA region and the localization of the cross-link site within that region. The simultaneous identification of two RNA fragments, or the identification of a sequence within a closed loop (the situations which arise in intra-RNA cross-linking experiments, as just discussed) can be achieved without difficulty, and 'ragged-ended' fragments present no problem. Furthermore, this method has the added advantage that any cross-contaminants or impurities in the fragment being analysed are impossible to overlook, and can be quantitatively assessed.

3. Summary of experimental techniques

In this section we give a brief resumé of the experimental procedures which we currently use for an intra-RNA or RNA–protein cross-link site analysis. The corresponding details of the procedures are given in Sections 4 and 5, respectively, with emphasis on the newer aspects. However, the application of this type of cross-linking technology to a given research problem is likely to vary considerably from one case to another. We therefore recommend the reader to bear in mind the general considerations discussed above (Section 2) when deciding to what extent the cross-linking methods described below (Sections 3, 4, and 5) can be incorporated into or adpated for his or her particular research programme.

3.1 Intra-RNA cross-linking

A typical intra-RNA cross-linking experiment with ribosomal subunits from *E. coli* can be sub-divided into the following steps:

Protocol 1. Steps in an intra-RNA cross-linking procedure

1. Grow ^{32}P-labelled *E. coli* cells, and isolate 30S and 50S ribosomal subunits.
2. Induce intra-RNA cross-links, by direct ultraviolet irradiation or treatment with nitrogen mustard.
3. Remove ribosomal proteins by proteinase K treatment and phenol extraction. Concentrate the RNA by ethanol precipitation.
4. Partially digest the RNA by hybridizing it with a chosen set of deoxyoligonucleotides, followed by hydrolysis with ribonuclease H. Repeat this step with a second set of deoxyoligonucleotides, to reduce the size of the rRNA fragments.

At this stage, the RNA fragments can be directly applied to a two-dimensional polyacrylamide gel, for separation of the individual intra-RNA cross-linked complexes. However, particularly in the case of fragments generated from the 50S subunit, the gel patterns tend to be very complex, and so a pre-fractionation step is advisable. This is carried out with the help of M-13 clones carrying DNA inserts which are complementary to selected regions of the rRNA (28). The single-stranded DNA from these clones is immobilized onto cellulose, and is used to select for rRNA fragments and cross-linked complexes with sequences corresponding to the particular rDNA insert in the clone.

5. Hybridize the RNA fragments from the ribonuclease H digest with immobilized DNA from a suitable M-13 clone. (The supernatant from this step can be used if required for successive further hybridizations with other clones).
6. Elute the hybridized RNA fragments and concentrate by precipitation.
7. Separate individual RNA fragments and cross-linked complexes by two-dimensional gel electrophoresis.
8. Extract cross-linked complexes from the gel, and subject aliquots to total digestion with ribonuclease T1 or ribonuclease A.
9. Make a 'finger-print' of oligonucleotides released, by two-dimensional thin-layer chromatography (cf. 29). If too many large oligonucleotides are present, the finger-print is repeated with a further aliquot of the cross-linked complex, using a different chromatographic elution system.
10. Extract individual oligonucleotides from the finger-print, and subject to secondary digestion with ribonuclease A or T1 (for oligonucleotides from T1 or A finger-prints, respectively).
11. Compare oligonucleotide sequences found with the known sequence of 16S or 23S RNA, and deduce the positions of the cross-link sites.

3.2 RNA–protein cross-linking

The corresponding steps in a typical RNA–protein cross-linking analysis are as follows:

Protocol 2. Steps in an RNA–protein cross-linking procedure

1. Prepare ^{32}P-labelled 30S and 50S ribosomal subunits, as in Section 3.1.
2. Induce RNA–protein cross-links by treatment with nitrogen mustard, 2-iminothiolane, or APAI. In the latter two cases, treatment with the reagent is followed by a brief ultraviolet irradiation to generate the cross-links. (Note that nitrogen mustard induces both intra-RNA and RNA–protein cross-links. Each type of cross-link can be analysed without interference from the other by following the appropriate procedure.)

3. Remove non-cross-linked ribosomal proteins by centrifugation of the ribosomal subunits through a sucrose gradient in the presence of SDS.
4. Collect the peak of RNA–protein cross-linked complexes (which runs together with 'free' RNA in the gradient), concentrate by precipitation, and hydrolyse with ribonuclease H and deoxyoligonucleotides as in *Protocol 1*.
5. Dilute the hydrolysate with high-salt buffer, and pass over a glass fibre filter. Free RNA fragments pass through the filter, whereas the RNA–protein cross-linked complexes are retained (cf., 30).
6. Elute the RNA–protein cross-linked complexes from the filter with SDS, and separate the individual complexes by two-dimensional gel electrophoresis in a high-salt system.
7. Extract the cross-linked complexes from the gel.

   The gel patterns at this stage (again, particularly those from cross-linked 50S subunits) tend to be very complex, and a final purification of the cross-linked complexes is achieved by affinity chromatography, using antibodies to individual ribosomal proteins bound to agarose via a second antibody. This step serves at the same time to give a positive identification of the proteins that are involved in the cross-links (31).

8. Bind the ribosomal protein antibodies on to agarose, and test aliquots of each isolated cross-linked complex with a selected set of antibodies, by measuring the amount of [^{32}P]-RNA radioactivity which binds to a particular agarose–antibody preparation.
9. Where a positive reaction has occurred, repeat the test with a larger aliquot, or with the entire cross-linked complex sample. (If more than one protein is found to be present in a given complex, the supernatant from the antibody–agarose selection step can be used for sequential binding to further antibody–agarose preparations).
10. Elute the RNA from the agarose by treatment with proteinase K in the presence of SDS.
11. Analyse the RNA by finger-printing, and deduce the positions of the cross-link sites, as in *Protocol 1*.

---

4. Detailed procedure for intra-RNA cross-linking

4.1 Cross-linking reactions

Protocol 3. Preparation of ribosomal subunits

1. The starting material for the experiment consists of 30S or 50S ribosomal subunits from *E. coli* strain MRE 600, labelled *in vivo* with

[^{32}P]-orthophosphate. Our procedure for isolating the labelled subunits has been described in detail elsewhere (3), and will not be repeated here. A typical preparation from a 40 ml *E. coli* culture yields 3–5 A_{260} units of 30S subunits in 0.5 ml of buffer, containing $1.0–1.5 \times 10^9$ counts/min of Cérenkov radioactivity, the corresponding yield of 50S subunits being about twice as much.

2. Dialyse the subunits into an appropriate buffer for the cross-linking reaction. For cross-linking with nitrogen mustard the buffer must be free of primary amines, and it is convenient to use the same buffer for cross-linking by direct ultraviolet irradiation. This buffer consists of 5 mM magnesium acetate (or chloride), 50 mM KCl and 25 mM triethanolamine–HCl (pH 7.8). Mercaptoethanol or other sulphur-containing compounds should not be included in the buffer.

Protocol 4. Reaction with nitrogen mustard (12)

1. Make a fresh 100 mM solution of nitrogen mustard (EGA Chemie) in the dialysis buffer (*Protocol 3*), and add a suitable aliquot to the ribosomal subunits so as to give a final reagent concentration of 1–3 mM. Incubate at 37 °C for 45 min.
2. Destroy the excess reagent by adding 1/20 volume of a 1 M solution of cysteamine (Merck) and incubate for a further 15 min at 37 °C.
3. Precipitate the subunits with ethanol (*Protocol 5*), and redissolve in 200 μl of 10 mM Tris–HCl (pH 7.8) containing 1 mM EDTA.
4. Destroy ribosomal protein by treatment with proteinase K followed by phenol extraction (*Protocol 6*).

Protocol 5. Ethanol precipitation of ribosomal subunits, RNA, or cross-linked fragment complexes

1. If the sample (see text) does not already contain salt ($\geqq$100 mM), add 1/10 volume of 1 M sodium acetate.
2. Add 2 vols ethanol, mix well and leave at −20 °C for 2 h.
3. Centrifuge at 10 000 r.p.m. for 30 min, using a swing-out rotor (or, in the case of Eppendorf tubes, a rotor where the sample is held horizontally in a suitable adaptor).
4. Carefully decant off the supernatant, wash the pellet with a little 80% ethanol, and centrifuge again (10 min).
5. Carefully decant the supernatant again, and allow the pellet to dry in air or under a slight vacuum. (Do not dry pellets under high vacuum, particularly if

proteins are present; this renders the precipitate very difficult to dissolve afterwards).

6. Resuspend the pellet in a low-salt buffer, appropriate to the next experimental step.

Protocol 6. Proteinase K digestion and phenol extraction

1. Dissolve the sample (see text) in 200 μl of 10 mM Tris–HCl (pH 7.8), 0.1% SDS, 1 mM EDTA, then add an equal volume of 1 M NaCl.
2. Add 1/10th volume of proteinase K (e.g., from Merck), 10 mg/ml, and incubate for 30 min at 37 °C.
3. Add a further 1/10th volume of proteinase K solution and repeat the incubation.
4. Add 2 μl of unlabelled tRNA carrier (10 mg/ml).
5. Add 500 μl of water-saturated phenol, and shake for 45 min at 4 °C.
6. Separate the phases by centrifugation (10 min at 10 000 r.p.m.) in a swing-out rotor.
7. Carefully pipette off the aqueous phase, and precipitate with ethanol as in *Protocol 5*.

Protocol 7. Cross-linking by direct ultraviolet irradiation (9)

1. Dilute the subunits with dialysis buffer (*Protocol 3*) to a concentration of 5 A_{260}/ml, and place in a shallow vessel (e.g. Petri dish) so as to give a solution depth of *c*., 1 mm.
2. Irradiate for 2–10 min. Our irradiation apparatus consists simply of four horizontally parallel-mounted Sylvania G8T5 germicidal lamps (Sylvania) set 5 cm apart from each other. Place the sample at a distance of 4–6 cm from the lamps, which at this distance deliver an energy of *c*., 25 Joules/m^2/sec.
3. Precipitate the subunits with ethanol (*Protocol 5*), and destroy ribosomal protein by proteinase K and phenol treatment (*Protocol 6*).

4.1.1 Cross-linking yield

We know of no way to estimate the overall yield of intra-RNA cross-linking in this type of reaction. The conditions just described are in both cases those found empirically to give reasonable yields of cross-linked products, without leading to any observable formation of subunit dimers or aggregates.

4.2 Partial digestion with ribonuclease H and oligodeoxynucleotides

4.2.1 Preparation of ribonuclease H

We recommend the method described by Darlix (32), with the exception that we omit the final purification step by DEAE–cellulose chromatography. After the first column separation on phosphocellulose (32), simply precipitate the enzyme from the appropriate column fractions with ammonium sulphate, and redissolve in 20 mM Tris–HCl (pH 7.8), 50 mM KCl, 100 mM NaCl, 0.1 mM EDTA, 0.1 mM dithiothreitol, and then add glycerol to 50%. The ribonuclease H prepared in this way has a very high activity, with negligible levels of contaminating ribonuclease activities, and the enzyme solution is stable for several years at −20 °C. (It should be added here that in our hands commercial preparations of ribonuclease H do not lead to efficient digestion of the rRNA, using the procedure described below.)

4.2.2 Preparation of oligodeoxynucleotides

Oligodeoxynucleotides with a chain length of ten bases are ideal for use in conjunction with ribonuclease H. These decanucleotides are long enough to form unique stable hybrids which can compete out the secondary structure of the RNA, but are at the same time short enough that the hydrolysis products generated by the ribonuclease H treatment are reasonably homogeneous. (Ribonuclease H tends to cut at one or two preferred positions with this type of short RNA–DNA hybrid, cf. reference 23.) Our decanucleotides are made on an Applied Biosystems 381A DNA synthesizer, with automatic removal of the 5′-terminal trityl group after the final cycle of synthesis. The protecting groups are removed by hydrolysis with ammonia according to the manufacturer's recommendation, and the solution is evaporated to dryness. The oligodeoxynucleotides are then taken up in water or 10 mM Tris–HCl (pH 7.8) and adjusted to a concentration of 100 A_{260} units/ml. They can be used for ribonuclease H hydrolysis of the rRNA without further purification.

We find it advantageous to work with two sets of oligodeoxynucleotides, in combination with a two-step ribonuclease H digestion. The first set of decanucleotides is designed to cut the rRNA into fragments 50–70 bases long, and the second set induces a further cut within each of these fragments. *Figure 1* shows as an example the sequence and secondary structure of *E. coli* 23S RNA (33, 34, 27), and indicates the positions of the two sets of decanucleotides that we have chosen to use within this structure. Since in practice not every decanucleotide position is cut quantitatively by the ribonuclease H, and other positions may be blocked by the cross-links themselves, the resulting hydrolysis pattern shows a satisfactory spread of RNA fragments.

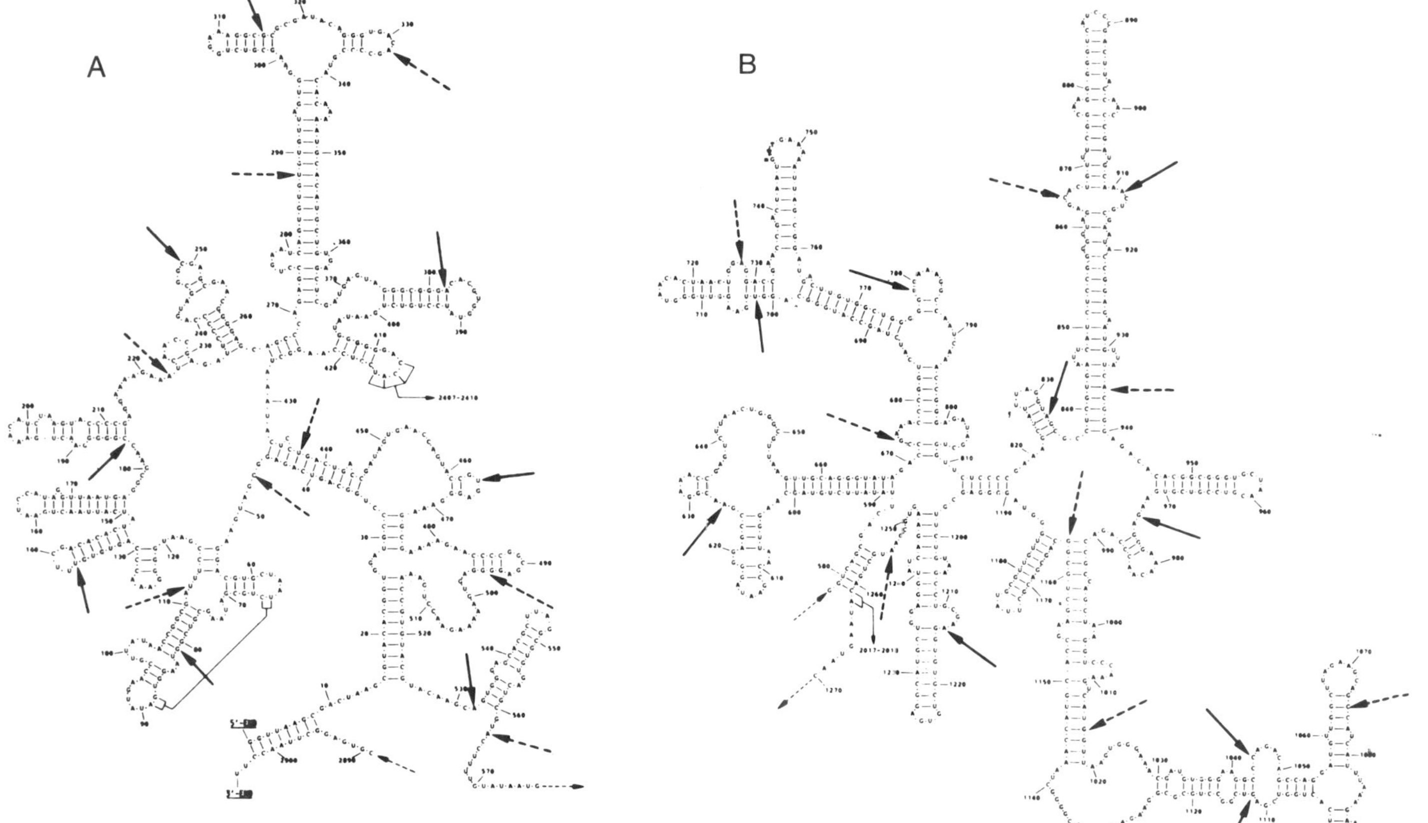

Figure 1. The secondary structure of *E. coli* 23S RNA (34, 27), showing the positions to which complementary oligodeoxynucleotides are hybridized, in order to create a substrate for ribonuclease H (Section 4.2.2). The arrows indicate the central position of each decanucleotide sequence within the four regions of the secondary structure, arrows with a solid tail denoting the oligonucleotide set used for the first ribonuclease H digestion step (*Protocol 9*), those with a dotted tail denoting the set used for the second hydrolysis step. (The smaller dotted arrows merely indicate the 5'- and 3'-termini of the four regions of the structure.)

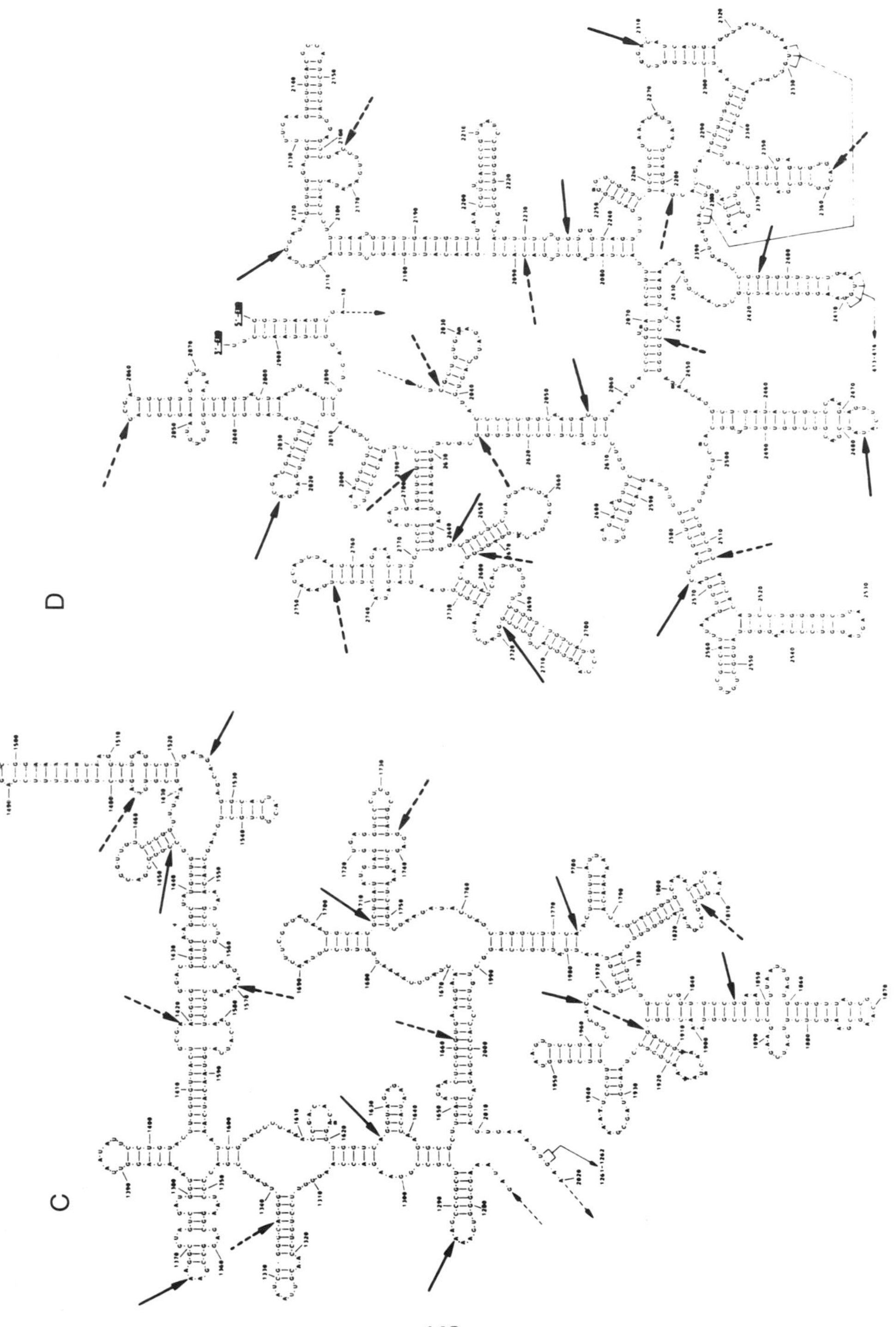
D
C

4.2.3 Ribonuclease H digestion

Protocol 8. Preparation of equimolar oligodeoxynucleotide-rRNA mixtures

1. Mix the cross-linked and deproteinized ^{32}P-labelled RNA sample (from *Protocol 4* or 7) with a set of decanucleotides (Section 4.2.2) so that each decanucleotide is present in an approximately 1:1 molar ratio to the 16S or 23S RNA (i.e. for 16S RNA (*c*., 1500 nucleotides long) take 0.006 A_{260} units of each decanucleotide per A_{260} unit of RNA, and for 23S RNA (*c*., 3000 nucleotides long) 0.003 A_{260} units of decanucleotide per A_{260} unit of RNA). As an example, if 30 different decanucleotides are to be used in the ribonuclease H hydrolysis, then mix together equal volumes (e.g. 2 μl) of each oligonucleotide from the 100 A_{260} units/ml stock solutions (Section 4.2.2); in the resulting solution each of the 30 decanucleotides will contribute 3.3 A_{260}/ml to the optical density, and 1 μl of the solution will therefore contain 0.0033 A_{260} units of each oligonucleotide. If the oligonucleotides are complementary to 23S RNA, then 1 μl is the correct dosage of oligonucleotide solution per A_{260} unit of 23S rRNA, whereas for 16S rRNA the corresponding dosage is 2 μl.
2. Carry out the two-step ribonuclease H digestion, as described in *Protocol 9*. (The optimal quantity of enzyme should first be evaluated by preliminary tests with unlabelled rRNA.)

Protocol 9. Digestion of RNA/oligodeoxynucleotide complexes with ribonuclease H

1. This recipe is for 1 A_{260} unit of rRNA; adjust volumes according to amount actually present.
2. Dissolve the sample to be digested (after ethanol precipitation) in 20 μl of 10 mM Tris–HCl (pH 7.8), 6 mM 2-mercaptoethanol.
3. Add 2 μl of 4 M NaCl, and 1 – 2 μl (see Section 4.2.2 and *Protocol 8*) of the first set of oligodeoxynucleotides (cf., *Figure 1*).
4. Place the sample tube in a beaker of water, warm to 70 °C, and keep at this temperature for *c*., 1 min.
5. Remove heat, and allow to cool to 37 °C over a period of *c*., 10 min, then incubate at 37 °C for 10 min.
6. Add 4 μl of 50 mM Mg acetate, 1 mM dithiothreitol, followed by a suitable volume (e.g., 6 μl) of ribonuclease H solution (32).
7. Incubate at 37 °C for 45 min.
8. Add 1–2 μl of the second set of oligodeoxynucleotides (cf., *Figure 1*).
9. Repeat steps 4 and 5.

10. Add a further 6 μl of ribonuclease H and incubate at 37 °C for 45 min.
11. Add EDTA to a final concentration of 10 mM.

4.3 Hybridization of rRNA fragments to M-13 clones

4.3.1 Preparation of M-13 clones, and immobilization of DNA

Restriction fragments of *E. coli* ribosomal DNA are prepared from plasmid pKK 3535 (35), and are cloned into M-13 by standard procedures (36, 37, cf. also reference 28). The length of the rDNA inserts which we use (e.g. reference 28) varies from *c.*, 50–500 bases, according to the pattern of restriction fragments available. Single-stranded DNA is isolated from M-13 clones grown in 500 ml cultures of *E. coli* JM103, again using standard procedures (37). The DNA is immobilized onto aminobenzyloxymethyl cellulose (Miles Chemical Corp.), using the method which has been described in detail by Goldberg *et al.* (38). The single-stranded DNA preparations typically contain 500–1000 μg, and in our hands the yield of the coupling reaction is 40–50% (28).

Protocol 10. Hybridization and elution of complementary RNA fragments

1. Pre-treat *c.*, 100 mg of an M-13 DNA aminobenzyloxymethyl cellulose preparation, carrying an appropriate rDNA insert (Section 4.3.1), by gently rotating at 42 °C for 1 h in 5 ml of 50% deionized formamide containing 400 mM NaCl, 40 mM trisodium citrate, 0.1% each of Ficoll, polyvinylpyrrolidone, and bovine serum albumin (5 × Denhardt's solution), 50 mM sodium phosphate (pH 6.8), 0.1% SDS and 0.1 mg/ml of unlabelled bulk tRNA.
2. Spin off the DNA-cellulose in a table centrifuge, and add the ^{32}P-labelled ribonuclease H-digested RNA fragments (*Protocol 9*) in the same milieu as just described, but with 1 × Denhardt's solution, 20 mM sodium phosphate (pH 6.8), and 0.2% SDS (cf., reference 28, 39). Incubate for 2 h at 42 °C.
3. Spin off the DNA-cellulose and wash five times for 10 min each at 42 °C with 5 ml of 50% deionized formamide containing 150 mM NaCl, and 15 mM trisodium citrate. (The first supernatant after the hybridization step can be used directly for hybridization with further DNA-cellulose preparations, containing DNA inserts complementary to other regions of the rRNA.)
4. Elute the RNA by washing the DNA-cellulose twice at 65 °C for 30 min with 5 ml of 99% formamide containing 0.1% SDS.
5. Combine the eluted RNA fractions, add sodium acetate to a concentration of 300 mM, and precipitate overnight at −20 °C by addition of 2 volumes of isopropanol. Centrifuge off the precipitated RNA and wash with 80% ethanol as in *Protocol 5*.

4.4 Separation of cross-linked RNA fragments by two-dimensional gel electrophoresis

The RNA fragment mixtures isolated from the DNA-cellulose (*Protocol 10*) will consist of non-cross-linked RNA fragments, and intra-RNA cross-linked complexes. These are separated by two-dimensional polyacrylamide gel electrophoresis in a system in which the first dimension is an acrylamide gradient

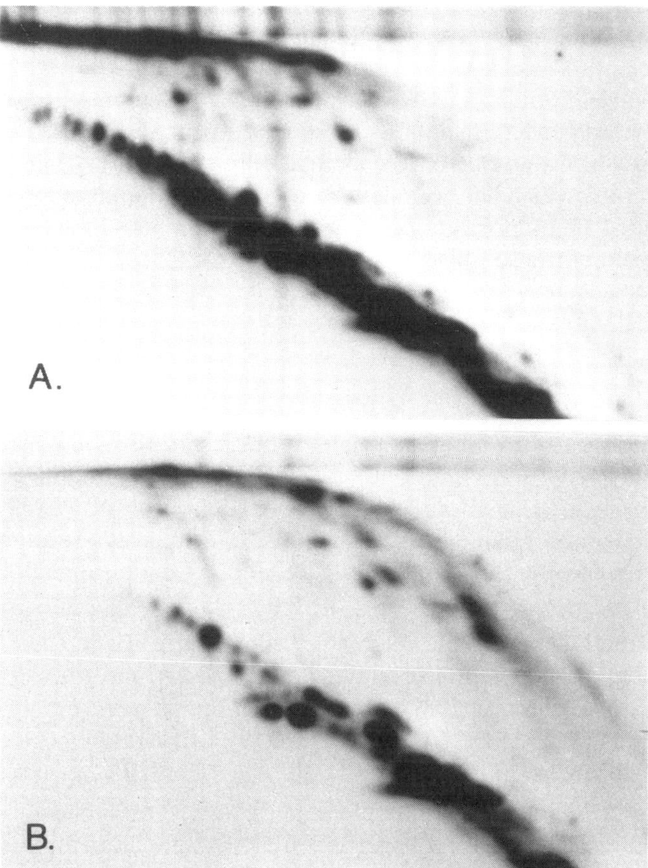

Figure 2. Autoradiograms of two-dimensional gels of intra-RNA cross-linked complexes (Section 4.4). The gel pattern is derived from a ribonuclease H hydrolysis of 50S subunits cross-linked by direct ultraviolet irradiation (*Protocol 7*). A: Mixture of digestion products after the first ribonuclease H digestion step (*Protocol 9*). B: Corresponding gel pattern after the second ribonuclease H digestion step. Direction of electrophoresis is from left to right (first dimension) and from top to bottom (second dimension). The strong 'diagonal' running from upper left to lower right consists of free RNA fragments, the cross-linked complexes being the spots lying above this diagonal. The pattern of complexes in gel B is clearly shifted to substantially lower molecular weight products, as compared to the pattern of gel A. See reference 28 for examples of gel patterns obtained after hybridization of the fragment mixtures to M-13 clones (Section 4.3).

gel containing both SDS and urea, whereas the second dimension is of constant acrylamide concentration and contains urea but no SDS; the gel system has been described in full detail in reference 3. In these two-dimensional gels, the non-cross-linked RNA fragments form a 'diagonal', and the cross-linked complexes appear as discrete spots above the diagonal. Typical examples, from 50S subunits cross-linked by direct ultraviolet irradiation (*Protocol 7*), are shown in *Figure 2*. *Figure 2A* is the autoradiogram of a gel obtained from the complete mixture of RNA fragments after the first ribonuclease H digestion step (*Protocols 8* and *9*) and *Figure 2B* shows the corresponding gel pattern after the second ribonuclease H digestion step. Examples of the simplification of the gel patterns which are observed after selective hybridization of the digestion mixtures to M-13 clones (Section 4.3) can be found in reference 28. Radioactive spots of interest are cut out from the gels and the RNA is extracted in the presence of SDS and phenol, again as described in reference 3. Radioactive yields of the individual cross-linked complexes at this stage vary from a few thousand up to several million counts/min of ^{32}P-radioactivity.

4.5 Oligonucleotide analysis

4.5.1 Total enzymatic digestion of cross-linked complexes

Protocol 11. Hydrolysis with ribonucleases T1, or A

1. After extraction from the two-dimensional gel (reference 3, and Section 4.4), precipitate the isolated intra-RNA cross-linked complexes with ethanol (*Protocol 5*), and take up the pelleted RNA in a small volume (20–50 μl) of 10 mM Tris–HCl (pH 7.8), 0.1% SDS.
2. Divide the sample into 10 μl aliquots for digestion with either ribonuclease T1 or ribonuclease A.
3. For each digestion add 2 μl unlabelled carrier tRNA (10 mg/ml) and then 5 μl of ribonuclease T1 (Sankyo, Tokyo, 1 mg/ml) or 5 μl of ribonuclease A (Sigma, 1 mg/ml). (The tRNA can, if preferred, be added after extraction from the gel in 1 above.)
4. Incubate for 20 min at 37 °C, and then raise the water bath temperature to 60 °C, so that the samples slowly warm up to this temperature during a further 15 min incubation period.
5. Lyophilize the samples, and take up in 7 μl of water containing a little xylene cyanol dye.

4.5.2 Thin-layer chromatography—'fingerprinting'

The oligonucleotides released by digestion with ribonuclease T1 or A are separated by two-dimensional chromatography on polyethyleneimine cellulose thin-layer plates (Macherey and Nagel), using the method of Volckaert and Fiers (29), with minor variations. Buffers for the various chromatographic steps are listed in *Table 1*.

Table 1. Buffers for two-dimensional thin-layer chromatography on polyethyleneimine cellulose (cf. reference 29).

Process	Buffer
Buffer for pre-washing plates	2 M formic acid adjusted to pH 2.2 with pyridine
'Normal' finger-print first dimension	7.3 M formic acid containing 5 M urea
'Normal' finger-print second dimension	1 M formic acid containing 4 M urea, adjusted to pH 4.3 with pyridine
'Alternative' finger-print first dimension	14.6 M formic acid containing 5 M urea
'Alternative' finger-print second dimension	2 M formic acid containing 4 M urea, adjusted to pH 4.3 with pyridine
Secondary digestion first dimension	5.8 M formic acid
Secondary digestion second dimension	1 M formic acid adjusted to pH 4.3 with pyridine

Protocol 12. Thin-layer chromatography of primary digestion products

1. Pre-wash the thin-layer plates (20 cm × 20 cm) with buffer (*Table 1*), rinse with water and allow to dry.
2. Spot the digested sample (Section 4.5.1) onto one corner of the plate, and again allow to dry.
3. Pre-run each plate in water until the water front is *c*., 1 cm above the sample application point, and then develop the plates (without drying) in first dimension buffer (*Table 1*), until the xylene cyanol dye marker has risen approximately 7 cm.
4. Wash the plates in flat trays containing 70% ethanol to remove urea, and allow to dry.
5. Pre-run each plate with water as in 3 above (in a direction at 90° to the first dimension), and then develop in second dimension buffer (*Table 1*), until the dye marker has risen *c*., 7 cm.
6. Wash again in 70% ethanol, dry, and subject to autoradiography overnight at −80 °C, using Kodak XAR 5 film (35 × 43 cm sheets) and intensifying screens. The thin-layer plates can be cut so that six plates can be autoradiographed under a single film. Mark the plates with radioactive ink to enable the radioactive spots to be located accurately afterwards.
7. If the X-ray film of the fingerprint shows a strong spot at the starting point (indicating that several large oligonucleotide digestion products are present), then the chromatography can be repeated with a further digestion aliquot (*Protocol 11*), using the 'alternative' buffer system of *Table 1*. Examples of

ribonuclease T1 fingerprints in the 'normal' and 'alternative' buffer systems are shown in *Figure 3*.

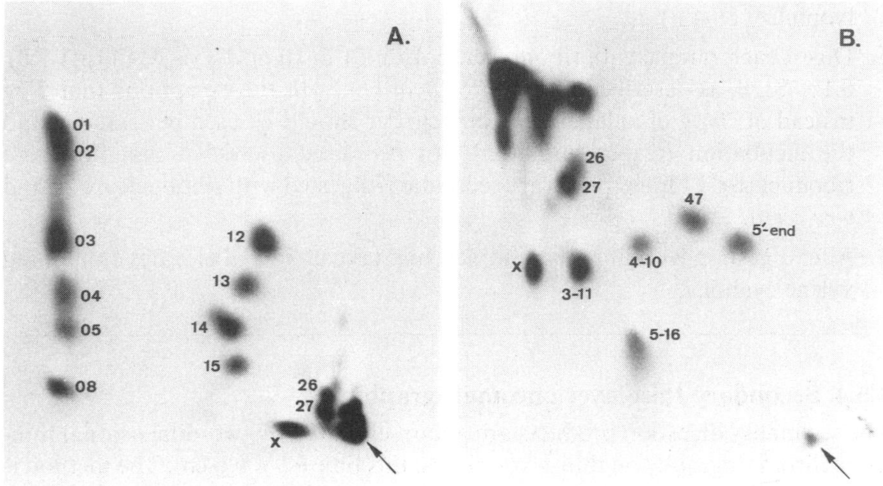

Figure 3. Typical ribonuclease T_1 finger-prints (Section 4.5). A: Finger-print of an intra-RNA cross-linked complex in the 'normal' solvent system (*Table 1*). B: Corresponding finger-print in the 'alternative' system (*Table 1*), to separate the larger oligonucleotides. The direction of the chromatography is from right to left (first dimension), and from bottom to top (second dimension), the arrow denoting the sample application point. The oligonucleotides are numbered according to the system of reference 40, in which the first digit gives the number of U-residues present in the oligonucleotide, and the second number the chain length. X denotes the cross-linked oligonucleotide. It can be seen that in finger-print B, the large oligonucleotides become separated at the expense of the separation of the smaller ones (cf., finger-print A).

4.5.3 Secondary digestion of oligonucleotides (cf. reference 29)

Protocol 13. Elution of oligonucleotides from fingerprints, and secondary digestion

1. Mark the positions of the oligonucleotides on the thin-layer plates (*Protocol 12*). To do this, lay the X-ray film on the chromatography plate (with the help of the radioactive ink markers), and circle the spots on the film firmly with a ballpoint pen; the circles will be clearly visible on the thin-layer plate.
2. Scratch out the polyethyleneimine cellulose containing each spot into a drawn-out capillary tube attached to a vacuum pump. (First suck a little cellulose powder into each capillary, so as to 'block' the end).
3. Elute the oligonucleotide with *c.*, 50 μl of 30% triethylamine adjusted to pH 9.5 with CO_2 (41), by attaching a syringe containing this solution to the

capillary tube. (If a small amount of the cellulose powder is co-eluted, this does not present any problem in the subsequent steps.)

4. Evaporate the eluates to dryness overnight, redissolve in 100 μl of water, and lyophilize (3–4 h).
5. Digest each sample with ribonuclease A or T1 in 10 mM Tris–HCl (pH 7.8), 0.1% SDS, as described above (*Protocol 11*), with the exceptions that 5 μg instead of 20 μg of unlabelled tRNA carrier should be used per sample, and the incubation temperature need not be raised to 60 °C. Samples from ribonuclease T1 fingerprints are secondary-digested with ribonuclease A, and *vice versa*.
6. After digestion lyophilize the samples, and take up in 5 μl of water containing xylene cyanol.

4.5.4 Secondary thin-layer chromatography

The secondary digestion products are again separated by two-dimensional thin-layer chromatography on thin-layer plates, this time 9.9 × 6.6 cm. The method is again that of Volckaert and Fiers (29), and these authors have described in detail the apparatus for this chromatographic step, as well as the method itself.

Protocol 14. Thin-layer chromatography of secondary digestion products

1. Spot the digested samples onto the thin-layer plates.
2. Pre-run the plates in water as in *Protocol 12*, and then develop in secondary digestion, first dimension buffer (*Table 1*) until the buffer front reaches the top of the thin-layer plate. Dry the plates (without washing in 70% ethanol).
3. Pre-run the plates in secondary digestion, second dimension buffer diluted 1:10 with water, and then develop in second dimension buffer (*Table 1*), again until the solvent reaches the top of the plate.
4. Dry the plates and autoradiograph for one week at −80 °C, using Kodak XAR 5 film and intensifying screens. Twenty secondary digestion plates will fit under a single 35 × 43 cm film.
5. The secondary digestion data are very simple to interpret, since only oligonucleotide products of the type $(Ap)_nCp$, $(Ap)_nUp$ and $(Ap)_nGp$ will be present, from both ribonuclease A and T1 fingerprints. The pattern of separation of these products can be found in *Figure 2* of reference 29.

4.6 Assessment of results

The advantages of the fingerprint method for this type of cross-linking experiment have already been pointed out in Section 2.4. It should further be noted that, with suitable organization, the method can be used to process very large numbers of samples in a relatively short time. Up to 50 fingerprints (*Protocol 12*) can be run simultaneously in a single large thin-layer chromatography tank (e.g. from Desaga, Germany), and the corresponding secondary digestions (Sections 4.5.3 and 4.5.4) can be handled in two or three days with four or five sets of the apparatus described by Volckaert and Fiers (29). A radioactivity of 3000 counts/min in an RNA fragment containing 100 bases is enough to allow a fingerprint and the corresponding secondary digestions to be made.

To analyse the data, it is advisable to prepare charts of the ribonuclease T1 digestion products which can be expected from the sequence(s) concerned. These charts should be ordered with regard to the chain length of the oligonucleotides and the number of U-residues each contains, so as to correspond to the pattern of separation of the oligonucleotides on the fingerprint (cf reference 40 and *Figure 3*). Similar charts can be prepared for the ribonuclease A digestion products (see reference 29 for pattern of separation). The position of each oligonucleotide on the fingerprint is used in conjunction with the secondary digestion data, and with the help of the chart of possible digestion products, to identify the oligonucleotides which are in fact present. The oligonucleotides are then fitted into the known sequence, which in our case is of course either that of the *E. coli* 16S (42) or 23S (33) rRNA.

It must be remembered when interpreting the fingerprint data that ribonuclease H digestion generates a 5′-phosphate group and a free 3′-hydroxyl group (23), so that the 5′-ends of the RNA fragments will give rise to oligonucleotides of the type 'pXpYpZp' on the fingerprints, whereas the 3′-ends will give 'XpYpZ', ('normal' oligonucleotides being of the form 'XpYpZp'). An extra 5′-phosphate group will move an oligonucleotide considerably to the right on the fingerprint (cf., *Figure 3*), and a 'missing' 3′-phosphate group correspondingly shifts the oligonucleotide to the left.

The cross-link site itself should be visible as a pair of 'missing' oligonucleotides within the fragment sequence present on the fingerprint, together with an extra anomalous cross-linked oligonucleotide spot whose analysis should correspond with the pair of missing oligonucleotides. The secondary digestion data of this cross-linked spot should also allow the cross-link site to be pin-pointed. Often the two components of the cross-link site will, however, only be unambiguously identifiable if the data from both the ribonuclease A and T1 fingerprints are considered together. In the case of nitrogen mustard cross-links (*Protocol 4*), the site of reaction is usually (but not always) a G-residue, which cannot then be digested by ribonuclease T1. This has the result that each 'gap' in the ribonuclease T1 data at the cross-link site will normally span two T1-oligonucleotides, and the cross-linked oligonucleotide will be composed of two pairs of (i.e. four) T1-oligonucleotides. Examples of individual cross-link site analyses can be found in any of our publications (e.g. 9, 12, 20). Upwards of 40 intra-RNA cross-links have so far been identified by this procedure in 16S and 23S RNA.

5. Detailed procedure for RNA–protein cross-linking

5.1 Cross-linking reactions

5.1.1 Reaction with nitrogen mustard (43)

Treat the ^{32}P-labelled 30S or 50S ribosomal subunits (*Protocol 3*) with nitrogen mustard, exactly as described in *Protocol 4*, proceeding up to the incubation with cysteamine and the ethanol precipitation step. Note, however, that the optimal reagent concentration may be different for RNA–protein as opposed to intra-RNA cross-linking.

Protocol 15. Reaction with 2-iminothiolane (14)

1. Dialyse the ^{32}P-labelled ribosomal subunits into 5 mM Mg acetate, 50 mM KCl, 25 mM triethanolamine–HCl (pH 7.8), 6 mM 2-mercaptoethanol. [In this case it is important that the buffer contains no primary amines, and that a reducing agent is present (cf. *Protocol 3*).]
2. Prepare a fresh solution of 500 mM 2-iminothiolane (Pierce Biochemicals, USA) in 500 mM triethanolamine base (cf. 13), and immediately add an aliquot to the dialysed ribosomal subunits, so as to give a final reagent concentration of 20 mM.
3. Incubatc for 20 min at room temperature (22 °C), then precipitate the subunits with ethanol (*Protocol 5*).
4. Resuspend the subunits in the dialysis buffer (minus 2-mercaptoethanol) to a concentration of 5 A_{260}/ml, and irradiate with ultraviolet light for 2 min under the conditions described in *Protocol 7*.
5. Add 2-mercaptoethanol to a concentration of 3%, and incubate for 30 min at 37 °C, to destroy any S–S bridges which might have formed. Precipitate again with ethanol (*Protocol 5*).

Protocol 16. Reaction with methyl-azidophenyl acetimidate (APAI) (15, 16)

1. Dialyse the ribosomal subunits into the buffer described in *Protocol 15*. (The buffer requirements here are the same as in the case of 2-iminothiolane cross-linking.)
2. Dissolve the APAI reagent in 100 mM triethanolamine–HCl (pH 9.0), 50 mM KCl, 5 mM Mg acetate, and immediately add to an equal volume of the dialysed ribosomal subunits. The solutions must be kept dark, to avoid premature reaction of the azide group. It should be noted that APAI is not commercially available; its synthesis is described in reference 15.

3. Incubate for 30 min at 30 °C, then precipitate the subunits with ethanol (*Protocol 5*).
4. Resuspend the subunits and subject to ultraviolet irradiation, exactly as in *Protocol 15*, step 4 above. Precipitate again with ethanol (*Protocol 5*). (In this case there is no need to incubate with 2-mercaptoethanol.)

5.1.2 Cross-linking yield

The yield in an RNA–protein cross-linking reaction can best be determined in preliminary experiments using double-labelled subunits, that is to say, subunits labelled with [^{3}H] in the RNA moiety and with ^{14}C in the proteins. The cross-linked subunits are run into a polyacrylamide gel containing SDS, and the yield of cross-linking is estimated from the amount of [^{14}C]-protein radioactivity co-migrating with the [^{3}H]-RNA peak. This procedure has been described in detail in reference 3. Under the conditions outlined above (Sections 5.1.1, *Protocols 15* and *16*) a cross-linking of 3–10% of the total ribosomal protein to the RNA should be observed, without any concomitant dimerization or aggregation of the subunits.

5.2 Removal of non-cross-linked protein and partial digestion of the RNA

Protocol 17. Sucrose gradient centrifugation in SDS

1. After the final ethanol precipitation of the cross-linked subunits (Sections 5.1.1, *Protocols 15* and *16*) suspend the pellet in 0.5 ml of 25 mM Tris–HCl (pH 7.8), 0.1% SDS, 2 mM EDTA, 6 mM 2-mercaptoethanol, and apply to a 7.5–30% sucrose gradient in the same buffer. A rotor such as the Beckman SW27 or SW 40 is suitable.
2. Centrifuge the gradients at 25 000 r.p.m. for 20 h at 10 °C. Pump out the gradient into *c*., 20 fractions and measure the radioactivity in a 2 μl aliquot of each fraction. The non-cross-linked protein (which is of course not visible in the ^{32}P-labelled samples) remains at the top of the gradient, whereas the RNA–protein cross-linked complexes run together with the peak of ^{32}P-labelled 16S or 23S RNA.
3. Pool the appropriate fractions in a 15 ml centrifuge tube, and precipitate with ethanol (*Protocol 5*). Resuspend the pellet in a small volume (*c*., 200 μl) of the same buffer as in 1 above, and transfer to a 1.5 ml Eppendorf tube. Repeat the ethanol precipitation, and take up the pellet in a small volume (20 μl per A_{260} unit of RNA) of 10 mM Tris–HCl (pH 7.8), 6 mM 2-mercaptoethanol.

5.2.1 Hydrolysis with ribonuclease H

Add the appropriate set of oligodeoxynucleotides to the mixture of RNA and RNA–protein complexes (Section 5.2), so that the oligonucleotides are present in a 1:1 molar ratio with respect to the rRNA, as described in *Protocol 8*. Carry out the ribonuclease H two-step digestion, exactly as instructed in *Protocol 9*.

5.3 Separation of non-cross-linked RNA fragments by glass-fibre filtration

---

Protocol 18. Glass-fibre filtration

1. After the second ribonuclease H digestion (Section 5.2.1), dilute the sample to 1.3 ml with 10 mM Tris–HCl (pH 7.8), 1 mM EDTA, 500 mM NaCl, and 6 mM 2-mercaptoethanol. Warm to 45 °C.
2. Pass the solution slowly over a glass-fibre filter disc (Whatman GF/C, 25 mm diameter), without applying a vacuum. Rinse the filter thoroughly with the same buffer used to dilute the sample in 1. Most of the radioactivity, corresponding to free RNA fragments, will have passed through the filter (cf. reference 30).
3. Roll up the glass filter (to which now only the RNA–protein cross-linked complexes are adsorbed), and place it in a small (0.6 ml) Eppendorf tube containing a little cellulose powder. Pierce the bottom of the tube with a fine needle, and place it in a larger (2.5 ml) Eppendorf tube.
4. Elute the cross-linked complexes from the filter with a solution containing 0.1% SDS, 170 mM trisodium citrate, 1 mM EDTA, 5 M urea, 24 mM 2-mercaptoethanol and 10 mM Tris–citric acid buffer (pH 8.0), together with a little bromophenol blue dye. To do this, add successive 50 μl aliquots of the elution buffer to the rolled-up filter, allow to stand for 1 min, and then centrifuge briefly in a table centrifuge. The eluate passes through the cellulose in the small Eppendorf tube and is collected in the large tube. 5×50 μl aliquots are sufficient to elute the cross-linked complexes.
5. The three eluate fractions containing the most ^{32}P-radioactivity are warmed to 70 °C for 5 min and then loaded directly on to a polyacrylamide gel for separation of the individual RNA–protein cross-linked complexes.

5.4 Separation of RNA–protein cross-linked complexes by two-dimensional polyacrylamide gel electrophoresis

The two-dimensional gel system for separation of the individual complexes eluted from the glass filter (*Protocol 18*) has been described in full detail in reference 3. Briefly, the principle of the system is to use a first dimension gel containing SDS, urea, and a high concentration of salt, with a view to separating the components of the mixture on a size basis. The high salt is essential to reduce electrostatic interactions between the RNA–protein complexes, and the polyacrylamide concentration is also high, to achieve maximum separation. In the second dimension the complexes are separated on a charge basis, in gels containing urea and (again) high salt, together with the zwitterionic detergent 'CHAPS', 3-[(3-cholamidopropyl)dimethyl ammonio]-1-propane sulphonate, (Serva). In this second gel dimension the polyacrylamide concentration is low (lower than that of the first dimension), and in consequence the first dimension gel must be fractionated into slices, and each slice eluted for application to the second dimension system (3).

A typical separation of RNA–protein cross-linked complexes (in this case from 30S subunits cross-linked with 2-iminothiolane, *Protocol 15*) is illustrated in *Figure 4*. It can be seen that the relatively small amounts of non-cross-linked RNA remaining after the glass fibre filtration step form a 'diagonal' (cf. *Figure 2*), with the cross-linked complexes lying above this diagonal, in well-defined rows according to the protein each complex contains. The radioactive spots of interest are cut from the two-dimensional gels, and extracted in the presence of SDS exactly as described in reference 3. Yields of the individual cross-linked complexes at this stage vary from a few thousand up to several hundred thousand counts/min of ^{32}P-radioactivity.

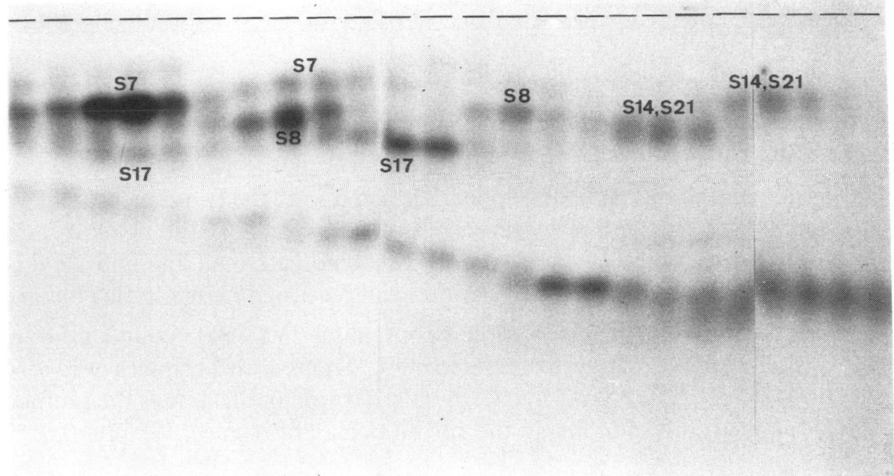

Figure 4. Autoradiogram of part of a set of two-dimensional gels of RNA–protein cross-linked complexes (Section 5.4, and reference 3). The direction of the first-dimension gel is from left to right, and the second-dimension gel slots (running from top to bottom with the dotted line indicating the top of the gel) contain the eluates from consecutive slices of the first dimension gel. The RNA–protein cross-linked complexes, numbered according to their protein content (see Section 5.5), can be seen as well-ordered rows of spots lying above the 'diagonal' of residual free RNA fragments. This relatively simple gel pattern is from 30S subunits cross-linked with 2-iminothiolane (*Protocol 15*).

5.5 Final purification of cross-linked complexes by antibody affinity chromatography

Protocol 19. Preparation of agarose–antibody complexes

1. The starting materials for this procedure are antigoat IgG raised in rabbits and immobilized on agarose (Sigma Chemical Corp, cat. no. 6903), and antisera to the individual ribosomal proteins, raised in sheep (44, 31).
2. Shake the agarose-bound antibody preparation to form a slurry, and pipette aliquots into Eppendorf tubes so that each aliquot contains *c.*, 40 μl of agarose, after it has settled out at the bottom of the tube.
3. Wash twice for 5 min by gently rotating with 1 ml of 'antibody buffer' [20 mM Tris–HCl (pH 7.8), 0.9% NaCl, 0.05% Tween 20 (Sigma)]. Spin off the agarose briefly in a table centrifuge and pipette off the supernatant after each wash.
4. Add 20 μl of appropriate ribosomal protein antibody (31, 44) together with 160 μl of antibody buffer, and rotate gently for 2 h at room temperature.
5. Spin off the agarose as in 3, pipette off the supernatant, and wash five times for 5 min with 1 ml of antibody buffer. The anti-ribosomal protein antibody is now bound to the agarose, ready for use.

Protocol 20. Identification of the proteins in RNA–protein cross-linked complexes

1. Precipitate the cross-linked complexes isolated from the two-dimensional gel (Section 5.4) with ethanol (*Protocol 5*). Resuspend in 100 μl of 20 mM Tris–HCl (pH 7.8), 0.1% SDS, 6 mM 2-mercaptoethanol, and add 20 μl of 3 M NaCl to the solution. The samples can be stored frozen in this buffer.
2. Take aliquots of each sample containing 300–500 counts/min of ^{32}P-radioactivity, and warm briefly to 60 °C. Dilute with 9 volumes of 20 mM Tris–HCl (pH 7.8), 100 mM NaCl, 0.05% Tween 20; this brings the samples into a milieu corresponding to the antibody buffer (*Protocol 19*), plus 0.01% SDS.
3. Dilute each aliquot further with antibody buffer plus 0.01% SDS to a final volume of 200 μl. Add these ^{32}P-samples to the appropriate 40 μl antibody–agarose preparations (*Protocol 19*), and rotate gently for 3 h at room temperature. [It should be noted here that the separation pattern of the two-dimensional gel (Section 5.4 and *Figure 4*) gives *per se* a preliminary idea as to which proteins are likely to be present in a particular isolated complex.]
4. Wash the agarose twice for 5 min with antibody buffer (without SDS), and

then measure the agarose-bound ^{32}P-radioactivity. Positive counts indicate that the protein present in the isolated cross-linked complex corresponds to the antibody bound to the agarose.

Protocol 21. Preparative isolation of pure cross-linked complexes

1. If the test of *Protocol 20* reveals the presence of only a single protein in the cross-linked complex as isolated from the two-dimensional gel (Section 5.4), then that complex can be analysed directly as in Section 5.6 below.
2. If more than one protein is present, or if the complex was from a smeared or otherwise 'overcrowded' region of the two-dimensional gel (cf. *Figure 4*), then simply repeat the procedure of *Protocol 20*, using the whole of the radioactive sample in a diluted volume of up to 1 ml (cf. *Protocol 20*, step 3) to bind to a single agarose-antibody preparation.
3. After incubation of the sample with the antibody–agarose preparation, use the supernatant directly for binding to further antibody–agarose preparations, corresponding to the other proteins that were found to be present in the cross-linked complex in the test reactions of *Protocol 20*.

5.6 Oligonucleotide analysis, and assessment of results

Protocol 22. Removal of ribosomal protein, and fingerprinting

1. Treat the agarose-bound RNA–protein complex (or the whole of the non-bound sample if only a single protein was detected; see *Protocol 21*) with proteinase K in order to release the RNA, and then extract with phenol. The procedure is the same as that of *Protocol 6* except that the proteinase K digestion is repeated twice, and the agarose then washed once with 100 μl of the proteinase K buffer. Precipitate the [^{32}P]-RNA from the combined agarose eluates with ethanol (*Protocol 5*).
2. Hydrolyse the RNA with ribonuclease T1 or ribonuclease A, and carry out the oligonucleotide fingerprint analysis, just as described in Section 4.5.

5.6.1 Assessment of results

The fingerprint data are interpreted in the same way as that outlined in Section 4.6. In this case, however, only a single RNA sequence region should be present, and the cross-link site is indicated by the absence from the fingerprint of a ribonuclease T1 or ribonuclease A oligonucleotide from the sequence region concerned, together with the appearance of a 'new' spot. This spot should represent the oligonucleotide–oligopeptide residue resulting from the combined nuclease plus proteinase digestion, and it should give secondary digestion products which correspond to those of the missing oligonucleotide. Due to the basic character of the peptide residue, these oligonucleotide–oligopeptide spots usually run well to the left in the fingerprints (cf. *Figure 3*). As with the intra-RNA cross-linking data, numerous examples of individual cross-link site analyses can be found in our publications (e.g. 14, 16, 43). The total number of RNA–protein cross-link sites that have so far been localized on 16S and 23S RNA by this method stands at over fifty.

Acknowledgements

Thanks are due to Dr G. Stöffler for providing us with antibodies to the ribosomal proteins, to Dr H. G. Wittmann for his continued support, and to many ex-members of our research group who laid the foundations for the methodologies described here.

References

1. Brimacombe, R., Atmadja, J., Stiege, W., and Schüler, D. (1988). *J. Mol. Biol.*, **199**, 115.
2. Schüler, D. and Brimacombe, R. (1988). *EMBO J.*, **7**, 1509.
3. Brimacombe, R., Stiege, W., Kyriatsoulis, A., and Maly, P. (1988).In *Methods in Enzymology*, Vol. 164 (ed. H. F. Noller and K. Moldave), p. 287. Academic Press Inc, San Diego.
4. Ehresmann, C., Baudin, F., Mougel, M., Romby, P., Ebel, J. P., and Ehresmann, B. (1987). *Nucleic Acids Res.*, **15**, 9109.
5. Stern, S., Wilson, R. C., and Noller, H. F. (1986). *J. Mol. Biol.*, **192**, 101.
6. Peattie, D. and Gilbert, W. (1980). *Proc. Nat. Acad. Sci. USA*, **77**, 4679.
7. Moazed, D. and Noller, H. F. (1986). *Cell*, **47**, 985.
8. Stern, S., Changchien, L. M., Craven, G. R., and Noller, H. F. (1988). *J. Mol. Biol.*, **200**, 291.
9. Steige, W., Glotz, C., and Brimacombe, R. (1983). *Nucleic Acids Res.*, **11**, 1687.
10. Stiege, W., Atmadja, J., Zobawa, M., and Brimacombe, R. (1986). *J. Mol. Biol.*, **191**, 135.
11. Geiduschek, E. P. (1961). *Proc. Nat. Acad. Sci. USA*, **47**, 950.
12. Atmadja, J., Stiege, W., Zobawa, M., Greuer, B., Oßwald, M., and Brimacombe, R. (1986). *Nucleic Acids Res.*, **14**, 659.
13. Traut, R. R., Bollen, A., Sun, T. T., Hershey, J. W. B., Sundberg, J., and Pierce, L. R. (1973). *Biochemistry*, **12**, 3266.
14. Wower, I., Wower, J., Meinke, M., and Brimacombe, R. (1981). *Nucleic Acids Res.*, **9**, 4285.
15. Fink, G., Fasold, H., Rommel, W., and Brimacombe, R. (1980). *Anal. Biochem.*, **108**, 394.
16. Oßwald, M., Greuer, B., Brimacombe, R., Stöffler, G., Bäumert, H., and Fasold, H. (1987). *Nucleic Acids Res.*, **15**, 3221.

17. Turner, S. and Noller, H. F. (1983). *Biochemistry*, **22**, 4159.
18. Rinke, J., Meinke, M., Brimacombe, R., Fink, G., Rommel, W., and Fasold, H. (1980). *J. Mol. Biol.*, **137**, 301.
19. Expert-Bezançon, A. and Wollenzien, P. L. (1985). *J. Mol. Biol.*, **184**, 53.
20. Zwieb, C. and Brimacombe, R. (1980). *Nucleic Acids Res.*, **8**, 2397.
21. Vassilenko, S. K. and Ryte, V. C. (1975). *Biokhimiya*, **40**, 578.
22. Vassilenko, S. K., Carbon, P., Ebel, J. P., and Ehresmann, C. (1981). *J. Mol. Biol.*, **152**, 699.
23. Donis-Keller, H. (1979). *Nucleic Acids Res.*, **7**, 179.
24. Ehresmann, C. and Ofengand, J. (1984). *Biochemistry*, **23**, 438.
25. Hui, C. F. and Cantor, C. R. (1985). *Proc. Nat. Acad. Sci. USA*, **82**, 1381.
26. Dams, E., Hendriks, L., van de Peer, Y., Neefs, J. M., Smits, G., Vandenbempt, I., and De Wachter, R. (1988). *Nucleic Acids Res.*, **16**, r87.
27. Gutell, R. R. and Fox, G. E. (1988). *Nucleic Acids Res.*, **16**, r175.
28. Stiege, W., Kosack, M., Stade, K., and Brimacombe, R. (1988). *Nucleic Acids Res.*, **16**, 4315.
29. Volckaert, G. and Fiers, W. (1977). *Anal. Biochem.*, **83**, 228.
30. Thomas, C. A., Saigo, K., McCleod, E., and Ito, J. (1979). *Anal. Biochem.*, **93**, 158.
31. Gulle, H., Hoppe, E., Oßwald, M., Greuer, B., Brimacombe, R., and Stöffler, G. (1988). *Nucleic Acids Res.*, **16**, 815.
32. Darlix, J. L. (1975). *Eur. J. Biochem.*, **51**, 369.
33. Brosius, J., Dull, T. J., and Noller, H. F. (1980). *Proc. Nat. Acad. Sci. USA*, **77**, 201.
34. Stiege, W. and Brimacombe, R. (1985). *Biochem. J.*, **229**, 1.
35. Brosius, J., Ulrich, A., Racker, M. A., Gray, A., Dull, T. J., Gutell, R. R., and Noller, H. F. (1981). *Plasmid*, **6**, 112.
36. Maniatis, T., Fritsch, E. F., and Sambrook, J. (1982). *Molecular Cloning, A Laboratory Manual*, Cold Spring Harbor Press, USA.
37. Messing, J. (1983). In *Methods in Enzymology*, Vol. 101 (ed. R. Wu, L. Grossmann, and K. Moldave), p. 20. Academic Press Inc. San Diego.
38. Goldberg, M. L., Lifton, R. P., Stark, G. R., and Williams, J. G. (1979). In *Methods in Enzymology*, Vol. 68 (ed. R Wu), p. 206. Academic Press Inc, San Diego.
39. Anderson, M. L. M. and Young, B. D. (1985). In *Nucleic Acid Hybridization, A Practical Approach* (ed. B. D. Hames and S. J. Higgins), p. 73. IRL Press, Oxford, UK.
40. Uchida, T., Bonen, L., Schaup, H. W., Lewis, B. J., Zablen, L., and Woese, C. R. (1974). *J. Mol. Evol.*, **3**, 63.
41. Brownlee, G. G. (1972). *Determination of Sequences in RNA*, North Holland Press, Amsterdam.
42. Brosius, J., Palmer, M. L., Kennedy, P. J., and Noller, H. F. (1978). *Proc. Nat. Acad. Sci. USA*, **75**, 4801.
43. Greuer, B., Oßwald, M., Brimacombe, R., and Stöffler, G. (1987). *Nucleic Acids Res.*, **15**, 3241.
44. Gulle, H., Brimacombe, R., Stöffler-Meilicke, M., and Stöffler, G. (1987). *J. Immunol. Methods*, **102**, 183.

8

Reconstitution of ribosomes

KNUD H. NIERHAUS

1. Introduction

Take 21 different proteins and one RNA molecule, mix them and incubate them, and the result is a homogeneous population of complexes which contain exactly one copy of the various components. This sounds more like an alchemist's recipe than a description of a serious biochemical experiment, but yet this is precisely what happens in the reconstitution of the small subunit of ribosomes from *E. coli*. The situation is even more unbelievable in the case of the large subunit, since here 36 different components are mixed together. In this case a two-step incubation has to be performed, in which the Mg^{2+} concentration and incubation temperature are changed for the second step. Again, the end result is a particle with one copy of each of the different components, the only exception being the protein L7/L12, which is found in four copies per particle. Not only is it surprising that the process works at all, but also that it does so with a remarkably high efficiency. About 50–100% of the input material is assembled into active particles, if the rules outlined in the next section are carefully considered.

The fact that ribosomal subunits can be totally reconstituted with relative ease indicates that both the prescription for this complicated assembly process and the information necessary to form the quaternary structure of the ribosome resides completely in the primary sequences of the proteins and rRNAs taking part.

When discussing the total reconstitution of ribosomes we usually refer to *E. coli*, since only the ribosomes from this organism have been extensively analysed by the reconstitution technique. Highlights of the progress in this field are the unravelling of assembly features of the small (1) and large subunits (2, see also Section 4.1), the assembly maps of both subunits (3 and 4, respectively), single-component omission experiments identifying the proteins that are candidates for the peptidyltransferase function (5), analysis of assembly mutants, where the phenotypic features of the mutant could be traced back to assembly defects (6), and the usage of the reconstitution technique to assess the arrangement of the proteins within the small subunit by means of neutron scattering (7). Accordingly, the main part of this chapter deals with the reconstitution of *E. coli* ribosomes (Sections 2–5), and following these sections the efforts to reconstitute ribosomes from other organisms of the eubacterial, archaebacterial, or eukaryotic kingdoms are briefly outlined (Section 6). The chapter closes with a short description of future applications and experimental strategies of the reconstitution method.

2. Prerequisites for a successful reconstitution

2.1 General rules

The chief concern and basic rule is to be aware of and to avoid ribonuclease contamination and activity. This rule has a number of precautionary consequences:

(a) Use an *E. coli* strain which is deficient in RNases, e.g. *E. coli* K12, strain A19 or D10 (8); CAN20-19E (9); MRE600.

(b) Cells grown in rich medium (see *Protocol 2*) should be harvested in the log phase at a concentration of 0.5 A_{650} units per ml. This represents an optimum, since at this stage the ribosomes are very active. At the later semi-log phase stage an RNase appears which sticks to ribosomes and impairs the reconstitution efficiency (10), and at the even later stationary phase the ribosomes are not optimally active.

(c) Since RNases are secreted from our hands, it is essential to wear gloves throughout the preparation of ribosomes.

(d) The use of disposable materials and apparatus is recommended whenever possible. Sometimes, however, it is impossible to avoid the use of non-disposable equipment. For example, 'clean' centrifuge tubes for pelleting zonally centrifuged 50S subunits must be washed with distilled water carefully and extensively. Other non-disposable instruments should be sterilized before use. Equipment should be reserved solely for use with these 'RNase-free' experiments. Such equipment should be cleaned and treated separately from other communal glassware, and so on.

(e) Use RNase-free sucrose for sucrose gradients (for example, ultrapure sucrose, BRL).

(f) Perform all preparations at temperatures $\leqq 4$ °C.

(g) There is generally no problem in preparing 30S subunits with intact 16S rRNA. However, it has been claimed that it is almost impossible to isolate 50S subunits with intact 23S rRNA. Following the rules outlined here should allow the isolation of 50S subunits with intact 23 rRNA (*Figure 1*, see also reference 12). If particles are not active following reconstitution, the intactness of 23S rRNA should be checked. Ideally, this should be done both before and after the reconstitution process in order to test for the presence of RNases. This check is made by SDS-gel electrophoresis (*Protocol 9*), where isolated rRNA as well as intact subunits or assembled particles can be directly applied to the gel.

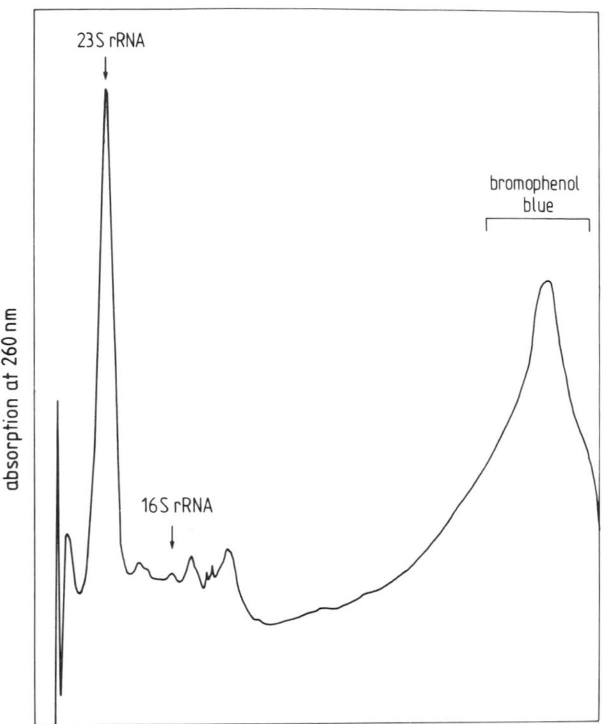

Figure 1. Profile of an A_{260}-scan of an SDS–RNA gel (*Protocol 9*) One A_{260} unit of 50S subunits was applied to the gel.

Protocol 1. Buffers used for the preparation of ribosomal subunits and in reconstitution experiments

1. Buffer 1 (ionic milieu for activity tests): 20 mM Hepes–KOH (pH 7.6 at 0 °C), 3 mM $MgCl_2$, 150 mM NH_4Cl, 2 mM spermidine, 0.2 mM spermine, 5 mM 2-mercaptoethanol.
2. Buffer 2 (subunit dissociation condition): 20 mM Hepes–KOH (pH 7.6 at 0 °C), 1 mM $MgCl_2$, 300 mM NH_4Cl, 2 mM spermidine, 0.2 mM spermine, 5 mM 2-mercaptoethanol.
3. Buffer 3 (association condition; buffer used for preparation of tight-couple ribosomes): 20 mM Hepes–KOH (pH 7.6 at 0 °C), 4 mM $MgCl_2$, 30 mM NH_4Cl, 2 mM spermidine, 0.2 mM spermine, 5 mM 2-mercaptoethanol.
4. Buffer 4 (previously termed Rec-4 buffer): 20 mM Tris–HCl (pH 7.4 at 37 °C), 4 mM Mg acetate, 400 mM NH_4Cl, 0.2 mM EDTA, 5 mM 2-mercaptoethanol.
5. Buffer 5 (previously termed Rec-4-6U buffer): same as buffer 4 except that

6 M urea (p.a., Merck) is present. The buffer is mixed with bentonite-SF (1 g per litre; Serva, Heidelberg) stirred at 4 °C for 1 h, and filtered through two layers of Selecta filters (Schleicher and Schüll, 595 1/2), and stored at 4 °C for up to 3 weeks. Alternatively, the 6 M urea solution may be passed through a DOWEX ion-exchange column (5 × 30 cm, DOWEX 1-X8, 20–50 mesh, Serva), and the salts added subsequently.

6. Buffer 6 (previously termed Rec-20 buffer): 20 mM Tris–HCl (pH 7.4 at 37 °C), 20 mM Mg acetate, 400 mM NH_4Cl, 1 mM EDTA, 5 mM 2-mercaptoethanol.
7. Buffer 7: 10 mM Tris–HCl (pH 7.6 at 0 °C), 4 mM Mg acetate.
8. Buffer 8: 10 mM Tris–HCl (pH 7.6 at 0 °C), 50 mM KCl.
9. Buffer 9: 110 mM Tris–HCl (pH 8.0 at 0 °C), 4 mM Mg acetate, 4 M NH_4Cl (note: 4 molar NH_4Cl), 0.2 mM EDTA, 20 mM 2-mercaptoethanol.
10. Buffer 10: 10 mM Tris–HCl (pH 7.6 at 0 °C), 10 mM Mg acetate.
11. Buffer 11: 20 mM Tris–HCl (pH 7.8 at 0 °C), 11 mM Mg acetate, 160 mM NH_4Cl, 4 mM 2-mercaptoethanol.

2.2 Isolation of 70S ribosomes and ribosomal subunits

The method given in *Protocol 2* is a well-tried and reliable procedure. The only special point is that buffer 3 is used, which contains 4 mM Mg^{2+} and polyamines which allows for the isolation of highly active ribosomes. If tight-couple ribosomes are to be prepared, then use the same buffer throughout until the zonal run, which is performed at reduced speed (at 21 000 r.p.m. for 18 h in a Beckman B15 Ti rotor). Next, pellet the 70S ribosomes (at 22 000 r.p.m. for 24 h in a 45 Ti rotor), and remove the residual 50S subunits in a second zonal run under the same conditions. With the exception of the buffer recommended above, a detailed description of the preparation of tight-couple ribosomes can be found in reference (13). See also Chapter 1.

In contrast to some prescriptions in the literature, we do not recommend washing the ribosomes or the ribosomal subunits with 0.5 M or 1 M NH_4Cl. This washing procedure partially removes some ribosomal proteins (S1, S5, S6, S8, S16, L1, L3, L6, L10, L11, L7/12, L24, L29, L30, L32/33; Fehner and Nierhaus, unpublished observation), and thus generates a heterogeneous ribosome population which is disadvantageous for functional studies.

Protocol 2. Isolation of 70S ribosomes, ribosomal subunits, and S-150 enzymes

1. Grow cells in rich medium, e.g. L.-H_2O medium [reference 14, 1 litre L-H_2O medium contains 10 g Bacto-Tryptone (Oxoid), 5 g yeast extract (Oxoid), 5 g

NaCl, 1 ml 1 M NaOH, 10 ml 20% glucose (w/v)], harvest at 0.5 A_{650} units per ml, and store pelleted cells (yields about 1–1.5 g per litre) at −80 °C.

2. Thaw, for example, 100 g cells in 200 ml buffer 3, pellet them at 16 000 *g* for 10 min (e.g., in a Sorvall GSA rotor), mix them with 200 g Alcoa A-305 (Serva), and grind them at 4 °C in a Retsch mill for 25 min.
3. Homogenize the cell paste with 200 ml buffer 3 for 15 min in the mill and subject it to two low-speed centrifugations, at 16 000 *g* for 10 min (e.g., using a Sorvall GSA rotor) to remove Alcoa, and at 30 000 *g* for 45 min to remove cell debris. The result is the S-30 supernatant (S-30 for supernatant obtained by a centrifugation at 30 000 *g*).
4. Centrifuge the S-30 at 50 000 *g* for 15–18 h (if tight-couple ribosomes are to be prepared: at 30 000 *g* for 18 h) e.g., in a Beckman 45 Ti rotor.
5. Resuspend the pellet of ribosomes in 30 ml buffer 2 per tube and clarify by low-speed centrifugation. The supernatant is used as a source of enzymes and elongation factors.[a]
6. Subject about 7000 A_{260} units to a zonal centrifugation in 0–40% sucrose in buffer 2 at 21 000 r.p.m. for 18 h in a Beckman Ti15 rotor.
7. Pool the fractions containing 30S and 50S ribosomal subunits, respectively, and pellet them at 100 000 *g* for 24 h in a 45 Ti rotor). Resuspend the pellets in buffer 3 (300–700 A_{260} units/ml), clarify by low-speed centrifugation, and store in small portions at −80 °C.

[a] Centrifuge the supernatant at 150 000 *g* for 3 h then take the upper two-thirds of the supernatant for use as S-150 enzymes (S for supernatant, 150 for 150 000 *g*).

3. Partial reconstitution with split proteins and core particles

3.1 General remarks

Ribosomal proteins can be split off the ribosome with salt solutions in such a way that the higher the ionic strength, the more proteins are washed off. This splitting process is almost an 'all-or-nothing' phenomenon for most of the proteins; they appear together in groups in the split fractions when distinct values of the ionic strength are reached. This indicates that the splitting process is co-operative. At least four co-operative protein families can be distinguished within the 50S subunit, comprising those which are split off at about 1.0 M LiCl (L1, L6, L7/L12, L9, L10, L11, L16, L28), those split off at 2.0 M LiCl (L5, L15, L18, L25, L27, L31, L33), those at 3.5 M LiCl (L14, L19, L24, L30, L32), and those proteins (L3, L4, L13, L17, L20, L21, L22, L23, L29 and L34) which remain quantitatively attached on the resulting '3.5 core' (obtained with 3.5 M LiCl). Protein L2 splits off over a broad range of ionic strength.

The co-operative behaviour provides a convenient means for fractionating the ribosomal proteins, and we take advantage of it when we isolate individual ribosomal proteins (Section 4.2). Interestingly, the sequence in which the ribosomal proteins can be washed off roughly reflects the inverse order of the incorporation of proteins during the course of ribosome assembly (4). Accordingly, the existence of co-operatively split families might also reflect the step-wise nature of ribosome assembly; in other words, the *in vivo* assembly of the 50S subunit might occur in three to five major steps.

The resulting core particles represent defined complexes of ribosomal proteins and rRNA. These complexes can be used for the reconstitution of active subunits by complementing them with the corresponding split fractions or even by adding total proteins. Partial reconstitution can be advantageous in the case of 50S subunits if one of the many L-proteins is to be replaced by a modified protein. In the case of 30S subunits it is convenient to apply the total reconstitution procedure, since most of the S-proteins can be separated by a single HPLC-run. Therefore, we consider here only washing procedures with 50S subunits, although the procedures described can be readily extended to 30S subunits as well.

The most effective salt for splitting off ribosomal proteins is LiCl. Defined cores lacking only a few proteins can also be obtained by the NH_4Cl/EtOH procedure.

3.2 The NH_4Cl/EtOH-split procedure (15)

A straightforward procedure, described in *Protocol 3*, yields the proteins (L1), L5, L6, L7/12, L10, L11, L16, (L25), L31, and L33 (proteins in parentheses are not quantitatively obtained). Protein L7/12 is preferentially washed off the ribosome if the 50S subunits are treated as described in *Protocol 3*, with the exceptions that 1 M NH_4Cl is present and the ethanol treatment is performed at 0 °C.

Protocol 3. Washing 50S ribosomal subunits with NH_4Cl/EtOH

Split proteins

1. Dilute, for example, 3500 A_{260} units of 50S subunits in 100 ml buffer containing 10 mM imidazole–HCl (pH 7.4), 1.5 M NH_4Cl, 20 mM $MgCl_2$, and 1 mM 2-mercaptoethanol (final concentration: 35 A_{260} units/ml). Warm up to 37 °C.
2. Add 50 ml pre-warmed ethanol and shake gently for 10 min at 37 °C.
3. Add a further 50 ml volume of pre-warmed ethanol and shake gently for 5 min at 37 °C.
4. Centrifuge the mixture at 16 000 *g* for 30 min (e.g. in a Sorvall GSA rotor).
5. Withdraw the supernatant, add 300 ml (3 volumes) acetone, and store at −20 °C overnight.

6. Collect the precipitated proteins by centrifugation at 16 000 *g* for 30 min.
7. Dissolve the protein pellet in a buffer suitable for the next step. If the split proteins are to be used as one fraction in reconstitution assays, dissolve the proteins in 10 ml of buffer 5.
8. Dialyse overnight against 100 volumes of the same buffer at 4 °C, and then three times for 45 min against 100 volumes of buffer 4.
9. Measure the A_{230} absorption, store these split proteins designated as SP37(1.5) in small aliquots at −80 °C.

If single proteins are to be isolated by reversed phase HPLC, treat the pellet obtained in step 6 as described in *Protocol 8*.

Core particles

1. Resuspend the pellet obtained in step 4 (above) in about 7 ml buffer 4.
2. Dialyse overnight against 10 volumes of the same buffer, clarify by a low-speed centrifugation.
3. Measure the concentration in A_{260} units/ml and store the resulting subunit core particles designated P37(1.5) in small aliquots at −80 °C.

3.3 The LiCl-split procedure (16)

An example using 3.5 M LiCl for the splitting procedure is given in *Protocol 4*, but any other LiCl concentration can be used correspondingly. It is important that the desired LiCl concentration is established before the 50S subunits are added. Adjust a stock-solution of 5 M LiCl in buffer 10 to the desired LiCl concentration with the same buffer and then add the 50S subunits to a final concentration of 20 A_{260} units/ml.

When isolating L-proteins we treat $P_{37(1.5)}$ cores (see *Protocol 3*) as described for 50S subunits in *Protocol 4*, and purify the individual proteins from the differential split fraction $SP_{3.5}$ via HPLC (Section 4.2).

Protocol 4. Washing 50S ribosomal subunits with LiCl (3.5 M)

Split proteins

1. Dilute 50S subunits in buffer 10 containing 3.5 M LiCl to a final concentration of 20 A_{260} units/ml (e.g. 800 A_{260} units of 50S subunits in 40 ml).
2. Incubate for 5 h at 0 °C, shake gently once every hour.
3. Centrifuge at 40 000 r.p.m. for 5 h.
4. Withdraw the supernatant, dialyse it against water, and lyophilize the sample.

5. Add 4 ml 66% acetic acid in 0.1 M Mg acetate and stir for 45 min at 4 °C in order to remove traces of rRNA.
6. Centrifuge at 30 000 *g* for 30 min.
7. Take the supernatant and add 20 ml (5 volumes) acetone and keep at −20 °C overnight (or at least for 2 h).
8. Centrifuge at 30 000 *g* for 30 min.
9. Take the pellet and remove the residual acetone by 5 min lyophilization or by placing for 30 min in a desiccator. (If the proteins of the SP3.5 fraction are to be isolated via HPLC continue with *Protocol 8*.)
10. Resuspend the pellet in 3 ml buffer 5 and dialyse overnight against the same buffer.
11. Dialyse three times each for 45 min against 10 volumes of buffer 4.
12. Centrifuge at 12 000 *g* for 5 min; measure absorption at 230 nm and store the resulting 3 ml of the SP3.5 fraction in small aliquots at −80 °C.

Core particles

Resuspend the pellet obtained in step 3 (above) in 2 ml of buffer 6, measure the concentration (A_{260} units/ml), and store the 3.5 core particles (3.5 c) at −80 °C in small aliquots.

3.4 Partial reconstitution procedure

In *Protocol 5* a partial reconstitution is described using LiCl derived 3.5 cores and the split proteins '$SP_{3.5}$', prepared from 50S subunits according to *Protocol 4*. The same ionic conditions can also be used for the partial reconstitution with components derived from 30S subunits, but in this case the incubation should be performed at 40 °C for 20 min.

It is relatively easy to assess the optimal quantities to add for the reconstitution of single proteins as well as of total proteins (see Section 4.2). This is in contrast to the case of the split fractions, where the relative molarities of the individual proteins may vary. In the latter case we recommend the following estimation procedure for, e.g. the $SP_{3.5}$ fraction, which is obtained according to *Protocol 4*. In this example 3 ml $SP_{3.5}$ were obtained from an input of 800 A_{260} units of 50S subunits. If there were no losses during the preparation, then one equivalent unit (e.u.) of the $SP_{3.5}$ fraction would be present in 3.75 μl (one e.u. is that amount of a protein or group of proteins which is present on one A_{260} unit of 50S subunits). With the reasonable assumption that the actual recovery of the proteins is about 60%, then one e.u. should be present in about 6 μl, and the 3 e.u. required for the partial reconstitution described in *Protocol 5* would thus be present in about 20 μl. Therefore, one would make test reconstitutions with 2.5 A_{260} units of 3.5c core particles and with varying amounts of the prepared $SP_{3.5}$ fraction in the range of 20 μl.

Protocol 5. Partial reconstitution of 50S ribosomal subunits

1. Take 2.5 A_{260} units of 3.5 M LiCl derived core particles (see *Protocol 4*) in 25 μl of buffer 6. Add about 3 equivalent units (e.u.) of $SP_{3.5}$ proteins (in buffer 4) and add buffer 4 to a final volume of 100 μl per aliquot. Add 3 μl of 0.4 M Mg acetate to adjust the Mg^{2+} concentration to 20 mM.
2. Incubate for 90 min at 50 °C.
3. Use 40 μl for activity measurements as described for the total reconstitution in *Protocol 13*.

---

4. Total reconstitution with total proteins and rRNA

4.1 General remarks

Both subunits of *E. coli* ribosomes can be separated into their protein and rRNA moieties, and reconstituted again to fully active subunits. These total reconstitutions are the basic methods for *in vitro* studies of the ribosomal assembly, culminating in the 'assembly maps' for both subunits (*Figure 2*).

The total reconstitution of the 30S subunit (19) is a one-step procedure: the total proteins of the 30S subunit (TP30) and the 16S rRNA are mixed in a stoichiometric ratio under defined conditions (buffer 6), and after 10–20 min at 40 °C 30S subunits are formed in high yield (i.e. 50–100% of the input material; see Section 4.5). In contrast the total reconstitution of the 50S subunit (10) is a two-step procedure. The reason for this is that during early and late assembly, respectively, conformational changes occur which require different ionic milieus and incubation temperatures. The total proteins (TP50) and (23S + 5S) rRNA are mixed stoichiometrically, and after an incubation at 44 °C in buffer 4 the Mg^{2+} concentration is raised to 20 mM, and a second incubation is performed at 50 °C (Section 4.4). The ionic conditions of the second step are identical both to the milieu for the 30S reconstitution and to that for the partial reconstitution of both subunits (Section 3.4). This fact enables a total reconstitution procedure for both subunits to be carried out in one test tube (20; Section 4.6).

The two-step procedure for the 50S subunit allows a clear distinction to be drawn between early and late assembly reactions, which has facilitated the elucidation of the assembly features of this subunit (*Protocol 6*; for review see reference 4). In the next three sections the isolation of TP50 (Section 4.2), the isolation of 23S and 5S rRNA (Section 4.3), and the procedure for the total reconstitution of the 50S subunit are described (Section 4.4). The reconstitution

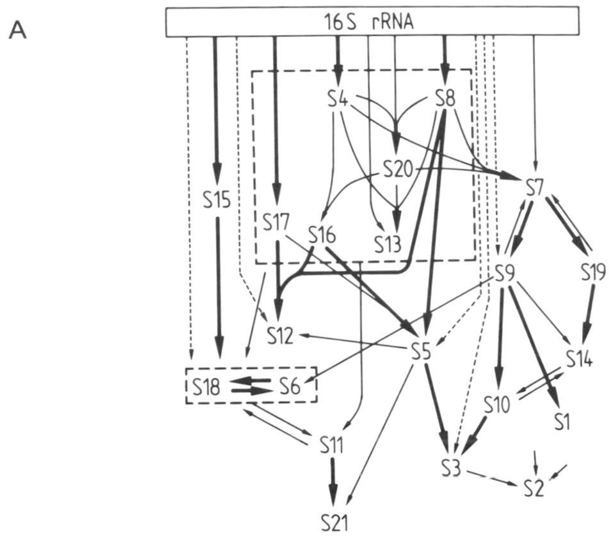

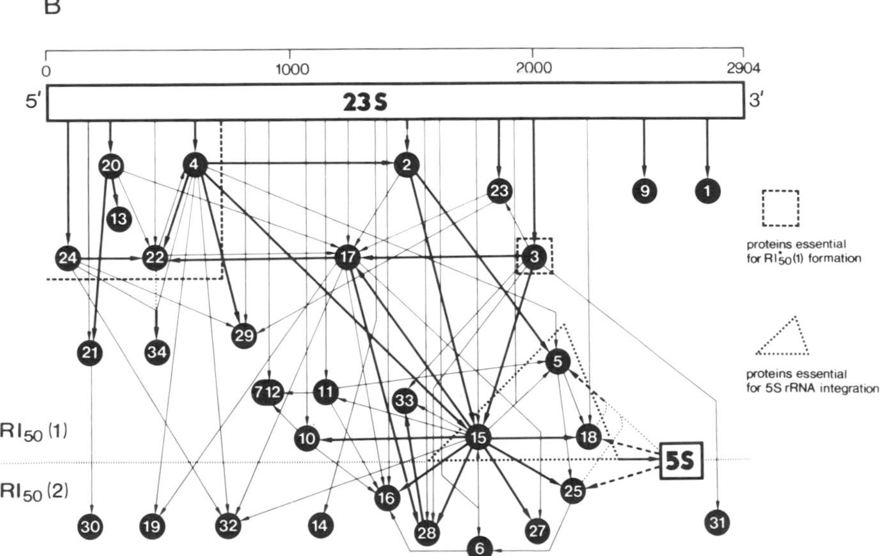

Figure 2. Assembly maps of the subunits from *E. coli* ribosomes. A, 30S subunit (3); supplemented with the results from reference 17 (broken arrows) and from reference 18 (S1). B, 50S subunit (4). The proteins boxed-in with a broken line are essential for the early assembly. $RI_{50}(1)$ and $RI_{50}(2)$ are reconstitution intermediates of the 50S assembly. (See information at end of Section 4.1.

methods for the 30S subunit (Section 4.5) and the 70S ribosome (Section 4.6) follow.

Protocol 6. Assembly features of the *E. coli* 50S ribosomal subunit

1. *In vitro*: There are three intermediates:

$$\underset{(33S)}{RI_{50}(1)} \rightarrow \underset{(42S)}{RI^{*}_{50}(1)} \rightarrow \underset{(48S)}{RI_{50}(2)} \rightarrow 50S$$

   In vivo: There are three precursors:

$$\underset{(32S)}{p_1 50S} \rightarrow \underset{(43S)}{p_2 50S} \rightarrow \underset{(\sim 50S)}{p_3 50S} \rightarrow 50S$$

2. The assembly starts with only two proteins (assembly initiator proteins), namely L24 and L3.
3. The initiator protein L24 can be replaced by L20 at temperatures below 36 °C, although L20 is not as efficient as L24.
4. The early assembly reactions depend on only five proteins: L4, L13, L20, L22, and L24 (early assembly proteins). L3 stimulates but is not essential.
5. The five early assembly proteins bind exclusively to the 5′ end of the 23S rRNA: this is the 'assembly gradient'.
6. At least two of the early assembly proteins, L20 and L24, are mere assembly proteins. They are not involved in either late assembly or function of the 50S subunit.

4.2 Isolation of the total proteins from 50S subunits (TP50)

The method of preparation for TP50 described in *Protocol 7* is equally well suited for the isolation of TP30 or TP70. The acetic acid extraction (21) is followed by a transient urea dialysis, which causes the ribosomal proteins to become reversibly denatured. In the course of the following dialysis steps the proteins refold and readily adopt a conformation which enables the assembly of active particles. The transient urea dialysis allows the preparation of reproducibly active protein fractions. The final protein solution should be clear and free of urea. Store the proteins in small aliquots at −80 °C and thaw them only once before use, since the activity of the proteins decays with repeated freezing and thawing.

If single proteins are to be isolated via HPLC techniques, the initial protein source can be the acetone pellet of a total protein preparation (*Protocol* 7, step 6), or that of an NH_4Cl/EtOH split preparation ($SP_{37(1.5)}$, see *Protocol 3*, steps 7–9), or a LiCl-split preparation (SP3.5, see *Protocol 4*, step 9). For the isolation of L-proteins we follow the scheme shown in *Figure 3*. The acetone pellets from the $SP_{37(1.5)}$, SP3.5, and 3.5cHAc (see below) preparations are dissolved and treated as described in *Protocol 8* before being applied to HPLC columns (P. Nowotny, personal communication). The 3.5cHAc fraction represents the

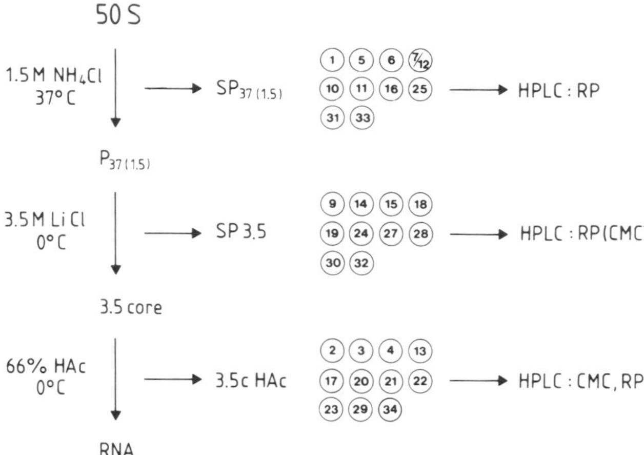

RP (reversed phase): Nucleosil 300-5 C_4
Gradient A: 0.1% TFA in water
B: 0.1% TFA in 2-propanol or acetonitrile

CMC (ion exchange): TSK 535 CM
Gradient A: 5M urea, 10 mM HEPES, 2 mM methylamine
B: + 0.5 M KCl (pH 7.0)

Figure 3. Outline for isolation of L-proteins from the 50S subunit. For details see the text, *Protocols 3, 4 and 8* and reference 22.

proteins of the 3.5 cores (*Protocol 4*) after extraction with acetic acid as described in *Protocol 7* (up to step 6). Details of the HPLC procedures can be found in reference 22.

Protocol 7. Preparation of total 50S ribosomal subunit proteins (TP50)

1. Take, for example, 300 A_{260} units of 50S subunits in 1 ml buffer 3, add 0.1 ml (0.1 volume) 1 M Mg acetate and 2.2 ml (2 volumes) acetic acid.
2. Stir for 45 min at 0 °C.
3. Centrifuge at 10 000 *g* for 30 min.
4. Take the supernatant and add 16.5 ml (5 volumes) acetone and keep at −20 °C overnight (or at least for 2 h).
5. Centrifuge at 10 000 *g* for 30 min.
6. Take the pellet and remove the residual acetone by 5 min lyophilization or by placing for 30 min, in a desiccator.
7. Resuspend the pellet in 1 ml buffer 5, yielding about 300 e.u./ml, and dialyse overnight against the same buffer.
8. Dialyse three times each for 45 min against 100 volumes of buffer 4.

9. Centrifuge at 5000 *g* for 5 min.
10. Measure the absorption at 230 nm (1:100 dilution); store in small aliquots at −80 °C.

Useful approximations for TP50: 1 A_{230} unit = 220 μg = 10 e.u.
for TP30: 1 A_{230} unit = 220 μg = 8 e.u.
for TP70: 1 A_{230} unit = 220 μg = 10 e.u.

Note that 1 A_{260} unit of 50S = 36 pmol = 1 e.u. TP50
1 A_{260} unit of 30S = 72 pmol = 1 e.u. TP30
1 A_{260} unit of 70S = 24 pmol = 1 e.u. TP70

Protocol 8. Treatment of acetone pellets of ribosomal proteins prior to HPLC purification

The preparation of the acetone pellets of SP37(1.5), SP3.5, and 3.5cHAc are described in *Protocols 3, 4, and 7*, respectively. Treat the acetone pellet for a reversed-phase chromatography (Nucleosil 300–5 C_4) as follows:

1. Take the acetone pellet and remove the residual acetone by 3–5 min lyophilization.
2. Resuspend in buffer 5 to a concentration of 200–500 e.u./ml.
3. Apply to the reversed phase column (Nucleosil) and start the gradient (see *Figure 3*).

For a CMC-column (TSK 535 CM):

1. Take the acetone pellet and remove the residual acetone by 3–5 min lyophilization.
2. Resuspend in a buffer containing 10 mM Hepes (pH 7.0 at 4 °C), 2 mM methylamine, and 5 M urea (the buffer has to be treated before as described for buffer 5 in *Protocol 1*) to a concentration of 200–500 e.u./ml.
3. Apply to the CMC column and start the gradient (see *Figure 3*).

If single ribosomal proteins are used in a reconstitution assay, it is of the utmost importance for the activity of the reconstituted particle that the protein is added in near stoichiometric amounts with respect to the rRNA. Our standard reconstitution assay contains 2.5 A_{260} units of rRNA (see Section 4.4), and therefore the optimal amount of a protein is 3 e.u. per assay. We have determined the molar extinction coefficient at 230 nm ε_{230} for each of the L-proteins (except L31, L34, L35, and L36; reference 22), and have used it to calculate a so-called *y*-value for each of the L-proteins. If the *y*-value for a particular protein is divided by the actual A_{230}/ml value of the respective protein solution, one obtains the volume in μl of the protein solution which contains 3 e.u. of that protein. This

Table 1. *y*-values of the L-proteins from the 50S subunit

Protein	*y*-value	Protein	*y*-value
L1	16.3	L18	6.4
L2	23.8	L19	5.2
L3	14.6	L20	11.1
L4	11.5	L21	7.0
L5	13.6	L22	6.6
L6	13.3	L23	5.8
L7/12	7.04 (×4)	L24	3.9
L9	8.4	L25	7.7
L10	11.0	L26=S20	5.0
L11	6.3	L27	6.2
L13	12.8	L28	6.6
L14	8.3	L29	3.8
L15	8.1	L30	3.0
L16	11.5	L32	5.0
L17	9.1	L33	3.5

$y/(A_{230}/ml) = \mu l$ of single protein/3 e.u.

The *y*-value of a protein L*x* divided by the concentration expressed as A_{230}/ml of the L*x* solution gives the volume in μl which contains 3 e.u. of L*x* (see the text for details, Section 4.2).

volume must be used in the reconstitution assay (see *Table* 1): $y/(A_{230}/ml) = \mu l$ of the solution containing 3 e.u. of the single protein concerned.

4.3 Isolation of 23S and 5S rRNAs from 50S subunits

Before isolating rRNA one has to confirm that at least 85% of the corresponding subunits contain intact rRNA. This can be done by SDS-gel electrophoresis (23), where native ribosomes and subunits can be applied directly, and it is not necessary to use purified rRNA. Each sample is loaded on to a gel in a plastic tube (4 mm i.d., 120 mm length), and after electrophoresis the gel is transferred to a quartz cuvette (5 mm i.d., 185 mm length), immersed in distilled water, and scanned at 260 nm. The procedure is presented in *Protocol* 9, a typical scan-pattern being shown in *Figure 1*.

Protocol 9. SDS gel-electrophoresis of rRNA

Buffers and solutions

10 × TEB: Dissolve 108 g Tris (p.a., Merck) in 500 ml distilled water. Add 9.3 g Na_2EDTA, 55 g boric acid, and distilled water to a total volume of 1000 ml. Store at 4 °C.

Electrophoresis buffer

Take 100 ml of 10 × TEB, add 2 g SDS and distilled water to a total volume of 1000 ml.

3.1% acrylamide solution

Take 30 g sucrose (RNase free), add 31 g acrylamide (Serva), 1.8 g methylene*bis*-acrylamide (Serva), and 1.8 g SDS, 100 ml 10 × TEB, and distilled water to a total volume of 1000 ml. Store at 4 °C.

Preparation of gels

1. Each tube gel requires 2 ml of 3.1% acrylamide solution.
2. De-gas 10 ml of the acrylamide solution (enough for 5 gels).
3. Add 5 μl TEMED and 200 μl 10% ammonium persulphate.

Preparation of samples and electrophoresis

1. Mix 0.5–1 A_{260} unit of RNA or ribosomal particles in about 20 μl with 1/10th volume of 10% SDS.
2. Incubate for 2 min at 70 °C.
3. Add 10 μl of 0.1% bromophenol blue in 60% sucrose.
4. Pre-run the gels for 10 min at 2 mA per gel before loading the samples.
5. Perform electrophoresis for 2 h at 0.5 mA per gel, then for 2 h at 1.5 mA per gel.
6. Scan the gels at A_{260} nm or, alternatively, stain with toluidine blue (stain in a solution containing 0.1% toludine blue, 1% acetic acid, and 30% methanol; destain in 10% acetic acid and 20% methanol).

Protocol 10 describes the preparation of rRNA from 50S subunits. The same procedure can also be used for the rRNA from 70S ribosomes, whereas for 30S subunits one phenol extraction instead of three is usually sufficient.

Protocol 10. Isolation of (23S + 5S) rRNA from 50S subunits

The whole procedure is performed at 4 °C. Wear gloves and use sterile equipment.

1. Take, for example, 250 A_{260} units of 50S subunits in 1 ml buffer 3. Do not use higher concentrations. Add 0.1 ml (0.1 volume) of 10% (w/v) SDS (Bio-Rad), 0.05 ml (0.05 volume) of 2% (w/v) bentonite-SF (Serva), and 1.2 ml (1.0 volume) of 70% phenol (freshly distilled phenol should be stored in 10 ml portions at −20 °C).
2. Shake vigorously for 8 min (Vortex) in sterile 15 ml-Corex tubes.
3. Centrifuge at 10 000 *g* for 10 min.
4. Take the aqueous phase (upper phase) and add 1 ml (1 volume) 70% phenol.
5. Shake vigorously for 5 min (Vortex) in sterile 15 ml-Corex tubes.

6. Centrifuge at 10 000 *g* for 10 min.
7. Repeat steps 4–6.
8. Take the aqueous phase (at this stage avoid any contamination with the phenol phase) and add 2 ml (2 volumes) ethanol, and keep at −20 °C for 2 h in sterile 15 ml-Corex tubes.
9. Centrifuge at 10 000 *g* for 45 min.
10. Wash the pellet with about 0.5 ml (0.5 volume) ethanol in order to remove traces of phenol (Vortex).
11. Centrifuge at 10 000 *g* for 30 min.
12. Resuspend the pellet in 1 ml buffer 7. Measure the concentration (A_{260}/ml) and store in small aliquots at −80 °C.

Separation of 23S and 5S rRNA can be readily achieved by a sucrose gradient centrifugation (22) or via HPLC (Nowotny, personal communication). Both techniques are described in *Protocol 11*.

Protocol 11. Separation of 23S and 5S rRNA

Sucrose gradient method

1. Take 1–2 ml of the aqueous phase (*Protocol 10*, step 8) containing 200–600 A_{260} units of (23S + 5S) rRNA, and layer it on to a sucrose gradient (10–30% sucrose in buffer 8 with 1% methanol).
2. Centrifuge at 20 000 r.p.m. for 17 h (e.g. using a Beckman SW27 rotor).
3. Fractionate the gradient and monitor absorbance at 290 nm. Pool fractions containing 23S and 5S rRNA, respectively, add 2 volumes of ethanol and keep the solutions at −20 °C overnight.
4. Centrifuge at 10 000 *g* for 45 min.
5. Resuspend the pellet of 23S rRNA in 1 ml buffer 7 and that of 5S rRNA in 0.2 ml buffer 7. Measure the concentrations (A_{260}/ml), convenient concentrations are 200 and 20 A_{260} units/ml for 23S and 5S rRNA, respectively. Store at −80 °C in small portions.

HPLC technique

1. Take 1 ml of the aqueous phase (*Protocol 10*, step 8) containing not more than 300 A_{260} units of (23S + 5S) rRNA and apply to the column (Nucleogen DEAE 500–10, 150 × 4 mm; Macherey and Nagel), and perform the rRNA separation at room temperature.
2. Start the gradient (A: 20 mM Hepes–KOH (pH 7.5), 5% methanol; B: 20 mM Hepes–KOH (pH 7.5), 5% methanol, 2 M KCl).
3. Monitor the absorbance at 260 nm and pool fractions containing 5S and 23S

rRNA, respectively. Add 2 volumes of ethanol and keep the solutions at −20 °C overnight.

4. Continue as described above in steps 4 and 5.

4.4 Total reconstitution of 50S subunits and activity measurements

The procedure for the total reconstitution of the 50S subunit is presented in *Protocol 12*. Exactly 2.5 A_{260} units of (23S + 5S)rRNA are mixed with 3 e.u. of total proteins or individual proteins in 100 µl, and subjected to a two-step incubation (10). If 23S and 5S rRNA are added separately, 2.5 A_{260} units of 23S rRNA should be mixed with 0.1 A_{260} units of 5S rRNA. After the two-step incubation one 40 µl aliquot per reconstitution assay is tested for activity in the poly(Phe) synthesis system (*Protocol 13*). A volume of 40 µl containing 1 A_{260} unit of reconstituted 50S particles, and 1 A_{260} unit of 30S subunits (molar ratio 50S : 30S = 1 : 2) is added, yielding up to 36 pmol of 70S ribosomes. The poly(Phe) system is adjusted so that the specific activity of the [^{14}C]Phe is 10 c.p.m./pmol. Hence, the number of Phe molecules incorporated, statistically, per 70S ribosome can be determined easily. The ionic conditions of the poly(Phe) system are those of buffer 11 containing 1.5 mM ATP, 0.05 mM GTP, 80 µM [^{14}C]Phe (10 c.p.m./pmol), and 5 mM phosphoenolpyruvate. Each aliquot contains in addition 3 µg pyruvate kinase, 40 µg $tRNA^{bulk}$ (*E. coli*), 100 µg poly(U), and about 25 µl of S-150 enzymes. The yield of active reconstituted particles is usually 50–100% as compared with native 50S subunits.

Protocol 12. Total reconstitution of 50S subunits

1. Take 2.5 A_{260} units of (23S + 5S)rRNA in, for example, 10 µl buffer 7. Add 1 µl (0.1 volume) of buffer 9 in order to adjust the ionic conditions to those of buffer 4. If several assays are to receive rRNA make the buffer adjustment of the rRNA before distribution to the various reconstitution samples.
2. Add 3 e.u. of TP50 or of each of the individual proteins. The proteins are already in buffer 4.
3. Add buffer 4 to a total volume of 100 µl and incubate at 44 °C for 20 min (first-step incubation).
4. Add 4 µl of 0.4 M Mg acetate in order to increase the Mg^{2+} concentration to 20 mM. Incubate at 50 °C for 90 min (second-step incubation; ionic milieu: buffer 6).
5. Measure the phenylalanine incorporating activity using a 40 µl aliquot (*Protocol 13*). The sample can also be stored at −20 °C or −80 °C without loss of activity.

Protocol 13. Activity measurements of the reconstituted particles in the poly(Phe) system

Solutions

[^{14}C]Phe–poly(U) mix:

1. Take 2.4 ml of a 2.5 mM Phe solution and add 0.6 ml [^{14}C]Phe (50 μCi/ml; Amersham, CFB.70), 60 mg poly(U) (Boehringer), and distilled water to a final volume of 12 ml.
2. Store at -20 °C in 20×600 μl portions.

Energy mix

1. Prepare 1 M Tris–HCl (pH 8.0 at 0 °C), 1 M Mg acetate, 2 M NH_4Cl, 40 mM ATP–Na_2 (Boehringer; 484.2 mg in 20 ml), 3 mM GTP–Li_2 (Boehringer; 32.06 mg in 20 ml), 200 mM phosphoenolpyruvate (Boehringer; 828.4 mg in 20 ml).
2. Take 0.75 ml of 1 M Tris–HCl (pH 8.0 at 0 °C) and mix with 0.174 ml of 1 M Mg acetate, 0.96 ml of 2 M NH_4Cl, 2.70 ml of 40 mM ATP, 1.20 ml of 3 mM GTP, 1.80 ml of 200 mM phosphoenolpyruvate, and 20 μl of 2-mercaptoethanol (14.3 M).
3. Adjust pH to 7.8 (0 °C) with 1 M KOH (about 1–2 ml).
4. Add distilled water to 12 ml and store at -20 °C in 20×600 μl portions.

Procedure

1. To prepare a mixture for 30 assays: Take 600 μl (20 μl per assay) of the [^{14}C]Phe–poly(U) mix, and add 600 μl of energy mix, 1050 μl of buffer 3 containing an optimal amount of S-150 enzymes (*Protocol 2*) (usually an equivalent of 25 μl S-150 and 10 μl buffer per assay), 90 μl pyruvate kinase (1 mg/ml), 60 μl of tRNA^{bulk} from *E. coli* (20 mg/ml; Boehringer).
2. Add 30 A_{260} units of 30S subunits (if reconstituted 50S subunits are to be tested) or 30 A_{260} units of 50S subunits (if reconstituted 30S subunits are to be tested). Since the added subunits are highly concentrated ($\geqq 200$ A_{260}/ml) neglect the added volume.
3. Distribute 80 μl per assay and add 40 μl of the reconstitution mixture (*Protocol 12*, step 5) or as a control (minus reconstituted particle) 40 μl of buffer 6.
4. Incubate the 120 μl per assay for 45 min at 30 °C.
5. Add 1 drop of a 1% (w/v) bovine-serum-albumin solution and 2 ml of 5% (w/v) trichloroacetic acid. Mix the sample.
6. Incubate at 90 °C for 15 min.
7. Pass sample over a glass filter (e.g., Schleicher and Schüll, no. 6, 2.3 cm diameter).

8. Wash the filter three times with 3 ml of 5% (w/v) TCA and once with 3 ml ether/ethanol (1:1).
9. Add scintillation cocktail and count.

4.5 Total reconstitution of 30S subunits

The preparation of the total proteins of the 30S subunit (TP30) follows precisely the protocol given for TP50 (*Protocol 7*), that of 16S rRNA follows the protocol given for (23S + 5S)rRNA (*Protocol 10*) except that the second and third cycles of phenol extraction are omitted (*Protocol 10*, steps 4–7). The one-step reconstitution procedure is shown in *Protocol 14*, the ionic milieu buffer 6 being identical with the second step of the 50S reconstitution. The activity measurement can therefore be performed as described for the 50S reconstitution (*Protocol 13*). The incubation for poly(Phe) synthesis can be performed at 20 °C for 60 min instead of the normal conditions of 30 °C for 45 min, in order to prevent the occurrence of any further 30S assembly events during this incubation. The yield of active particles is again usually 50–100% of the input material.

Protocol 14. Total reconstitution of 30S subunits

1. Take 1.25 A_{260} units of 16S rRNA in, for example, 10 μl buffer 7. Add 1 μl (0.1 volume) of buffer 9 in order to adjust the ionic conditions to those of buffer 4. If several samples are to receive the same rRNA make the buffer adjustment of the rRNA before distribution to the various reconstitution samples.
2. Add 1.5 e.u. of TP30 or of each of the individual proteins. The proteins are in buffer 4.
3. Add buffer 4 to a total volume of 100 μl, and then 4 μl of 0.4 M Mg acetate to increase the Mg^{2+} concentration to 20 mM (ionic milieu: buffer 6).
4. Incubate for 20 min at 40 °C.
5. Measure the activity of a 40 μl aliquot in the polyphenylalanine synthesis assay (*Protocol 13*).

4.6 Total reconstitution of 70S ribosomes

30S subunits can be assembled completely under the conditions of the first step of the 50S reconstitution procedure (20), whereas 50S subunits require in addition the second-step incubation. Two counteracting effects determine the reconstitution optimum for 70S ribosomes in the second step, namely, the temperature sensitivity of the reconstituted 30S subunits on the one hand, and the temperature and Mg^{2+} requirements of the 50S reconstitution on the other. The superposition of both effects yields the optimal conditions for the 70S reconstitution. The procedure is identical to that of the 50S reconstitution except that the incubation temperature of the second step is reduced from 50 to 47 °C (20). Furthermore, TP70 (total proteins from 70S ribosomes) from tight-couple ribosomes (Section 2.2) should be used rather than a TP70 preparation derived from crude 70S ribosomes, in order to avoid contamination with RNases.

The methods of preparation of TP70 and TrRNA (total rRNA from 70S ribosomes) follow, essentially, the procedures given in *Protocols 7 and 10*, respectively. The reconstitution procedure is given in *Protocol 15*. The activity measurement is described in *Protocol 13*, omitting the addition of either subunit. As already mentioned, the incubation for the poly(Phe) synthesis can be performed at 20 °C for 60 min instead of the normal conditions of 30 °C for 45 min, in order to prevent further 30S assembly events occurring during the reaction. The reconstitution efficiency for 70S ribosomes is typically 50–85%.

Protocol 15. Total reconstitution of 70S ribosomes

1. Take 3.75 A_{260} units of TrRNA (total rRNA from tight-couple 70S ribosomes) in, for example, 10 μl buffer 7. Add 1 μl (0.1 volume) of buffer 9 in order to adjust the ionic conditions to those of buffer 4. If several samples are to receive rRNA make the buffer adjustment of the rRNA before distribution to the various reconstitution samples.
2. Add 4.5 e.u. of TP70 (total proteins from tight-couple 70S ribosomes) or of each of the individual ribosomal proteins. Note that here 1 e.u. is that amount of proteins present on 1 A_{260} unit of 70S ribosomes. The proteins are already in buffer 4.
3. Add buffer 4 to a total volume of 100 μl and incubate at 44 °C for 20 min (first-step incubation).
4. Add 4 μl of 0.4 M Mg acetate in order to increase the Mg^{2+} concentration to 20 mM. Incubate at 47 °C for 120 min (second-step incubation; ionic milieu: buffer 6).
5. Measure the activity of a 40 μl aliquot in the polyphenylalanine synthesis assay (*Protocol 13*), without addition of either of the native subunits.

---

5. Applications of the reconstitution procedure

5.1 Single component omission experiments

In single component-omission experiments all the components except one from either the small or large ribosomal subunit are used to assemble the particle and test a specific function. If all the components of a subunit are analysed in this way, one can define a group of components, the omission of which, one by one, abolishes the function concerned. This group is essential for the assembly of the active centre under study, and it necessarily contains that or those component(s) actually executing the activity. The latter point represents an exclusive advantage of this experimental strategy. Other techniques such as affinity labelling, or protection experiments against small modifying reagents such as kethoxal, are topographical methods, i.e. they identify a component at or near the binding site of a ligand, but they cannot intrinsically distinguish between 'at' or 'near'. (See also Chapter 11).

An important example of a series of single component-omission experiments is the search for the peptidyltransferase, which is a specific function of the 50S subunit. With *E. coli* ribosomes L2, L3, L4, L15, L16, and 23S rRNA have been identified (5), whereas with ribosomes from *Bacillus stearothermophilus* the same group was found except for L15 (24), which is also lacking in an *E. coli* mutant (25). Thus, L2, L3, L4, L16, and 23S rRNA remain as the peptidyltransferase candidates, although most of the evidence argues for L2 and 23S rRNA (for discussion see reference 4).

The reconstitution procedure requires purified individual proteins, and their stoichiometric addition should be assessed with the help of *Table 1* in the case of L-proteins. The reconstitution procedures for 30S and 50S subunits, and 70S ribosomes can be found in *Protocols 14*, *12*, and *15*, respectively.

5.2 Replacement experiments

This strategy is applied if one or two modified ribosomal components are to be incorporated into the ribosome. Examples are components modified with a fluorescent dye for use in distance measurements (26), modified with a gold cluster as isomorphic replacements in ribosomal crystals (27), or as protonated components in a deuterated ribosomal matrix for structural analysis by means of neutron scattering (7, 28).

Whenever possible, one should use mutants lacking one or the other protein. About one third of the ribosomal proteins have been found to be separately missing in various mutants, including the proteins S1, S6, S9, S13, S17, S20, L1, L11, L15, L19, L24, L27, L28, L29, L30, and L33 (29). For example, the mutant lacking L11 (30) is particularly useful, since even tight-couple 70S ribosomes from this mutant are able to bind purified L11 quantitatively during the second step of the two-step reconstitution procedure for the 50S subunit (*Protocol 12*, Rheinberger and Nierhaus, unpublished observation).

The next best approach in the case of 50S subunits is to use a washing procedure with relatively low concentrations of LiCl (e.g. 1.5 M LiCl, *Protocol 4*), or the NH_4Cl/EtOH split procedure (*Protocol 3*). There are two reasons for preferring these methods. First, only a relatively small number of proteins are washed off, so that the individual proteins can be relatively easily purified (*Protocol 8* and *Figure 3*). Second, under these conditions proteins at the ribosomal surface are most likely to be removed. These will represent late-assembly proteins and thus can be reincorporated even as modified proteins often without affecting the assembly process.

5.3 Assembly mapping experiments

Assembly mapping experiments have unravelled the inter-relationships between the individual ribosomal components during the course of the assembly process of both ribosomal subunits. The results are compiled into 'assembly maps' (*Figure 2*). Two proteins have not yet been integrated in the 50S assembly map, namely, the recently detected L35 and L36 (31).

The outline of the assembly mapping experiments is depicted in *Figure 4*, and *Protocol 16* presents the procedure for 50S subunits. The reconstitution intermediate RI_{50} (2), which already contains all the 50S components, can be formed in the first-step condition of the two-step procedure (10). Therefore, the assembly mapping experiments are performed under the first-step condition, i.e. using buffer 4. The sucrose gradient is made in the presence of buffer 6 in order to stabilize the rRNA–protein interactions during the centrifugation. For 30S subunits the same protocol can be used except that the incubation milieu for the 16S rRNA–protein mixture is buffer 6.

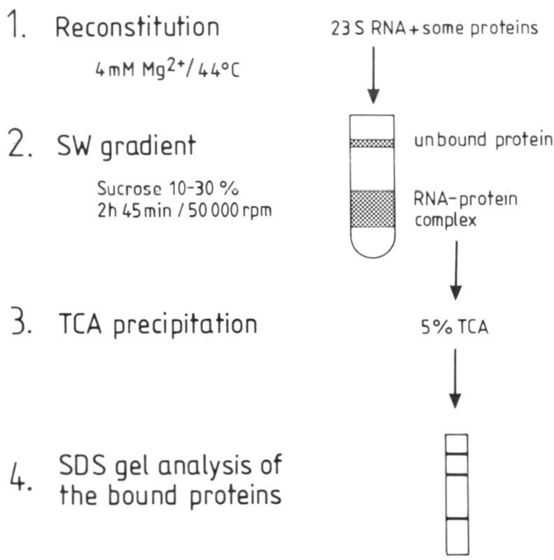

Figure 4. Outline of assembly mapping experiments. The procedure is given in *Protocol 16*.

Protocol 16. Assembly mapping experiments for 50S subunits

1. Mix five A_{260} units of 23S rRNA with 10 e.u. of each large ribosomal subunit—protein to be tested for assembly in 200 μl of buffer 4.
2. Incubate the mixture for 20 min at 44 °C (first-step condition).
3. Layer the sample on to a sucrose gradient (10–30% sucrose made up in buffer 6) and centrifuge at 250 000 *g* for 2 h and 45 min (e.g., in a Beckman SW60 rotor).

4. Fractionate the gradient and monitor the absorbance at 290 nm, and pool the fast-running 'heavy' half of the ribonucleoprotein peak.
5. Add 0.1 volume of 50% (w/v) TCA to the pooled fractions and keep the mixture at 4 °C overnight.
6. Centrifuge at 5000 *g* for 10 min.
7. Resuspend the pellet in 30 μl of a buffer containing 0.1 M Tris (unbuffered), 1 M 2-mercaptoethanol, 1.5% SDS, 15% (v/v) glycerol, and 0.001% (w/v) bromophenol blue.
8. Incubate at 90 °C for 5 min (for destruction of the rRNA).
9. Apply to a standard SDS-slab gel (for example: gel dimensions 140 × 140 mm, thickness 1.2 mm, consisting of 20% acrylamide, 0.5% bisacrylamide, 0.1% SDS in the presence of 0.4 M Tris–HCl (pH 8.8). Fill the upper 30 mm with a spacer gel consisting of 6% acrylamide, 0.15% bisacrylamide, 0.1% SDS in 0.1 M Tris–HCl (pH 6.8)).
10. Perform electrophoresis at 60 V for 70 min, followed by 150 V for 4 h. (Electrophoresis buffer: 0.05 M Tris–glycine (pH 8.8), 0.1% SDS).
11. Stain with Coomassie (0.4% Coomassie Brilliant Blue R250, 10% acetic acid, 50% methanol) at 60 °C for 4 h.
12. Destain with a solution containing 10% acetic acid and 25% isopropanol at room temperature.

5.4 Identification of assembly-initiator and assembly-leader proteins

The detailed description of the assembly process as demonstrated in the assembly maps is not yet sufficient to explain the enormous efficiency of the assembly process. For example, the 50S assembly map (*Figure 2B*) implies that 20 L-proteins bind directly to 23S rRNA. However, only two of these, L3 and L24, can initiate the assembly process (32). Only those 23S rRNA molecules which bind both of these L-proteins will be assembled to an active particle. The 30S assembly starts likewise with two initiator proteins (33).

An initiator protein is defined as a protein which binds without co-operativity to 23S rRNA (i.e. its binding cannot be stimulated by other proteins) and which is essential for the formation of an active particle.

After initiation of assembly the subsequent process seems to occur in three to four 'jumps', each of which is governed usually by one and not more than two proteins, the 'assembly leader proteins'. The assembly initiator and leader proteins probably trigger the major conformational transitions through which the assembly has to proceed in order to form active particles. The identification of these proteins is of basic importance for an understanding of the assembly process.

The following equation plays a central role in the identification of these crucial proteins:

$$A = E^{1-n}, \qquad (1)$$

where A is the fraction of total proteins (TP) which appears in active particles, E is the molar ratio of rRNA to TP (E for rRNA excess), and n represents the number of initiator (or assembly-leader) proteins. Assume, for example, that there is only one initiator protein. If the molar ratio rRNA:TP is E, then the probability of finding the initiator protein on one distinct rRNA molecule is E^{-1}, whereas for n initiator proteins the probability is E^{-n} (independent events). Since the probability is the same for each rRNA molecule, the total probability of obtaining a complete initiation complex (i.e. a complex of rRNA and all n initiator proteins) is $E \times (E^{-n}) = E^{1-n}$. This is identical with the fraction of TP, (A) which appears in active particles, since only complete initiation complexes will form active particles. Hence, $A = E^{1-n}$, and

$$\ln A = (1 \quad n)\ln E. \qquad (2)$$

Equation 2 gives us directly the experimental strategy to use for the elucidation of the number of initiator proteins. One keeps the level of TP50 constant and increases the input of (23S + 5S)rRNA. The reconstitution is then performed and the output of active particles, A, determined. A double logarithmic plot of $\ln A$ versus $\ln E$ should give a straight line, the slope of which is equal to $(1-n)$, and thus gives n. A slope of 1 was found with both subunits, clearly indicating that the number of initiator proteins, n, is two for both subunits (32, 33).

It is important to equalize the rRNA content of the various samples before measuring the activity in the poly(Phe) system, in order to normalize the inhibitory effects of rRNA excesses on poly(Phe) synthesis. For example, if the molar ratio of TP50 to rRNA is varied from 1:1 up to 1:10 during the series of reconstitution tests, then rRNA should be added after the reconstitution procedure so that all samples now contain the same high amount (i.e. 10-fold excess) of rRNA.

If an initiator protein ($n = 2$) is to be identified, one also takes advantage of Equation 1, which predicts a low activity (small A) if the initiator proteins are confronted with a large excess E of rRNA. One incubates rRNA with a selected group of purified proteins under stoichiometric conditions, then the rRNA content is strongly increased (e.g. five-fold), total proteins are added and the standard reconstitution incubation follows. Only if the selected group of proteins contains both initiator proteins will stoichiometric amounts of initiation complexes be formed, thus leading to a high yield of active particles during the standard reconstitution. In contrast, if the selected group of proteins contains only one or neither of the initiator proteins, then the initiator proteins, which are present at least in the total proteins, are faced with the large excess of rRNA during the standard reconstitution procedure, which will lead to formation of only a small number of complete initiation complexes and thus to only a low yield of active particles.

If the assembly-leader proteins, which, e.g., dictate the assembly in the step just after initiation complex formation, are to be identified, the same strategy is applied, except that now complete initiation complexes are added in excess over TP50 rather than an excess of rRNA.

This concept works as long as the value of n for assembly-leader proteins is greater than 1. If $n = 1$, which is the case for later assembly stages, the activity A becomes independent of E according to Equation 1. Here a chasing strategy (32) has to be applied. For example, core particles are incubated with [^{14}C]TP (e.g., ratio of cores to [^{14}C]TP = 1:0.5), and at various times during the reconstitution incubation an excess of [^{3}H]TP is added (e.g., ratio of cores to [^{3}H]TP = 1:2.5), and the incubation is continued. The reconstituted particles are isolated by centrifugation, then the proteins are extracted by acetic acid, separated by 2-dimensional gel electrophoresis and the [^{14}C]/[^{3}H]-ratio determined for each protein. That protein with the highest [^{14}C]/[^{3}H]-ratio (lowest chasing efficiency) is the assembly-leader protein. For details concerning 50S subunits see (32), for 30S subunits (33).

5.5 Analysis of assembly mutants

Mutants defective in ribosomal assembly can give important clues as to the *in vivo* assembly. Examples are a mutant defective in the early 50S-assembly which lacks ribosomal protein L24 (6), and one defective in the late 50S-assembly which lacks protein L15 (F. Franceschi and K. H. Nierhaus, submitted). Before starting the assembly analysis one should ascertain whether or not the mutant is an RNase deficient strain amenable to a reconstitution analysis. If not, one should first transduce the assembly lesion into the genetic background of one of the strains mentioned in the information listed in Section 2.1, point (a). A standard P1 transduction can be used. If the P1 phage does not grow very well on one of the strains, the more aggressive phage T4 can be chosen for transduction. Rules for the analysis of an assembly mutant are compiled in *Protocol 17*.

Protocol 17. Rules for the analysis of assembly mutants

1. Ensure that the mutant is an RNase deficient strain amenable to reconstitution analysis. If necessary, transduce the assembly lesion into the genetic background of one of the strains mentioned in Section 2.1, point (a).
2. Determine the growth rate during logarithmic growth in comparison to wild type.
3. Subject the S-30 (*Protocol 2*, step 3) to a sucrose-gradient centrifugation, and compare the A_{260} profile with that of the wild type, in order to determine the subunit ratio and the presence of ribosomal precursors.
4. Isolate the subunit with an assembly defect and determine both intactness of the rRNA (SDS-gel electrophoresis, *Protocol 9*) and the protein content (2-dimensional gel electrophoresis).

5. Complement with the appropriate wild-type subunit and analyse the association profile in buffer 3 via sucrose-gradient centrifugation.
6. Complement with the appropriate wild-type subunit and measure activity in the poly(Phe) system (*Protocol 13*).
7. Separate protein (*Protocol 7*) and rRNA moieties (*Protocol 10*), and perform a reconstitution (*Protocols 12, 14*, and *15*).
8. If necessary, modify the reconstitution procedure according to the nature of the assembly defect (early or late assembly, compare with the assembly map) and to the phenotypic features of the mutant. (For example, if the mutant is defective in the early 50S-assembly, and heat sensitive, then a reconstitution with the first-step incubation at lower temperature, below 44 °C, should be tested.)

---

6. Reconstitution of ribosomes in organisms other than *E. coli*

6.1 Eubacteria

Both subunits of ribosomes from *Bacillus stearothermophilus* can be totally reconstituted. The reconstitution of 50S subunits from this organism was in fact the first reconstitution of a large ribosomal subunit (34). It was a one-step procedure at 60 °C under ionic conditions similar to buffer 6, but with KCl replacing NH_4Cl. Only a relatively low yield of up to 30% was obtained. However, this technique did not become as widely used as the corresponding method for *E. coli*. It was reported (10) that the two-step procedure of the *E. coli* method (*Protocol 12*), with 60 instead of 50 °C as the second step temperature, leads to significantly higher yields of active particles. Nowadays, this technique (4 mM Mg^{2+}, 44 °C→20 mM Mg^{2+}, 60 °C) is also standard for the total reconstitution of 50S subunits from *B. stearothermophilus* (e.g. reference 35).

6.2 Archaebacteria

Total reconstitutions have only been achieved with the large ribosomal subunits from the extremely thermophilic organism *Sulfolobus solfataricus* (36) and from the halophilic *Halobacterium mediterranei* (Londei, personal communication). In the first case a two-step procedure is also required. TP50 and whole-cell RNA are incubated at 65 °C for 45 min in a buffer containing 20 mM Tris–HCl (pH 7.2 at 37 °C), 20 mM Mg acetate, 300 mM KCl and 10 mM of the polyamine thermine. The Mg^{2+} concentration is raised to 40 mM, and the second incubation is performed at 80 °C for 60 min. For the 50S subunit of *H. mediterranei* a one-step procedure is obviously sufficient in the presence of extremely high salt concentrations near to the corresponding intracellular values, namely 3.1 M K^+ and 60 mM Mg^{2+}; incubation is performed at 42 °C for 120 min. This technique is of particular interest since it allows an analysis of the important problem as to how specific protein–nucleic acid interactions can occur under these extreme salt conditions.

6.3 Eukaryotes

Reconstitution procedures with eukaryotic ribosomes are still in their infancy. Only a partial reconstitution of the large 60S subunit of rat liver ribosomes has been achieved. About 30% of the total proteins can be removed from the ribosome by treatment with the modifying reagent dimethylmaleic anhydride. This reagent modifies the amino groups at pH 8.2 but the modifying residue can be easily removed at pH 6.0 (37).

The total reconstitution of eukaryotic ribosomes starting with mature rRNA and total proteins could be extremely difficult for two reasons:

(a) The assembly gradient (i.e. the coupling between rRNA transcription and ribosome assembly in the nucleoli; reference 4) could play a more fundamental role in assembly than is the case in prokaryotes, and thus a total reconstitution using mature rRNAs might not be possible.

(b) A core particle has been derived from the 40S subunits of wheat germ by two successive washing procedures with 4 M LiCl (*Protocol 4*). This particle was thus certainly free of any cytoplasmic and non-ribosomal proteins, but it nevertheless contained at least one protein which efficiently degraded 18S rRNA under prokaryotic reconstitution conditions (Sikorsky and Nierhaus, unpublished observation).

7. Prospects

In addition to the analysis of assembly mutants (Section 5.5), the research into ribosome assembly will concentrate on two main points in the near future. First, the identification of the 'assembly-leader proteins' (Section 5.4) will significantly contribute to our understanding of the principles of ribosomal assembly. Second, we assume that the 'assembly gradient' (i.e. the coupling of rRNA synthesis and ribosomal assembly; reference 4) is one of the central features of assembly *in vivo*, and that this explains the ease of the assembly process inside the cell in contrast to the harsh conditions which have to be applied *in vitro*. Modern genetic techniques can test this assumption. The 16S rRNA gene under the control of a T7 promoter has already been transcribed by a T7 RNA polymerase with high efficiency, and the transcript was successfully incorporated into 30S particles during a subsequent reconstitution experiment (38). It must be remembered, however, that processing (methylation and tailoring of the transcript) and assembly still have to be coupled to the rRNA transcription *in vitro*, which is by no means a trivial enterprise.

As a method, reconstitution has at least two major applications. The first application is in replacement experiments: for example, one ribosomal component of a fully deuterated ribosome can be replaced by a protonated one. If one now binds a protonated tRNA to either the A, P, or E site, then the distance from the tRNA to the protonated ribosomal component can be determined, and thus the centre of gravity of the three tRNA-binding sites mapped within the ribosome. The second application is that reconstitution of ribosomes from extreme halophilic bacteria (Section 6.2) will possibly unravel features of protein–RNA recognition under extreme salt conditions.

Acknowledgements

I thank Drs H. G. Wittmann and R. Brimacombe for support and advice, P. and V. Nowotny and F. J. Franceschi for discussions and many suggestions, and J. Belart and M. Rühl for assistance.

References

1. Nomura, M. and Held, W. (1974). In *Ribosomes* (ed. M. Nomura, A. Tissières, and P. Lengyel), p. 193. Cold Spring Harbor Laboratory Press, NY.
2. Nierhaus, K. H. (1980). In *Ribosomes* (ed. G. Chambliss, G. R. Craven, J. Davies, K. Davis, L. Kahan, and M. Nomura), p. 267. University Park Press, Baltimore.
3. Held, W. A., Ballou, B., Mizushima, S., and Nomura, M. (1974). *J. Biol. Chem.*, **249**, 3103.
4. Herold, M. and Nierhaus, K. H. (1987). *J. Biol. Chem.*, **262**, 8826.
5. Schulze, H. and Nierhaus, K. H. (1982). *EMBO J.*, **1**, 609.
6. Herold, M., Nowotny, V., Dabbs, E. R., and Nierhaus, K. H. (1986). *Mol. Gen. Genet.*, **203**, 281.
7. Capel, M. S., Kjeldgaard, M., Engelman, D. M., and Moore, P. B. (1988). *J. Mol. Biol.*, **200**, 65.
8. Gesteland, R. F. (1966). *J. Mol. Biol.*, **16**, 67.
9. Zaniewski, R., Petkaitis, E., and Deutscher, M. P. (1984). *J. Biol. Chem.*, **259**, 11651.
10. Dohme, F. and Nierhaus, K. H. (1976). *J. Mol. Biol.* **107**, 585.
11. Amils, R., Matthews, E. A., and Cantor, C. R. (1978). *Nucleic Acids Res.*, **5**, 2455.
12. Nowotny, V., Rheinberger, H.-J., Nierhaus, K. H., Tesche, B., and Amils, R. (1980). *Nucleic Acids Res.*, **8**, 989.
13. Rheinberger, H.-J., Geigenmüller, U., Wedde, M., and Nierhaus, K. H. (1988). In *Methods in Enzymology*, Vol. 164 (ed. H. F. Noller and K. Moldave), p. 658. Academic Press, San Diego.
14. Lennox, E. S. (1955). *Virology*, **1**, 190.
15. Hamel, E., Koka, M., and Nakamoto, T. (1972). *J. Biol. Chem.*, **247**, 805.
16. Homann, H. E. and Nierhaus, K. H. (1971). *Eur. J. Biochem.*, **20**, 249.
17. Hochkeppel, H.-K., Spicer, E., and Craven, G. R. (1976). *J. Mol. Biol.*, **101**, 155.
18. Laughrea, M. and Moore, P. B. (1978). *J. Mol. Biol.*, **122**, 109.
19. Traub, P. and Nomura, M. (1969). *J. Mol. Biol.*, **40**, 391.
20. Lietzke, R. and Nierhaus, K. H. (1988). In *Methods in Enzymology*, Vol. 164 (ed. H. F. Noller and K. Moldave), p. 278. Academic Press, San Diego.

21. Hardy, S. J. S., Kurland, C. G., Voynow, P., and Mora, G. (1969). *Biochemistry*, **8**, 2897.
22. Nowotny, P., Nowotny, V., Voß, H., and Nierhaus, K. H. (1988). In *Methods in Enzymology*, Vol. 164 (ed. H. F. Noller and K. Moldave), p. 131. Academic Press, San Diego.
23. Ceri, H. and Maeba, P. Y. (1973). *Biochim. Biophys. Acta*, **312**, 337.
24. Auron, P. E. and Fahnestock, S. R. (1981). *J. Biol. Chem.*, **256**, 10105.
25. Lotti, M., Dabbs, E. R., Hasenbank, R., Stöffler-Meilicke, M., and Stöffler, G. (1983). *Mol. Gen. Genet.*, **192**, 295.
26. Deng, H.-Y., Odom, O. W., and Hardesty, B. (1986). *Eur. J. Biochem.*, **156**, 497.
27. Yonath, A., Leonard, K. R., Weinstein, S., and Wittmann, H. G. (1987). *Cold Spring Harbor Symp. Quant. Biol.*, **52**, 729.
28. Nierhaus, K. H., Lietzke, R., May, R. P., Nowotny, V., Schulze, H., Simpson, K., Wurmbach, P., and Stuhrmann, H. B. (1983). *Proc. Nat. Acad. Sci. USA*, **80**, 2889.
29. Dabbs, E. R. (1986). In *Structure, Function and Genetics of Ribosomes* (ed. B. Hardesty and G. Kramer), p. 733. Springer-Verlag, NY.
30. Stöffler, G., Hasenbank, R., and Dabbs, E. R. (1981). *Mol. Gen. Genet.* **181**, 164.
31. Wada, A. and Sako, T. (1987). *J. Biochem. (Tokyo)*, **101**, 817.
32. Nowotny, V. and Nierhaus, K. H. (1982). *Proc. Nat. Acad. Sci. USA*, **79**, 7238.
33. Nowotny, V. and Nierhaus, K. H. (1988). *Biochemistry*, **27**, 7051.
34. Nomura, M. and Erdmann, V. A. (1970). *Nature*, **228**, 744.
35. Vogel, D. W., Hartmann, R. K., Bartsch, M., Subramanian, A. R., Kleinow, W., O'Brien, T. W., Pieler, T., and Erdmann, V. A. (1984). *FEBS Lett.*, **169**, 67.
36. Londei, P., Teixidò, J., Acca, M., Cammarano, P., and Amils, R. (1986). *Nucleic Acids Res.*, **14**, 2269.
37. Conquet, F., Lavergne, J.-P., Paleologue, A., Reboud, J.-P., and Reboud, A.-M. (1987). *Eur. J. Biochem.*, **163**, 15.
38. Krzyzosiak, R., Denman, R., Nurse, K., Hellmann, W., Boublik, M., Gehrke, C. W., Agris, P. F., and Ofengand, J. (1987). *Biochemistry*, **26**, 2353.

9

Coupled transcription–translation of ribosomal proteins

GEORGE A. MACKIE,
B. CAMERON DONLY, and PHILIP C. WONG

1. Introduction: transcription and translation as tools

The synthesis of ribosomal components *in vitro* has been fundamental to the mapping of the corresponding genes and in elucidating such processes as the feedback control of ribosomal protein expression in prokaryotes (reviewed in 1 and 2). In outline, a fragment of DNA is transcribed by RNA polymerase, or is used to select mRNAs by hybridization, and the RNA product is employed as a template in relatively crude extracts for the synthesis of ribosomal proteins whose identity can be verified. Correlating gene products with a physical or restriction map of the DNA permits the mapping of genes for ribosomal proteins or RNAs to specific regions of the genome in the absence of either classical genetic markers or of extensive primary structural data. Patterns of expression observed *in vivo* can be reproduced by systematically altering the conditions of synthesis *in vitro*, thereby defining mechanisms which control ribosomal gene expression. *In vitro* synthesis of ribosomal components also facilitates the synthesis of altered RNAs and proteins from templates constructed with the aid of the full range of contemporary technologies. A strategy like this one permits the synthesis of ribosomal components defective in assembly which might not otherwise accumulate *in vivo*.

The basic requirements for the *in vitro* synthesis of ribosomal components are first, a template, DNA or RNA, and second, a source of enzymes and cofactors, usually a crude extract, capable of catalysing the synthesis of these components. The aim of mapping experiments is to use the synthesis of a known ribosomal component as an assay for its template. In such experiments, the ribosomal product must be identifiable beyond doubt. In other experiments, the template is available and characterized, usually in the form of a cloned DNA, and the goal is the synthesis of its product, either RNA or protein. The purpose of this chapter is to outline methods which should enable the synthesis of ribosomal proteins *in vitro* from prokaryotic or eukaryotic sources in yields sufficient to meet the needs of mapping, regulatory, and biochemical experiments.

2. *In vitro* synthesis of prokaryotic ribosomal proteins

2.1 Strategies and problems

The methods developed by Zubay and his collaborators (3) and by Gold and Schweiger (4) form the basis of many investigations of the synthesis of ribosomal proteins in *E. coli*, and are potentially applicable to ribosomal proteins in many other prokaryotic organisms. Briefly, a crude extract (the S-30) is prepared from a suitable strain of *E. coli* and is depleted of endogenous templates. When supplied with substrates, cofactors, and a DNA template, such extracts can catalyse the synthesis of a number of enzymes and bacterial ribosomal proteins. Originally, a product was assayed by its enzymatic activity (3). Ribosomal proteins are more conveniently identified by supplementing extracts with a radioactive amino acid and identifying the labelled product on polyacrylamide gels (see Section 2.5.1). Several groups have fractionated S-30 extracts into soluble supernatant, crude RNA polymerase, ribosomes, and initiation factors, (5, 6) and demonstrated the ability of the reconstituted extract to direct the synthesis of ribosomal proteins.

Synthesis of ribosomal proteins from organisms other than *E. coli* may be simplified by supplying a cloned DNA (or equivalent) from the organism in question to an extract prepared from *E. coli*, rather than attempting to develop an homologous coupled system. Several problems could arise with this approach. First, signals for initiation of transcription or translation may differ from the optimal signals in *E. coli*. Secondly, the frequency of codon usage may differ from that in *E. coli*, particularly in organisms whose DNA base composition differs significantly from that of *E. coli*. Thirdly, some organisms or organelles utilize a variation on the 'universal' genetic code. Solutions to the first problem can be sought by cloning the template in an expression vector functional in *E. coli* (7). Problems two and three may be circumvented, in part, by supplementing the *E. coli* extract with bulk tRNA from the organism in question, although heterologous tRNAs may not interact as well with elongation factor EF-Tu as the homologous tRNAs. It may also be possible to employ S-30 extracts from *E. coli* strains carrying a suppressor mutation which in effect mimics the non-universal code in the organism of interest (e.g., where UGA is read as Trp).

2.2 Growth of *E. coli* and preparation of extracts

2.2.1 Choice of strain

It is convenient to employ a *lac*$^-$ strain of *E. coli* for the preparation of extracts since the extract's activity can be assayed readily with any one of a number of templates carrying the *lacZ* gene (see Section 2.4.3 below). We have routinely used strain RD100 (8), and have also prepared active S-30 extracts from strains such as C600. There may be advantages to preparing extracts from protease-deficient strains or from *recB* mutants for the assay of linear templates (9).

Protocol 1. Buffers

Buffer A: 10 mM Tris–acetate (pH 7.8), 14 mM Mg acetate, 60 mM potassium acetate (this can be prepared as a 10-fold concentrated stock).

Buffer B: 10 mM Tris–HCl (pH 7.5), 50 mM NaCl, 1 mM EDTA.

Buffer C: 50 mM Tris–HCl (pH 7.5), 20 mM $MgCl_2$, 5 mM 2-mercaptoethanol.

Buffer D: 20 mM sodium phosphate (pH 7.0), 200 mM NaCl, 0.2 mM PMSF, 2% Triton X-100, 1% sodium deoxycholate, 0.2% SDS.

2.2.2 Growth of *E. coli*

Grow cultures as outlined in *Protocol 2* in *Z*-medium (3): 5.6 g KH_2PO_4, 28.9 g K_2HPO_4, 10 g yeast extract per litre with thiamine and glucose added to 10 mg/litre and 0.4%, respectively, after autoclaving. We occasionally add $MgSO_4$ to 2 mM. A total of 2.5 litre of culture is convenient to handle and avoids recourse to fermentors or continuous flow centrifugation (3). Good aeration is essential. To achieve this, use no more than 800 ml of medium in a 4 litre Erlenmeyer flask and apply vigorous shaking. Follow growth by monitoring the absorbance of the culture at 600 nm. Do not overgrow the culture; extracts prepared from cells grown to high densities may exhibit unacceptable levels of proteolytic activity (10). While the cells are growing, check to ensure that all the necessary rotors and materials for harvesting are chilled to 4 °C and are ready to use.

Protocol 2. Growth of *E. coli* for preparation of S-30 extracts

1. Inoculate a single colony into 20 ml of complete *Z* medium (Section 2.2.2) and grow at 35 °C to saturation (i.e., overnight).
2. Establish larger cultures by inoculation with 7.5 ml/litre of the saturated starter culture and grow at 35 °C.
3. Harvest the cultures while they are still in exponential growth, i.e., at absorbances (600 nm) of 0.9 to 1.2.
4. Chill the cultures quickly by swirling the flasks in a slurry of wet ice. Make every effort to keep the cell pellets or suspensions ice cold during all the following steps.
5. Harvest the cells by centrifugation at 6000 *g* for 15 min at 4 °C in a large capacity rotor (e.g., a Dupont/Sorvall GS-3, or a Beckman JA-10 rotor). Decant the spent culture medium.
6. Suspend the pellets by careful pipetting, or with a rubber policeman, with ice cold buffer A (*Protocol 1*) supplemented with 6 mM 2-mercaptoethanol (10–20 ml/litre of starting culture).

7. Transfer the cell suspension to chilled, weighed, 50 ml centrifuge tubes. Collect the suspended cells by centrifugation at 6000 *g* for 10 min at 4 °C (e.g., using a Beckman JA-20 rotor).
8. Resuspend the washed cell pellet in cold buffer A containing 6 mM 2-mercaptoethanol (10 ml per litre of culture).
9. Collect the cells again by centrifugation at 10 000 *g* for 10–15 min at 4 °C.
10. Determine the wet weight yield of cells. The cell pellet can be frozen quickly and stored at −70 °C; it is preferable, however, to process the cells immediately (e.g., see *Protocol 3*).

Protocol 3. Preparation of S-30 extracts

1. Suspend the 'wet' cell pellet prepared as described in *Protocol 2* with buffer A (*Protocol 1*) containing 1 mM dithiothreitol and 0.5 mM PMSF (e.g., from Eastman) using 1.3 ml buffcr per g wet weight of cells.
2. Stir the buffer into the packed cell pellet with a heavy glass rod. Then suspend the cells evenly by pipetting the suspension up and down. Keep the cell suspension cold at all times. Large cell pellets may require recourse to mechanical homogenizers for even resuspension (3, 11).
3. Load the cell suspension into a clean, cold French pressure cell and rupture the cells in a single pass at about 3500 p.s.i. Collect the lysate directly into a chilled ultracentrifuge tube containing 10 μl 0.1 M dithiothreitol per ml of cell suspension. Do not add DNase.
4. Centrifuge the lysate immediately for 30 min at 4 °C in a chilled Beckman Ti50 rotor or equivalent at 30 000 g.
5. Decant the clear amber supernatant into a suitable graduated container and keep it on ice. Discard the pellet containing unbroken cells and debris.
6. The S-30 extract is 'pre-incubated' to complete nascent polypeptides and to destroy endogenous mRNA and DNA fragments as follows:
   (a) Supplement the S-30 with the ingredients given in *Table 1*.
   (b) Transfer the fortified S-30 to a small Erlenmeyer flask, cover the flask to protect it from light, gas it with nitrogen or argon, stopper the flask, and incubate it for 80 min with gentle swirling at 35–37 °C.
7. Dialyse the S-30 at 4 °C for a total of 8 h against two changes of at least 50 volumes of buffer A (*Protocol 1*) containing 0.2 mM dithiothreitol and 0.1 mM PMSF.
8. Dispense the dialysed S-30 into chilled 1.5 ml microcentrifuge tubes in convenient aliquots (e.g., 50 μl) and freeze them quickly in a dry ice-acetone bath or in liquid nitrogen. The aliquots can be stored at −70 °C for periods of several years.

Table 1. Ingredients for the pre-incubation of S-30 extracts

Component	Volume added per ml of crude S-30
1 M Tris–acetate (pH 7.8)	0.10 ml
0.14 M Mg acetate	0.02 ml
20 mM ATP (pH 7, adjust with KOH)	0.04 ml
75 mM trisodium PEP (made fresh in H_2O)[a]	0.12 ml
0.1 M DTT	0.01 ml
amino acid cocktail I[b]	0.02 ml
pyruvate kinase[c]	5 μg

[a] Phosphoenol pyruvate (e.g., from Calbiochem, cat. no. 52494; other grades or other sources may not be acceptable)
[b] A solution which is 500 μM in each of the 20 common amino acids
[c] Calbiochem cat. no. 5506

There is considerable divergence in the literature in the conditions used for lysis. Reported pressures range from 1800–10 000 p.s.i. Extracts prepared at higher pressure may be considerably less active in the synthesis of β-galactosidase or other enzymes (12). As an alternative to using a French press, it is possible to prepare S-30 extracts by grinding with levigated alumina (13). We have found that such extracts are much less active than those prepared as described above.

2.3 Templates for *in vitro* protein synthesis

2.3.1 Alternative templates

Templates for the synthesis of ribosomal proteins have included specialized transducing phages (14, 15), recombinant plasmids (6), linear restriction fragments (14), or even unfractionated *E. coli* DNA (5). Efficient synthesis requires that the template contain a promoter; nonetheless, linear fragments, even lacking a promoter, have been observed to direct detectable synthesis of ribosomal proteins. This may be sufficient for the purposes of gene mapping (14, 15). In most cases, however, it will be necessary to supply an efficient bacterial promoter 'upstream' of the sequences to be expressed when heterologous DNA is being used as template (e.g., after being cloned into a λ-based vector). The choice of suitable vectors is beyond the scope of this chapter, but the reader can consult reference 7 for suggestions.

2.3.2 Purification of templates derived from bacteriophage

High titre stocks of bacteriophage λ may be prepared by infection of a suitable host in liquid culture, by induction of a lysogen, or by confluent lysis on plates. We have relied on the second method in the past and have followed procedures in the manual by Miller (13). Alternative methods for preparing adequate stocks of recombinant λ phages may be found elsewhere in this series (16). At least 10^{12} pfu of phage are required to provide useful quantities of DNA for coupled transcription–translation. Regardless of the method employed to prepare a high-titre phage lysate, further purification of the phage, usually by buoyant density centrifugation, is essential.

We extract DNA from phage using the methods described in reference 13, which are summarized in *Protocol 4*. The advantages of this method are two-fold: it entails little handling of the DNA, and avoids the difficulties of suspending high molecular weight DNA after ethanol precipitation. The quality of the DNA preparation should be checked by determining the minimum quantity of a restriction enzyme necessary to achieve total digestion. Assuming the enzyme preparation is fully active, any need for more than 1–2 units of enzyme to digest 1 μg of DNA in 60 min implies the presence of contaminants in the DNA preparation such as SDS or residual agar (from plate lysates).

Protocol 4. Extraction of DNA from bacteriophage λ and its derivatives

1. Transfer the phage suspension to a glass centrifuge tube, adjust it to an absorbance of 10 (260 nm) with buffer B, and make it 0.5% in SDS with 0.05 volumes 10% SDS.
2. Heat this suspension for 10 min at 55 °C. The suspension should lose its opalescence and become clear.
3. Add 1/6 volume of 3 M KCl to the lysed phage suspension and chill it for 10–15 min on ice, during which time a heavy white precipitate of protein–detergent complexes will form.
4. Remove the precipitate by centrifuging the suspension at 5000 *g* for 5–10 min (e.g., in a Beckman JA-20 rotor).
5. Carefully remove the viscous supernatant with a wide bore pipette. To minimize handling, transfer the DNA-containing supernatant directly to dialysis tubing sealed at one end.
6. Dialyse the DNA against 3 changes of 500 ml sterile 10 mM Tris–acetate (pH 8.0) for 40–48 h at 4 °C. Lengthy dialysis is required to remove residual micelles of SDS in the crude DNA preparation.
7. Store the final preparation of DNA at 4 °C in a sterile tube. Estimate the concentration by its absorbance at 260 nm assuming an A_{260} of 20 is equivalent to 1.0 mg/ml. A preparation of λ spc2 (15) obtained by inducing 1 litre of culture will typically yield 500 μg of DNA.

2.3.3 Purification of plasmid templates

Effective methods for the preparation of plasmid DNA are described elsewhere in this series (11). We have successfully used templates prepared by modifications of either the cleared lysate procedure of Clewell and Helinski (17), or by the alkaline lysis procedure of Birnboim and Doly (18, 19). In either case, it is necessary to purify the plasmid by dye-buoyant density centrifugation. Crude preparations of plasmid can sometimes be cleaned up by spun column centrifugation. We have been unable to do so reproducibly, however. As in the case of bacteriophage

DNAs, plasmid DNAs should be tested for digestibility with minimal quantities of a suitable restriction enzyme.

2.3.4 Purification of linear DNA templates

Restriction fragments of bacteriophage DNA carrying genes for ribosomal proteins have been eluted from agarose gels before use as templates (14, 20). The original methods were usually harsh and of modest efficiency. We have more recently employed 'Geneclean' (obtained from Bio 101, California) to solublize agarose containing DNA fragments of interest. We have recovered the DNA fragments in high yield and found them to be competent templates for phage-encoded RNA polymerases. We presume that they would also function well in coupled systems.

2.4 Coupled transcription–translation with S-30 extracts

2.4.1 Preparation and storage of the reagents for coupled transcription–translation

We have used Zubay's procedure (3) as the basis for our work, and have modified it with the advice of colleagues. *Table 2* gives the ingredients in a model incubation. Further details are provided in *Table 3* and in the comments in *Protocol 5* below.

Much useful information regarding the preparation and storage of the components of the *in vitro* system is provided in the article by Pratt in this series (11). Prepare all reagents in sterile distilled or deionized water. Neutralize all nucleotides to pH 7 with KOH. Salt and buffer solutions can usually be autoclaved (see reference 11), but sterile filtration will often suffice. Store all the components in part A of the cocktail in *Table 3* at −20 °C. Most can be frozen and thawed repeatedly. Folinic acid develops a precipitate with time. Store stock solutions of this reagent in smaller portions and discard the working stock after

Table 2. A typical incubation for coupled transcription–translation *in vitro*

A. PEG-6000[a] (300 mg/ml)	2.5 μl
DNA: λ-derived	0.2–1.0 μg
plasmid-derived	0.1–0.5 μg
Labelled amino acid	as required[b]
Effectors (see the text)	as required
Sterile water	to 20 μl
B. Cocktail[c]	15 μl
C. S-30[d]	15 μl

[a] PEG-8000 can be substituted.

[b] The amount of radioactive label used depends on the goals of the experiment and must be determined empirically. We find that up to 20–25% of the added label is incorporated into protein in our experiments (see the text).

[c] Refer to *Table 3* for details.

[d] The S-30 is taken from storage and thawed on ice immediately prior to use. Discard any unused S-30; do not refreeze it.

Table 3. Components of the cocktail for coupled transcription–translation.[a]

Stock solution	Volume (μl)
A. 1 M Tris–acetate (pH 8.2)	60.5
5 M K acetate	11.5
0.1 M DTT	20.5
2 M NH_4 acetate	20
1 M Mg acetate	variable[b]
33 mM methionine	variable[c]
75 mM leucine	variable[c]
0.27% FAD	5
0.27% NADP	15
1.5% NAD	15[d]
0.27% pyridoxine	15
0.11% p-aminobenzoic acid	15
0.27% folinic acid	15[e]
10 mg/ml *E. coli* tRNA	15
0.1 M solution of CTP, GTP, UTP	8[f]
0.1 M ATP	33[f]
Amino acid cocktail II	150[g]
B. trisodium-PEP.7 H_2O	10.5 mg
1 M CaCl	variable[h]
sterile distilled water	to 450 μl[i]

[a] Mix the ingredients in the order shown. It is possible to prepare part A of the cocktail and freeze it for repeated use. Once the components of part B are added, the cocktail must be used promptly. A precipitate will form at this stage.

[b] The optimal concentration of Mg acetate needed must be determined empirically (see Section 2.4.2). Typically, it is 8–12 mM, corresponding to 12–18 μl of a 1 M solution.

[c] For labelling with [^{35}S]-methionine (typically 25 μCi) we also add 1.5 μl of 33 mM methionine (to 50 μM) and 4 μl of 75 mM leucine (to 200 μM).

[d] NAD is an addition to Zubay's original formulation and was suggested to us by Dr. Joel Kirschbaum (see also reference 21).

[e] We use folinic acid (calcium leucovorin) obtained from Gibco, Incorporated (catalog no. 850-3380).

[f] Neutralize ribonucleoside triphosphates to pH 7 with KOH.

[g] The 18 amino acids (minus leucine and methionine) are dissolved in sterile water in the presence of 10 mM Tris base and are neutralized with KOH as required to yield a solution which is 2.2 mM in each amino acid.

[h] The final concentration of CaCl must be determined empirically (see Section 2.4.2). It is typically about 2/3 that of Mg acetate.

[i] Total volume for parts A + B.

several uses. Take every precaution to avoid the introduction of bacteria, molds, or skin oils into any of the solutions or their containers.

It is convenient and practical to prepare part A of the cocktail (*Table 3*) in advance and to store it frozen in small portions. This eliminates repetitive pipetting each time an experiment is performed. The additions in part B of the cocktail must be made immediately prior to its use. A precipitate will form in the cocktail almost immediately and should be suspended before the cocktail is dispensed.

Protocol 5. The basic incubation for protein or RNA synthesis

1. Assemble the cocktails and the individual incubations in sterile 1.5 ml microcentrifuge tubes on ice. Add water, template, PEG, and radioisotope

first (step A in *Table 2*). If necessary, the radioisotope can be dried down in the tube prior to any additions. Make certain that the preparation of radioisotope does not contain potential inhibitors of protein or RNA synthesis (e.g., ethanol or HCl).

2. Prepare the cocktail (*Table 3*), separately and add 15 μl to each incubation tube.
3. Start the incubation by the addition of 15 μl of the S-30 and mix the components well by pipetting up and down.
4. Incubate samples at 30–40 °C for the desired time. In general, the rate of reaction increases with higher temperatures as would be expected. The spectrum of products synthesized may change, however (e.g. see reference 22). We usually incubate for no longer than 60 min, but overnight incubations have been reported.
5. Stop reactions by chilling (or freezing for longer storage).
6. Portions of the mixture can be withdrawn for the determination of acid-precipitable radioactivity. Pipette 2.5 μl of the chilled incubation into 0.5 ml cold water containing 1 drop of 0.5% bovine serum albumin as carrier.
7. Add an equal volume of 10% trichloroacetic acid and boil for 15 min.
8. Chill the tube and collect the precipitate by filtration on to a glass fibre filter (e.g. Whatman GF/C).
9. Wash the filter several times with 5% TCA, and once with acetone.
10. Dry the filter and count in a toluene-based scintillation fluid.
11. Determine the total radioactivity in the incubation by pipetting a second 2.5 μl portion directly on to a glass fibre filter, drying it, and counting it as above. More rigorous means of analysis are described in sections 2.5.1 and 2.5.2 below.

2.4.2 Optimization of the incubation conditions.

Each preparation of S-30 extract will exhibit a slightly different optimal requirement for Mg and Ca ions depending on the bacterial strain, the conditions of the cells at the time of harvest, the means used for cell breakage, and the quantity of endogenous template. The optimal concentrations must be determined empirically. In our experience, the Mg optimum for the synthesis of many ribosomal proteins is very close to that for β-galactosidase, typically 8–12 mM. Optimal concentrations of Ca ions are about 2/3 those for Mg. To determine these concentrations, prepare the cocktail in *Table 3*, omitting both Mg acetate and Ca chloride. Vary the final Mg concentrations in incubation mixtures (*Table 2*) from 5–12 mM in 1 mM increments at three Ca concentrations (e.g., 5, 7, and 9 mM). Employ either a template containing cloned ribosomal protein genes or pBR322, and label the products with a radioactive amino acid. Determine the yield of product(s) as outlined in Section 2.5.2 below. [^{35}S]-methionine is particularly convenient in the case of pBR322 since the major product is the 30 kDa precursor to β-lactamase which contains 10 methionine residues and is readily labelled (cf. *Figure 1a* below). Leucine may be a more suitable radioactive precursor for many ribosomal proteins whose content of methionine can be variable.

An alternative approach to optimization uses a *lacZ*$^+$ DNA such as λ gt11 (16) as template. In this case, include 3′,5′-cyclic AMP in the incubation at 1 mM, and assay for the synthesis of β-galactosidase enzymatically as described in reference 13. The advantage of the enzymatic assay is that the results are available within an hour. A good S-30, properly optimized, will direct the synthesis of 1 A_{420} unit/h of β-galactosidase at 37 °C with 1 μg λ plac5 DNA as template in a 50 μl incubation. We have recently employed a translational fusion between S20 and β-galactosidase (pGP9 in reference 23) as a convenient means of optimizing extracts. The fusion protein depends on the translational signals for S20 but is enzymatically active and thus easily assayed.

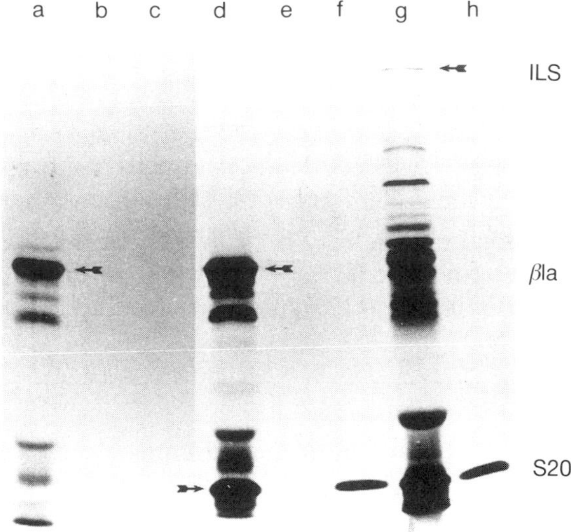

Figure 1. Immunoprecipitation of proteins synthesized *in vitro*. S-30 extracts were programmed with pBR322 (lanes a–c), λ dapB2 (21) which contains a 15–17 kb insert encoding ribosomal protein S20 of *E. coli* (lanes g and h), and pGM9 which contains a 2.9 kb fragment derived from λ dapB2 (lanes d–f). Incubations were prepared as outlined in *Tables 2* and *3* with [^{35}S] Met as label. Portions of the incubations were centrifuged to remove ribosomes (Section 2.5.1, part d) and subjected to indirect immunoprecipitation with non-immune (lanes b and e) or anti-S20 (lanes c, f, and h) sera, generously supplied by Dr. L. Kahan (University of Wisconsin, Madison). Equivalent amounts of sample (by original volume of the incubation) were applied to a 12.5% SDS–polyacrylamide gel, the proteins separated by electrophoresis, and the radioactive products visualized after fluorography. The arrows denote the positions of isoleucyl tRNA synthetase (ILS), β-lactamase (βla), and S20.

The concentration of template is also a variable. Template DNA in excess will yield the maximum incorporation of radioactivity, but may result in the accumulation of incomplete polypeptides. We have found that lower concentrations of DNA (e.g., 50–100 ng DNA in a 25–50 μl incubation) often yield much cleaner *in vitro* products than those made at saturating levels of template.

2.4.3 Effectors of ribosomal protein synthesis *in vitro*

Coupled transcription–translation has demonstrated that the synthesis of ribosomal proteins can be regulated *in vitro* by effectors such as guanosine 5′-diphosphate, 3′-diphosphate (MSI) (25, 26), by free rRNA (27), or by free ribosomal proteins (6, 28, 29). Most studies of autogenous repression of ribosomal protein synthesis *in vitro* have required a substantial concentration of free ribosomal protein to achieve repression, typically 1–4 μM. In one instance, only S20 purified by salt extraction (as opposed to acetic acid) was capable of exerting a specific, dose-dependent inhibition of its own synthesis (29). Ideally, the concentration of added ribosomal protein should be varied over a 10-fold range to ensure a true dose–response. It is important to include controls, such as a non-ribosomal template, to ensure specificity. A further test of specificity is to include free rRNA in the incubation at concentrations close to 1 μM (27) since autogenous repression of ribosomal protein synthesis can often be reversed by this means. Care should be taken to ensure that any observed effects on synthesis of a particular ribosomal protein are due to the effector directly, and not due to accompanying contaminants (such as urea and KCl). The yield of the particular ribosomal protein synthesized in the presence of an effector can be determined quantitatively by the methods suggested in section 2.5.2 below.

It is frequently necessary to 'uncouple' an *in vitro* incubation in order to ascertain whether an effector acts at the level of transcription or translation. This is done most conveniently by withholding amino acids for the first 20–30 min of incubation to allow mRNA to accumulate (with or without the effector). Although mRNA has rarely been assayed directly in such experiments, Section 2.5.3 below provides methods for so doing. In a second stage of the incubation, further transcription is inhibited by the addition of rifampicin to 25 μg/ml (or by destruction of the template with DNase) and translation of the accumulated mRNA proceeds in the presence of amino acids including a suitable labelled precursor and the effector or its control.

2.4.4 Trouble-shooting

Many cures for common problems with coupled transcription–translation systems are cited in articles by Zubay (3) and by Pratt (11). Frequently, difficulties can be traced to solutions whose concentrations were calculated improperly. Pay particular attention to the divalent cations. It is rarer that an S-30 is the source of problems. See Section 2.2.2 for relevant comments.

2.5 Analysis of products synthesized in coupled systems

2.5.1 Methods for identifying ribosomal proteins synthesized *in vitro*

Most ribosomal proteins can be identified with a fair degree of certainty by two-dimensional gel electrophoresis. Others, particularly in organelles, may be subject to post-translational modifications which may obscure the identity of the primary translation product. Confirmation of tentative identifications can be achieved by immunological means, by peptide mapping, or ideally by direct comparison of the primary sequence of a template or product, with a well-characterized, purified ribosomal protein. Details of several useful identification methods follow.

(a) *Polyacrylamide gel electrophoresis*. Several formulations for polyacrylamide gels capable of resolving mixtures of ribosomal proteins are reviewed in this volume and in an accompanying volume in this series (30). A sample containing 25 000 c.p.m. of [^{35}S]-labelled protein can usually be visualized with fluorography quite adequately after 24 h exposure. If the sample volume is large, or the incorporation is low, concentrate and de-salt the sample by precipitation with 5 volumes of cold acetone. Dissolve the pellet in sample buffer. We commonly analyse samples on a one-dimensional gel such as that shown in *Figure 1*. Two dimensional gels may be necessary if a mixture of ribosomal proteins is anticipated, or if the identity of the *in vitro* products cannot be established on a one-dimensional gel.

(b) *Chromatographic methods*. We have used chromatography on phosphocellulose to identify fMet-S20 synthesized *in vitro* (22). This is, tedious, however, and is limited to one sample at a time. Nonetheless, proteins can be recovered from the column eluant much more easily than from dried polyacrylamide gels. Advances in the separation of ribosomal proteins by HPLC hold considerable promise for the purification of ribosomal proteins synthesized *in vitro*.

(c) *Reconstitution* (*Protocol 6*). Essentially, rRNA is used as a specific ligand for the purification of ribosomal proteins from other *in vitro* products (5).

   i. Prepare a coupled reaction as outlined in *Tables 2* and *3* and incubate for 60 min at 35–37 °C.

   ii. Determine the acid-precipitable radioactivity (see *Protocol 5*, step 6).

   iii. Follow the procedure outlined in *Protocol 6* for binding a ribosomal protein to rRNA.

   An example of such an experiment is shown in *Figure 2*. In the absence of heating (to minimize ribonuclease activity) the particle sediments at 22S–24S as an intermediate of reconstitution (fractions 15–16 in *Figure 2*). Such conditions are suitable for proteins which bind directly to rRNA or 'early' in the assembly pathway. For further details concerning reconstitution and the identification of proteins bound to rRNA, see Chapter 8.

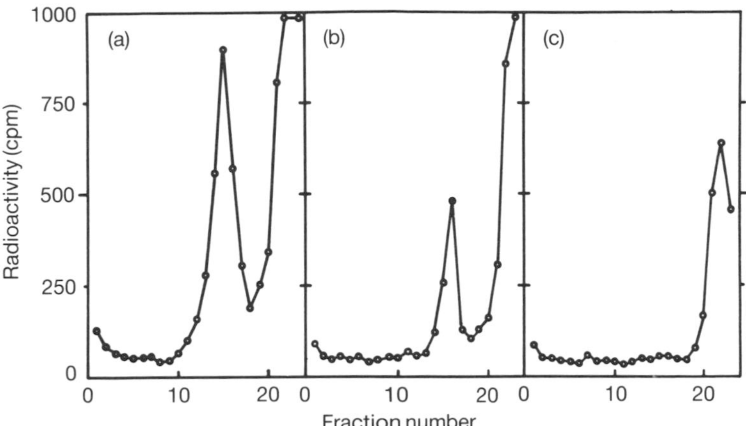

Figure 2. Recovery of ribosomal protein S20 synthesized *in vitro* from S-30 extracts by selective binding to 16S rRNA. S-30 extracts were programmed with pGM9 (a), λ dapB2 (b), or pBR322 (c), as in *Figure 1*, except that the incubation with pGM9 used [^{3}H]Met as label. Ribosomal proteins were extracted from the completed incubation as described in Section 2.5.1, part c, and *Protocol 6*. Samples containing 10 000 c.p.m. (panels a and c) or 28 000 c.p.m. (panel b) were incubated with 16S rRNA, and the complexes so formed resolved by sucrose gradient centrifugation. Fractions of 13 drops were collected and counted.

(d) *Immunoprecipitation* (*Protocol* 7). Ribosomal proteins synthesized *in vitro* may be identified by immunoprecipitation. To avoid competition with ribosomes in the S-30 extract, it is advisable to subject *in vitro* incubation mixtures to high speed centrifugation to remove ribosomes and other aggregates prior to immunoprecipitation. Dilute an *in vitro* incubation (25 μl) to 100 μl with 50 μl buffer A (*Protocol 1*) and 25 μl 2 M NH_4Cl. Centrifuge at 4 °C for sufficient time to pellet the ribosomes. Remove the supernatant and store on ice. Follow the procedure outlined in *Protocol 7*. *Figure 1*, lanes b, c, d, e, f, and h, illustrate a typical immunoprecipitation of ribosomal protein S20 synthesized from two templates using the procedure in *Protocol 7*.

(e) *Partial proteolytic digestion* (*Protocol 8*). Partial proteolytic digestion *in situ* as developed by Cleveland *et al.* (31) is particularly appropriate for comparing the peptide maps of an *in vitro* product with that of its authentic counterpart. Additional details and suggestions are provided in Chapter 6 of reference 30.

Protocol 6. Selection of ribosomal proteins synthesized *in vitro* by specific binding to rRNA[a]

1. Dilute an *in vitro* reaction mixture containing 2×10^5 c.p.m. (^{35}S) to 0.5 ml with 1 mg purified 30S ribosomal subunits (which act as carrier) and buffer A

(*Protocol 1*), then mix with 0.5 ml of 4 M LiCl in 8 M urea and leave on ice for 18 h.

2. Centrifuge at 12 000 *g* for 6 min in a microcentrifuge to remove precipitated RNA.
3. Dialyse the supernatant for a total of 24 h at 4 °C against two changes of buffer C (*Protocol 1*) containing 1 M KCl, 0.5 mM EDTA, and 0.1 mM PMSF.
4. Mix 50 μl of the dialysed LiCl–urea extract of ribosomal protein(s) synthesized *in vitro* (10 000–20 000 c.p.m.) with 50 μg 16S rRNA and 1–2 equivalents (30 μg) of total unlabelled 30S ribosomal protein as carrier (in addition to that added in step 1 above) in buffer C to a final volume of 125 μl and a concentration of K^+ of 0.4 M.
5. Incubate on ice for 30 min, then apply the mixture to a 3–15% sucrose gradient made up in buffer C and containing 0.35 M KCl.
6. Centrifuge for sufficient time that the rRNA would be no more than half way through the gradient (e.g., for *E. coli* 16S rRNA, suitable conditions are at 20 000 r.p.m. for 16 h at 4 °C in a Beckman SW40 rotor).
7. Collect fractions from the gradient and count the radioactivity in each.

[a] This outline employs 16S rRNA to select for small subunit ribosomal proteins. It could be adapted for the selection of large subunit ribosomal proteins by substituting 23S rRNA and 50S subunit proteins as carrier. See also Chapter 8 in this volume.

Protocol 7. Immunoprecipitation of ribosomal proteins synthesized *in vitro*

1. Mix portions of an *in vitro* reaction mixture with 50 μl of buffer D (*Protocol 1*) and sterile H_2O to a volume of 100 μl. Keep the concentration of SDS (if present) to 0.1% or less.
2. Add 1–2 μl of the appropriate antiserum and competitors (such as unlabelled antigen). Incubate on ice for 60 min (or longer, if convenient).
3. Prepare a 10% suspension of killed *Staphylococcus aureus* (available from Calbiochem as 'Pansorbin', catalog no. 507681). About 10 μl of a 10% suspension is required for each sample prepared in the previous step. Thaw or reconstitute this amount of *S. aureus*.
4. Collect the *S. aureus* cells by centrifugation (2 min at full speed in a microcentrifuge).
5. Suspend the cells at 1% (w/v) in buffer D diluted with an equal volume of sterile water (50% buffer D). Incubate for 20 min at room temperature.
6. Chill the cell suspension and harvest the cells by centrifugation as above. Discard the supernatant which contains cell fragments and soluble protein A.

7. Resuspend the pellet at 1% (w/v) in cold 50% buffer D and collect the cells by centrifugation.
8. Repeat step 7, twice.
9. Suspend the final cell pellet at 10% (w/v) in 50% buffer D on ice, and use without undue delay.
10. Add 10 μl of the washed *S. aureus* preparation to each tube from the first stage (i.e. steps 1 and 2) of the immunoprecipitation. Incubate 30 min on ice.
11. Collect the *S. aureus*–immune complexes by centrifugation at 10 000 g for 5 min.
12. Suspend the pellets with 0.5 ml cold 50% buffer D and repeat the centrifugation. Do this a total of five times. This set of washes serves to remove trapped proteins and unincorporated labelled amino acids from the antigen–IgG–*S. aureus* complex.
13. Drain the final pellet well. Suspend it in an appropriate sample buffer [e.g., 60 mM Tris–HCl (pH 6.8), 2% SDS, 0.1 M 2 mercaptoethanol (50 mM DTT is more effective as a reductant), 0.1 mM PMSF, 10% glycerol] and denature the complexes by heating for at least 2 min at 95 °C.
14. Centrifuge the samples briefly at 10000 g to remove debris.
15. Apply the samples to a suitable polyacrylamide gel. Run the gel (with appropriate standards), treat it for fluorography, dry it, and expose following suggestions in reference 31.

Protocol 8. Partial proteolytic digestion

1. Mix a sample of the *in vitro* synthesized product with 100 μg of total ribosomal protein, and separate by two-dimensional electrophoresis using a 0.75 mm thick gel for the second dimension.
2. Identify the proteins by staining the gel lightly with Coomassie Blue.
3. Excise the protein spot of interest and embed it in the sample well of a 15% SDS–polyacrylamide gel.
4. Follow the procedure of Cleveland *et al.* (31) for partial digestion and running the gel. The amount of protease to be used for digestion must be determined empirically.

The peptide map of the unlabelled ribosomal protein can be determined by staining the gel with silver. This pattern is recorded by photography after which the gel can be prepared for fluorography, dried, and exposed to reveal the pattern of the labelled protein synthesized *in vitro*. The two maps should be compared knowing that acidic residues which constitute the sites of cleavage by V8 protease are usually infrequent in ribosomal proteins. The sequence of the cloned cDNA or gene can be helpful in reconciling the map of the labelled protein with that of the carrier, since the distribution of methionine will determine which V8 peptides will be labelled.

2.5.2 Quantification of ribosomal proteins synthesized *in vitro*

Total incorporation of radioactivity is only a crude indicator of the synthesis of individual proteins. It is preferable to separate the actual products of *in vitro* synthesis by polyacrylamide gel electrophoresis (or analytical HPLC) and visualize them after autoradiography (see Section 2.5.1, part (a), above).

Protocol 9. Quantification of synthesized proteins—radioactivity determination

1. Excise the band corresponding to the product of interest from the dried gel.
2. Rehydrate the gel slice in 50 μl H_2O for 60 min in a glass scintillation vial with a teflon cap liner.
3. Add 10 ml 3% Protosol in Econofluor (NEN), seal the vial tightly, and incubate for 12–18 h at 37 °C.
4. Cool the sample and count it taking precautions to avoid photoluminescence.

Knowledge of the specific activity of the methionine in the incubation and the amino acid composition of the product permits a direct calculation of the yield of product in pmoles. Alternatively, densitometric scanning of autoradiograms will provide estimates of the relative yields of a particular product.

The absolute quantity of protein synthesized *in vitro* will depend on factors intrinsic to the template, such as promoter strength, codon usage, and stability of the mRNA synthesized *in vitro*, in addition to the concentrations of 'key' ions, substrates, and effectors in the coupled system. Despite these variables, it is possible for coupled systems to direct the synthesis of substantial quantities of protein. We have measured the yields of S20 as a function of promoter strength and incubation temperature (24). Up to 400 moles of S20 can be synthesized per mole of template at 37 °C at limiting concentrations of template (1 μg/ml) provided that both of the S20 promoters are functional. This value can be increased further by mutating the initiation codon from UUG to AUG, or by changing the sequences 5′ to the initiation codon (23). The high yields of S20 synthesized *in vitro* are sufficient to permit the measurement of biochemically useful parameters such as the affinity of the ribosomal protein for 16S rRNA (10).

2.5.3 Analysis of RNAs synthesized in coupled systems

Measurement of the synthesis of a particular mRNA in a coupled system is often necessary to demonstrate whether a particular effector of the synthesis of ribosomal proteins acts at the level of transcription or translation (23).

Protocol 10. Analysis of mRNA transcripts

1. Prepare parallel incubations, one containing a radioactive amino acid, the second without. After the incubation, save the first sample for a check of protein synthesis (e.g., Section 2.5.1, part (a)) and dilute the second with 4 volumes of 6.25 mM EDTA (pH 8.0) in 0.25% SDS.
2. Extract the diluted incubation twice with phenol–chloroform-isoamyl alcohol (25:24:1).
3. Precipitate the final aqueous phase with ethanol.
4. Recover the precipitated RNA by centrifugation (10 min at 10000 *g*), drain the pellet, and reprecipitate the RNA from 0.25 M sodium acetate (pH 5.8) with ethanol.
5. Wash the final pellet of RNA once with 80% ethanol, and dissolve it in the original volume of sterile H_2O.
6. Determine the recovery of RNA by measuring the absorbance of a 1:100 dilution.
7. Analyse a 5 μg portion of the extracted RNA for the mRNA of interest by Northern blotting as described (19, 32).
8. Visualize the RNA bands by autoradiography and determine the relative yields of RNA by densitometry. Include standards to demonstrate the linearity of the signal with increasing RNA.

3. *In vitro* synthesis of eukaryotic ribosomal proteins

3.1 Strategies: cDNA to mRNA to protein

In vitro synthesis of eukaryotic ribosomal proteins can be used to investigate the size and distribution of ribosomal protein mRNAs (33–36), to identify cloned cDNAs capable of encoding ribosomal proteins (37, 38), and to investigate post–translational modifications of ribosomal proteins. The technology can, in principle, also be employed to investigate the mechanisms of post-translational control of ribosomal protein synthesis. As in bacterial systems, such experiments require both template and a template-dependent 'system' for the synthesis of ribosomal proteins. Total polyadenylated mRNA from several eukaryotic cells, tissues, or organisms has been used as a template to demonstrate the synthesis of ribosomal proteins *in vitro* (33–36). Since the mRNA for any single ribosomal protein constitutes in the order of 0.1% of the total mRNA, it is clear that the synthesis of an individual ribosomal protein cannot be particularly efficient with an unfractionated template. Two effective methods for obtaining single templates for ribosomal proteins are hybrid selection and *in vitro* transcription. Extracts capable of directing the synthesis of ribosomal proteins *in vitro* have been prepared from a number of sources (33, 34). Translation 'kits' containing all the necessary ingredients for using the rabbit reticulocyte system (39) are available from a number of suppliers and have proven satisfactory in our experience.

3.2 Preparation of RNA templates

3.2.1 Preparation of total polyadenylated RNA

We have extracted substantially intact total RNA from cells in tissue culture using either a modification of the guanidine-HCl method of Strohmann *et al.* (40), or the guanidinium isothiocyanate/CsCl method (41; see also page 196 in reference 19). We have used affinity chromatography on oligo-dT cellulose (obtained from Collaborative Research, Inc.) to enrich the polyadenylated fraction of total RNA. The procedure is described by Maniatis *et al.* (page 197 in reference 19). Omit SDS from the final elution buffer since this detergent can be difficult to remove completely by ethanol precipitation. Its presence earlier in the procedure protects the applied RNA from nuclease activity. Expect the yield of polyadenylated RNA to be about 5% of the RNA applied to the column. Up to half the polyadenylated fraction will be contaminating rRNA, however.

3.2.2 Size selection of ribosomal protein mRNAs

Polyadenylated RNA can be further enriched for ribosomal protein mRNAs by size fractionation since most ribosomal proteins are less than 30 kDa and their mRNAs are proportionally sized (34–37). See *Protocol 11*.

Protocol 11. Size fractionation of mRNAs on sucrose gradients

1. Dissolve 100 μg polyadenylated RNA in 100 μl of 0.1 mM EDTA (pH 7.0).
2. Heat for 90 sec at 85–90 °C, and chill on ice.
3. Apply the RNA to a 14 ml 5–20% sucrose gradient made in 10 mM Tris–HCl (pH 7.5), 100 mM LiCl, 1 mM EDTA, 0.1% SDS (autoclave the components, except SDS, separately).
4. Centrifuge the samples in a Beckman SW40 rotor at 28 000 r.p.m. for 17 h at 7 °C. This is sufficient time to pellet the larger rRNA and most RNAs greater than 20S.
5. Fractionate the gradient taking care not to disturb pelleted RNA and determine the absorbance at 260 nm of each fraction.
6. Concentrate RNA in the fractions of interest by several cycles of ethanol precipitation, if necessary in the presence of 10 μg carrier calf liver tRNA (from Boehringer-Mannheim or Sigma).

The quality of the total or fractionated polyadenylated RNA should be tested in several ways. First, determine that it will stimulate translation in a cell-free translation system (see below). There should be roughly a linear response between 100 ng and 1 μg of RNA added to the assay. The polyadenylated RNA should not inhibit the standard template usually supplied with commercial translation kits in a mixing experiment. Most importantly, the products of translation should be resolved on a two-dimensional gel capable of separating most ribosomal proteins (see Section 2.5.1 above and reference 30) to ensure that mRNAs for ribosomal proteins are present in the RNA sample and are translatable.

3.2.3 Enrichment of ribosomal protein mRNAs by hybrid-selection

This procedure was first described by Ricciardi *et al.* (42). A DNA template (e.g., a cloned cDNA) is denatured, immobilized on a solid support, and annealed to a mixture of total polyadenylated mRNAs. The unannealed mRNA species are washed away from this immobilized hybrid and the specifically bound mRNA is eluted from the immobilized DNA by boiling in H_2O. Follow the procedures given on pages 330–333 of reference 19. It is possible to substitute nylon (e.g., HyBond N from Amersham, Inc.) for nitrocellulose. The selected RNA is subsequently employed as a template in a translational assay (Section 3.3 below).

3.2.4 Preparation of ribosomal protein mRNAs by transcription *in vitro*.

A cDNA containing a full-length coding sequence for the ribosomal protein of interest is subcloned into a vector containing a suitably positioned promoter for a bacteriophage-encoded RNA polymerase (7, 43). The recombinant template is linearized distal to the coding sequence and transcribed *in vitro* essentially as outlined in Section 4.3 in Chapter 10 by Zengel *et al.* in this volume, omitting the radioactive UTP. We make no effort to cap the transcript. If capping is desired, include m7GpppG (available from Pharmacia; cat. no. 27-4635) at 500 μm and reduce the concentration of GTP to 20 μM (44). The RNA product is translated directly, without extraction or purification, in a cell-free extract in the presence of a radioactive amino acid (Section 3.3 below). A major advantage of this method over hybrid selection is that much higher quantities of template RNA can be prepared. Moreover, the template can be manipulated *in vitro* to permit the synthesis of an altered ribosomal protein.

3.3 Translation of ribosomal protein mRNAs

We have routinely translated mRNAs prepared as described above in a nuclease-treated rabbit reticulocyte lysate (39) supplemented with [^{35}S]-Met. Not all ribosomal proteins will contain internal methionine residues and it is possible, therefore, to fail to detect an mRNA encoding a ribosomal protein. Procedures for translating the selected RNAs are provided in the instructions supplied with commercial translation kits (available from BRL, NEN, Amersham, and Promega Biotech).

Protocol 12. Translation of mRNAs with commercial kits

1. Dissolve selected mRNAs (Sections 3.2.1–3.2.4) in a small volume (typically 2.5 μl) of sterile H_2O compatible with the procedure and proceed as directed. Use 2.5 μl of an *in vitro* transcribed RNA directly. Include both a no RNA control and one with a 'standard' template known to be active.
2. Stop the incubations by the addition of 2 U ribonuclease T1 and incubate for 10 min at 37 °C, then chill or freeze.
3. Assay the protein(s) synthesized by one of the methods described in Section 2.5.1 above.

Acknowledgements

Work in the authors' laboratory has been supported by grants from the Medical Research Council of Canada. We thank our colleagues, particularly Drs Jack Greenblatt, Joel Kirschbaum, and Mark Pearson, who have supplied strains and advice which have assisted our research.

References

1. Nomura, M., Gourse, R., and Baughman, G. (1984). *Ann. Rev. Biochem.*, **53**, 75.
2. Lindahl, L. and Zengel, J. M. (1986). *Ann. Rev. Genet.*, **30**, 297.
3. Zubay, G. (1973). *Ann. Rev. Genet.*, **7**, 267.
4. Gold, L. M. and Schweiger, M. (1971). In *Methods in Enzymology*, Vol. 20 (ed. K. Moldave and L. Grossman), p. 537. Academic Press Inc., London.
5. Kaltschmidt, E., Kahan, L., and Nomura, M. (1974). *Proc. Nat. Acad. Sci USA*, **71**, 446.
6. Brot, N., Caldwell, P., and Weissbach, H. (1980). *Proc. Nat. Acad. Sci. USA*, **77**, 2592.
7. Rodriguez, R. L. and Denhardt, D. T. (1988). *Vectors, a Survey of Molecular Cloning Vectors and Their Uses.* Butterworths, Boston.
8. Dottin, R. D. and Pearson, M. L. (1973). *Proc. Nat. Acad. Sci. USA*, **70**, 1078.
9. Yang, H.-L., Ivashkiv, L., Chen, H.-Z., Zubay, G., and Cashel, M. (1980). *Proc. Nat. Acad. Sci. USA*, **77**, 7029.
10. Donly, B. C. and Mackie, G. A. (1988). *Nucleic Acids Res.*, **16**, 997.
11. Pratt, J. M. (1984). In *Transcription and Translation, A Practical Approach* (ed. B. D. Hames and S. J. Higgins), p. 179. IRL Press, Oxford.
12. Jacobs, K. A. and Schlessinger, D. (1977). *Biochemistry*, **16**, 914.
13. Miller, J. H. (1972). *Experiments in Molecular Genetics*, Cold Spring Harbor Laboratory, Cold Spring Harbor, NY.
14. Lindahl, L., Yamamoto, M., Nomura, M., Kirschbaum, J. B., Allet, B., and Rochaix, J.-D. (1977). *J. Mol. Biol.*, **109**, 23.
15. Nomura, M. (1976). *Cell*, **9**, 633.

16. Huyhn, T. V., Young, R. A., and Davis, R. W. (1985). In *DNA Cloning, Volume 1—A Practical Approach* (ed. D. M. Glover), p. 49. IRL Press, Oxford.
17. Clewell, D. B. and Helinski, D. R. (1969). *Proc. Nat. Acad. Sci. USA*, **62**, 1159.
18. Birnboim, H. C. and Doly, J. (1979). *Nucleic Acids Res.*, **7**, 1513.
19. Maniatis, T., Fritsch, E. F., and Sambrook, J. (1982). *Molecular Cloning, a Laboratory Manual.* Cold Spring Harbor Laboratory Press, Cold Spring Harbor, NY.
20. Mackie, G. A. (1979). *Gene*, **5**, 45.
21. Austin, S. (1974). *Nature*, **252**, 596.
22. Mackie, G. A. (1977). *Biochemistry*, **16**, 4497.
23. Parsons, G. D., Donly, B. C., and Mackie, G. A. (1988). *J. Bacteriol.*, **170**, 2485.
24. Mackie, G. A. and Parsons, G. D. (1983). *J. Biol. Chem.*, **258**, 7840.
25. Lindahl, L., Post, L., and Nomura, M. (1976). *Cell*, **9**, 439.
26. Wirth, P., Buckel, P., and Böck, A. (1977). *FEBS Lett.*, **83**, 103.
27. Wirth, R. and Böck, A. (1980). *Mol. Gen. Genet.*, **178**, 481.
28. Yates, J. L., Arfsten, A. E., and Nomura, M. (1980). *Proc. Nat. Acad. Sci. USA*, **77**, 1837.
29. Wirth, R., Littlechild, J., and Böck, A. (1982). *Mol. Gen. Genet.*, **188**, 164.
30. Hames, B. D. and Rickwood, D. (ed) (1981). *Gel Electrophoresis of Proteins. A Practical Approach.* IRL Press, Oxford.
31. Cleveland, D. W., Fischer, S. G., Kirschner, M. W., and Laemmli, U. K. (1977). *J. Biol. Chem.*, **252**, 1102.
32. Mackie, G. A. (1986). *Nucleic Acids Res.*, **14**, 6965.
33. Wu, B. C., Rao, M. S., Gupta, K. K., Rothblum, L. I., Mamrack, P. C., and Busch, H. (1977). *Cell Biol. Int. Rep.*, **1**, 31.
34. Hackett, P. B., Egberts, E., and Traub, P. (1978). *J. Mol. Biol.*, **119**, 253.
35. Nabeshima, Y.-I., Imai, K., and Ogata, K. (1979). *Biochim. Biophys. Acta*, **564**, 105.
36. Bollen, G. H. P. M., Mager, W. H., Jenneskens, L. W., and Planta, R. J. (1980). *Eur. J. Biochem.*, **105**, 75.
37. Meyuhas, O. and Perry, R. P. (1980). *Gene*, **10**, 113.
38. Bozzoni, I., Beccari, E., Luo, Z. X., and Amaldi, F. (1981). *Nucleic Acids Res.*, **9**, 1069.
39. Pelham, H. R. B. and Jackson, R. J. (1976). *Eur. J. Biochem.*, **67**, 247.
40. Strohman, R. C., Moss, P. S., Micou-Eastwood, J., Spector, D, Pryzbyla, A., and Paterson, B. (1977). *Cell*, **10**, 265.
41. Chirgwin, J. M., Pryzbyla, A. E., MacDonald, R. J., and Rutter, W. J. (1979). *Biochemistry*, **18**, 5294.
42. Ricciardi, R. P., Miller, J. S., and Roberts, B. E. (1979). *Proc. Nat. Acad. Sci. USA*, **76**, 4927.
43. Melton, D. A., Kreig, P. A., Rebagliati, M. R., Maniatis, T., Zinn, K., and Green, M. (1984). *Nucleic Acids Res.*, **12**, 7035.
44. Pelletier, J. and Sonenberg, N. (1985). *Cell*, **40**, 515.

10

A hybrid selection technique for analysing *E. coli* mRNA: applications to the study of ribosomal protein operons

JANICE M. ZENGEL, JOSEPH R. McCORMICK, RICHARD H. ARCHER, and LASSE LINDAHL

1. Introduction: advantages of the hybrid selection technique

One important parameter in analysing a gene's expression is the amount of translatable transcript representing the gene at any given time. The amount of transcript, in turn, is determined by both the rate of *de novo* synthesis and the rate of processing and decay of the mRNA. Since messages usually are unstable, these parameters are often difficult to measure accurately. Northern analysis and nuclease S1 mapping measure steady-state concentrations, but do not differentiate between synthesis and processing or decay. Moreover, the technique most often used for direct measurements of RNA synthesis, filter hybridization of pulse-labelled RNA, does not allow mapping of endpoints or analysis of the RNA integrity.

This chapter describes the procedures that we have used in order to analyse the synthesis of *E. coli* ribosomal protein messages. Our procedure combines the advantages of hybridization of labelled RNA with the fine-structure analysis capability of nuclease mapping techniques. This hybrid selection technique is based on a procedure originally reported by Hansen and Sharp (1) and modified by Lamond and Travers (2).

2. Labelling of cells and extraction of nucleic acids

When working with RNA all reagents, tubes, and so on, must be nuclease-free. We autoclave all glassware, glass pipettes, disposable tubes, and tips, and buffers for 1 h.

2.1 Reagents

(a) Nuclease-free water. Dissolve 1 ml diethylpyrocarbonate (Sigma) in 5 ml ethanol. Add the solution to 1 litre glass-distilled water. Stir overnight at 37 °C. Autoclave for at least 1 h.

(b) [5, 6-^{3}H] uridine, aqueous solution (1 mCi/ml, 40–50 Ci/mmol: Amersham, New England Nuclear, or ICN). Concentrate 5–10 fold by lyophilization.

(c) [^{32}P] orthophosphate (8 mCi/ml, carrier-free, acid-free, Amersham).

(d) Phenol, analytical grade. Distill phenol, collecting at vapour temperatures between 178 °C and 179 °C into about 100 ml distilled water per 500 g phenol. If necessary, slowly add additional water while mixing vigorously until a two-phase system forms. Add a few crystals (about 0.1%) of 8-hydroxyquinoline (a reducing agent) until the phenol phase (bottom phase) has a light yellow colour. The colour facilitates separating the phases during RNA purification. Store at −20 °C indefinitely or at 2–4 °C for several weeks.

(e) Yeast carrier RNA [Turola yeast ribonucleic acid (type VI), Sigma]. Dissolve 0.8 g in 40 ml distilled water. Extract three times with phenol, two times with phenol/chloroform–isoamyl alcohol, three times with chloroform–isoamyl alcohol and precipitate with ethanol, all as described below (Section 2.3) for purification of radioactive RNA. Re-dissolve in nuclease-free water at approximately 10 mg/ml (A_{260} = 200).

(f) Chloroform-isoamyl alcohol: mix in the volume ratio of 24:1.

(g) FPLC pure DNase I (Pharmacia).

2.2 Growth and radioactive labelling of cells

Cells for radioactive labelling should be grown in a minimal medium. For [^{3}H]-uridine labelling, we use either AB (5) or MOPS minimal medium (6) supplemented with a carbon source and appropriate requirements. The radioactive uridine should not be exhausted during the labelling period to ensure that radioactivity is not chased from the unstable message fraction into stable RNA species. At the isotope concentration (75 μCi/ml of cells) and cell density (0.5–1 × 10^8 cells/ml) we use, complete uptake requires more than 10 min at 37 °C. For $^{32}PO_4$-labelling, we use MOPS medium (6) with 0.2 mM phosphate, but in the overnight culture we use 1 mM phosphate, since phosphate depletion leads to a long lag period after dilution of the culture. The labelling is terminated by lysing the cells directly in hot buffer containing sodium dodecylsulphate. This procedure assures that incorporation of label is arrested instantly and helps prevent further processing and/or degradation of RNA. Details are given in *Protocol 1*.

Protocol 1. Labelling of cells with radioactive uridine or phosphate

1. Grow cells overnight in minimal medium supplemented with appropriate

amino acids, vitamins, and so on. For labelling with radioactive phosphate, adjust to 1 mM with non-radioactive phosphate.

2. Dilute the culture to an A_{450} of 0.05–0.1 (about 10^7 cells/ml). For labelling with radioactive phosphate, adjust the phosphate concentration to 0.2 mM, taking into consideration the phosphate contributed by the overnight culture (about 0.1 mM phosphate is taken up by a culture which has grown to a density of A_{450} of 1).
3. When the culture has grown to an A_{450} of 0.6–0.8, label with radioactive precursor:
   (a) Add 0.225 mCi [5, 6-^{3}H] uridine (40–50 Ci/mmol) to 3 ml of the culture.[a] After the desired labelling time (usually 1 min), mix the culture with 3 ml TSEI-SDS[b] (pre-heated to 90–100 °C)[c] containing 100 μg yeast RNA,

   or

   (b) Add 1.5 mCi carrier-free $^{32}PO_4$ to 3 ml of the culture. After the desired labelling time (usually 3–6 min), harvest the cells as above.
4. Keep the mixture at 90–100 °C for 2–3 min and then extract as described in *Protocol 2*.

[a] If the labelling is done in a 30-ml glass centrifuge tube, lysis with TSEI–SDS (see note *b*) as well as the following extractions (*Protocol 2*) can be done in the same vessel.

[b] TSEI–SDS is 0.02 M Tris–HCl (pH 7.4), 0.2 M NaCl, 0.04 M Na-EDTA, 1% SDS. Add SDS after autoclaving.

[c] Use a heat block rather than a water bath to avoid steam burns on hands.

2.3 Purification of RNA

Extraction and concentration procedures are given in *Protocol 2*. When labelling with [^{3}H]-uridine, relatively little radioactivity enters DNA during a pulse of 1–2 min, since conversion of uridine into thymidine goes through several intermediates resulting in a slow increase in the specific activity of the dTTP pool relative to the UTP and CTP pools. However, $^{32}PO_4$ rapidly enters DNA, carbohydrates, and lipids. The radioactivity in these components often generates an unacceptable background in the gels used for analysis of the RNA (see below). This problem can be eliminated by introducing a DNase step and several additional extractions with organic solvents (*Protocol 2*; see also Section 5.1).

Protocol 2. Extraction of RNA from labelled cells

A *Standard extraction*

1. Let radioactive lysates from *Protocol 1* cool at room temperature until lukewarm.[a]
2. Add 6 ml of water-saturated phenol. Cover tightly with parafilm and vortex for 10–15 sec. Repeat vortexing 2–3 times over the next 2 min.

3. Centrifuge at 8–10 000 r.p.m. for 10 min. Carefully remove and discard the phenol layer (bottom layer with yellow colour). Alternatively, the water phase can be transferred to a clean centrifuge tube.
4. Add 1.2 ml 10.5 M ammonium acetate or 0.25 ml 5 M NaCl.
5. Add 15 ml 95% ethanol, mix, and store at −20 °C for 2 h or longer.
6. Centrifuge at 10 000 r.p.m. for 15 min. Discard the supernatant and drain the pellet well.
7. Resuspend the pellet in 0.4 ml TE[b] and transfer to a 1.5 ml microcentrifuge tube.
8. Add 0.4 ml water-saturated phenol and vortex as in step 2. Centrifuge at 10–12 000 r.p.m. for 5 min. Discard the phenol phase as in step 3. Some precipitate will remain in the interphase between the phenol and water layers until the chloroform–isoamyl alcohol extractions (see the following steps).
9. Add 0.4 ml of chloroform–isoamyl alcohol mixture and vortex as in step 2. Centrifuge at 10–12 000 r.p.m. for 2 min. Discard the bottom organic phase as in step 3.
10. Repeat step 9.
11. Add 80 μl 10.5 M ammonium acetate or 16 μl 5 M NaCl.
12. Add 1.0 ml 95% ethanol, mix and store for 15 min or longer at −70 °C. Centrifuge at 10–12 000 r.p.m. for 15 min. Discard the supernatant and drain the pellet.
13. Redissolve in 0.4 ml TE. Perform two more ethanol precipitations as described in steps 11 and 12.
14. After the third ethanol precipitation, dry the pellet under vacuum for 10–15 min. It is preferable to use a vacuum chamber with a low speed centrifuge (such as a Savant Speed-Vac) to avoid the loss of pellets.
15. Redissolve the pellet in 100 to 200 μl TE.

B *Optional procedures (see also text)*

1. Incorporate 2–3 extractions (after step 8 above) with phenol–chloroform–isoamyl alcohol[c] and/or several additional extractions with chloroform–isoamyl alcohol (step 9 above).
2. Degrade the DNA by redissolving the final pellet in 50 μl DNase I buffer[d] containing 30 μg yeast RNA. Add 7.5 units of DNase I and incubate for 10 min at 37 °C. Inactivate the DNase by incubating for 15 min at 68 °C. Precipitate RNA by adding 12 μl 10.5 M ammonium acetate or 2 μl 5 M NaCl followed by 150 μl ethanol. Mix and incubate at −70 °C for 15 min or longer. Centrifuge and redissolve the RNA pellet in TE as in steps 12, 14, and 15 above.

[a] Samples can be extracted with phenol while they are still hot, but hot phenol dissolves parafilm. To avoid spills, first cover the tube with plastic film, then cover the plastic film tightly with parafilm.

[b] TE is 10 mM Tris–HCl (pH 7.4), 0.1 mM Na-EDTA.
[c] Add 1/2 volume of water-saturated phenol (relative to volume of the RNA solution) and 1/2 volume of chloroform–isoamyl alcohol. Mix and centrifuge as in step 9. Discard the bottom organic layer.
[d] DNase I buffer is 50 mM Tris–HCl (pH 7.4), 10 mM $MgCl_2$.

3. Hybrid selection of mRNA

3.1 Reagents

- Ribonuclease T1 (Bethesda Research Laboratories).
- Proteinase K (Beckman, Boehringer-Mannheim, or U.S. Biochemicals).
- Nitrocellulose filters (e.g., Schleicher and Schuell, BA 85, 0.45 μm, 25 mm discs).
- Single-stranded M13 DNA containing appropriate inserts from the gene studied.

3.2 Hybridization procedure

Total radioactive RNA is hybridized in solution to single-stranded DNA containing an insert complementary to the region of the transcript under study. We have used probes ranging in size from 60 bases to 890 bases. After the hybridization, all unhybridized RNA molecules as well as 5′ and 3′ flanking regions of the hybridized RNA which are not complementary to the probe are digested with RNase T1. The resulting RNA–DNA hybrids are then trapped on a nitrocellulose filter by using buffer conditions which allow binding of single-stranded DNA (and the protected RNA fragment hybridized to it) but not unhybridized RNA. The bound RNA is then released from the filter, concentrated, and analysed on a denaturing urea polyacrylamide gel. Details of the procedure are given in *Protocol 3*.

Protocol 3. Hybridization and RNase T1 treatment

1. Each 20 μl hybridization reaction contains an amount of labelled RNA equivalent to 0.1–1 ml cells, 20–30 μg yeast carrier RNA, 1–2 μg single-stranded M13 DNA probe and 4 μl 5 X hybridization buffer.[a]
2. Heat at 68 °C for 10 min, then incubate at 50–55 °C for 1–4 h.
3. Add 160 μl T1 buffer[b] and 30 units RNase T1. Incubate at 30 °C for 20–30 min.
4. Inactivate the RNase with 20 μl proteinase K (2 μg/μl) at 37 °C for 30 min.
5. Filter the sample slowly through a 25 mm nitrocellulose filter (pre-soaked in T1 buffer).
6. Rinse the sample tube and the filter with 3 × 300 μl T1 buffer.

7. Cut the filter with scissors into quarters; place all four pieces into a 1.5-ml microcentrifuge tube.
8. Add 500 μl 5 mM EDTA containing 10 μg yeast RNA.
9. Boil for 5 min (keep the tube lid open to avoid pressure build-up).
10. Chill rapidly. Transfer the supernatant to a new microcentrifuge tube containing 90 μl 10.5 M ammonium acetate or 20 μl 5 M NaCl.
11. Add 1 ml 95% ethanol and mix. Store at −70 °C for 30 min or longer.
12. Centrifuge for 15–30 min, carefully remove the supernatant, then dry the pellet.
13. Resuspend the RNA in 5–10 μl TE.[c]
14. Mix 2–4 μl of RNA with an equal volume of loading solution[d] just prior to gel electrophoresis (see *Protocol 4*).

[a] 5 X hybridization buffer is 3.75 M NaCl, 0.25 M Hepes–NaOH (pH 7.0), 5 mM Na-EDTA.
[b] T1 buffer is 10 mM Hepes–NaOH (pH 7.5), 200 mM NaCl, 1 mM Na-EDTA.
[c] TE is 10 mM Tris–HCl (pH 7.4), 0.1 mM Na-EDTA.
[d] Loading solution contains 95% deionized formamide, 10 mM Na-EDTA, 0.1% bromophenol blue, and 0.1% xylene cyanol.

4. Gel electrophoresis and quantitation

After hybrid selection and RNase T1 digestion, the protected RNA fragments are fractionated by electrophoresis through denaturing urea gels. The bands are then visualized by autoradiography or fluorography. If desired, the amounts of radioactivity in the relevant bands can be quantitated by scintillation counting or by densitometric scanning of autoradiograms.

4.1 Supplies

- Enhance (New England Nuclear) or Amplify (Amersham).
- Intensifying screen (Cronex Quanta III, Du Pont).
- X-ray film (Kodak, e.g., XAR5).
- T7 RNA polymerase (U.S. Biochemicals).
- Uridine 5′-α-[^{35}S]-thiotriphosphate ([^{35}S]-UTP), triethylammonium salt, 800–1200 Ci/mmol, 10 or 40 mCi/ml (Amersham).
- Nitrocellulose filters, precut, 0.45 μm (Bethesda Research Lab.).
- Catalase (product number C-40, Sigma).

4.2 Gel electrophoresis procedures

For ^{32}P-labelled RNA, we usually use large (30 × 40 cm) gels to maximize resolution. For ^{3}H-labelled RNA we use smaller (16 × 20 cm) gels, because they are easier to handle during the fluorography procedure. Details are given in *Protocol 4*. Examples of gels displaying RNase T1-protected ^{32}P- and ^{3}H-labelled RNA fragments are shown in *Figures 1* and *2*.

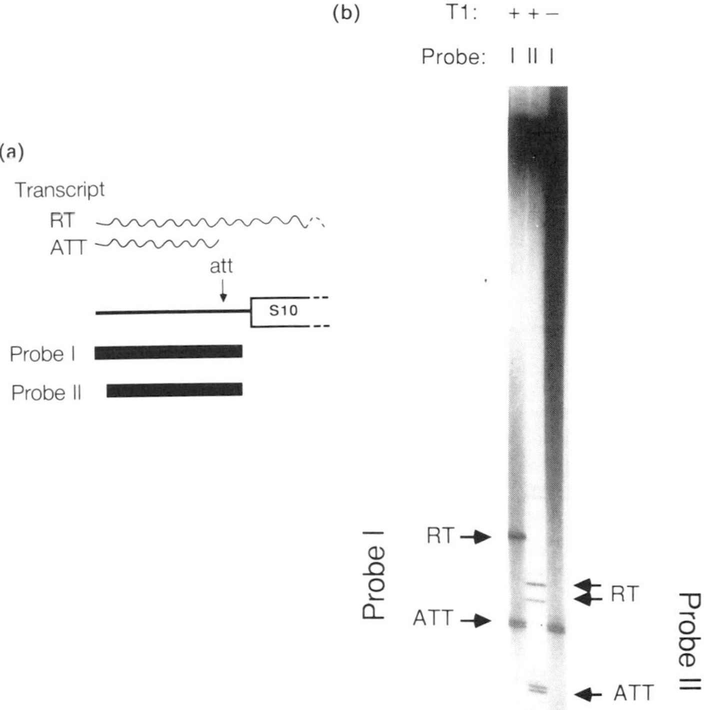

Figure 1. Analysis of RNA synthesized from the leader of the S10 ribosomal protein operon. Cells carrying a plasmid with the proximal portion of the S10 operon were pulse-labelled for 3 min with $^{32}PO_4$. The RNA was extracted and hybridized to Probe I or Probe II. The hybridization mixture was then treated with RNase T1 (half of the Probe I mix was subjected to a mock T1 treatment), filtered, electrophoresed on a denaturing gel, and analysed by autoradiography. (a) Schematic representation of the readthrough (RT) and attenuated (ATT) transcripts synthesized from the S10 promoter. Transcription of the S10 operon is regulated by an attenuator in the leader (10). The vertical arrow shows the position of the attenuation site. Note that Probe II has the same 3′ end as Probe I, but is 15 bases shorter at the 5′ end. (b) Autoradiogram of hybrid-selected RNA using Probes I and II, with (+) and without (−) nuclease T1 treatment. In this experiment, Probe II generated two RT fragments, probably because of incomplete nuclease cleavage at the 3′ side of the hybridized readthrough RNA. A comparison of the two Probe I lanes shows that the attenuated transcript (ATT) is not affected by T1 treatment, and hence is contained within the region protected by Probe I. In this case the double band results from transcription termination at two adjacent bases in the S10 leader. The minus T1 lane also illustrates the ability of the hybrid-selection technique to purify specific RNAs even in the absence of nuclease treatment.

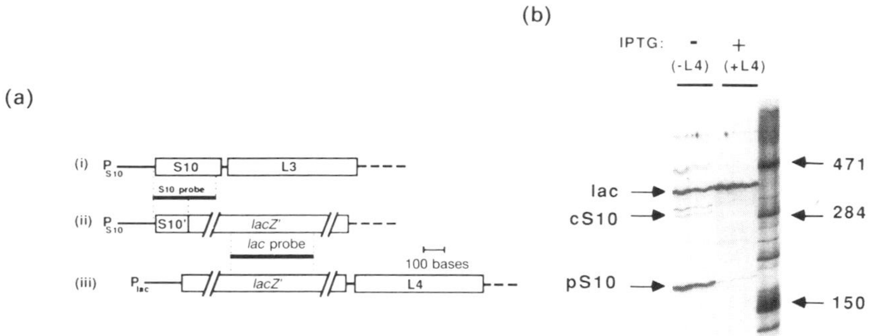

Figure 2. Analysis of RNA synthesis after induction of ribosomal protein L4. Cells carried the normal chromosomal copy of the S10 ribosomal protein operon (panel a, part i), and two plasmids, one containing the beginning of the S10 operon fused to the *lacZ* gene (panel a, part ii) and the other containing most of the *lacZ* gene and the complete r-protein L4 gene under control of the *lac* promoter (panel a, part iii). Cells were pulse-labelled for 1 min with [^{3}H]-uridine either before or 10 min after addition of isopropylthiogalactoside (IPTG). The RNA was extracted, hybridized to a mixture of a *lac*- and an S10-specific probe, treated with RNase T1, filtered, and electrophoresed on a denaturing gel. Panel (b) shows a fluorogram of the gel with the protected RNA fragments identified. Duplicate labellings were analysed. Transcripts corresponding to the chromosomal (cS10) and the plasmid (pS10) copies of the S10 gene can be distinguished because the S10 probe protects RNA which extends 119 bases beyond the junction of the S10'/*lacZ'* fusion. The addition of IPTG, which turns on expression of both *lacZ* and the L4 gene on the plasmid shown in (iii), results in an increase in the intensity of the *lac* band, and a decrease in the intensities of both the chromosomal and plasmid-derived RNA hybridized to the S10 probe. The decrease in S10 RNA synthesis results from L4-stimulated termination of transcription within the S10 leader (10). Note that the *lac* probe protects transcripts from both the S10'/*lacZ'* and *lacZ'*-L4 plasmids. After IPTG addition, S10'/*lacZ'* transcription decreases (data not shown), but this decrease is obscured by the increased synthesis of *lacZ* from the other plasmid. The right-most lane of the gel shows a mixture of ^{35}S-labelled markers synthesized using T7 RNA polymerase and several different linearized DNA templates.

Protocol 4. Gel electrophoresis

1. For a large (30 × 40 × 0.04 cm) gel, mix 80 ml Instagel Solution[a] with 0.9 ml 10% ammonium persulphate in a sidearm flask. [For a small (16 × 20 × 0.04 cm) gel, use 25% of the volumes listed in steps 1 and 3 of this *Protocol*.]
2. De-gas the solution by evacuating the flask for about 2 min.
3. Add 80 μl TEMED.
4. Mix briefly and then slowly inject the solution between glass plates using a syringe without a needle.
5. Insert a slot former and clamp the glass plates with three heavy duty spring-type clamps along the slot former.

6. Let polymerize for at least 1 h.
7. Remove the slot former and rinse the top of the gel with distilled water.
8. Install in the gel apparatus, add 1 × TBE buffer[b] to the buffer chambers, and pre-run the gel at 75 W (voltage begins at 1300–1600 V) for 45 min (the gel will get warm). Small gels are run at a constant voltage of 500 V.
9. Disconnect the electrodes and squirt running buffer into the slots to rinse out urea which has leached out from the gel during the pre-run.
10. Load the samples, reconnect the electrodes, and run at a maximum voltage of 1800–1900 V for 3–5 h.[c] Run small gels at 500 V for 2–3 h.[c]
11. Fix the gel in 5% acetic acid, 5% methanol, for 30–40 min.
12. For large gels, dry, and expose to X-ray film. Exposure times for ^{32}P radioactivity can be reduced about five-fold by using an intensifying screen at −70 °C.
13. After fixing small gels, impregnate with a fluor solution (Enhance or Amplify) according to the manufacturer's instructions, dry, and expose to a pre-flashed X-ray film[d] at −70 °C.

[a] Instagel Solution is made up as follows: Mix 160 ml 40% acrylamide, 160 ml 2% bisacrylamide, 80 ml 10 × TBE (see note *b*), and 336 g ultrapure urea. Bring the volume to 800 ml with distilled water. Filter through two sheets of Whatman No. 1 filter paper. Store at 2–4 °C.

[b] TBE is 90 mM Tris, 90 mM boric acid, and 2.5 mM Na-EDTA. 10 × TBE is a 10-fold concentrated stock solution. The buffer will have a pH of 8.1–8.3.

[c] The running time depends on the sizes of the RNA fragments to be analysed. A useful guide is that the xylene cyanol marker dye co-electrophoreses with RNA fragments about 70 bases long.

[d] To reduce the exposure time, the film should be activated by a limited exposure to visible light from an electronic flash (7). The appropriate exposure is determined empirically by covering the flash with layers of neutral density filters, exposed X-ray film or paper, and by varying the distance between the flash and the film. A correctly activated piece of X-ray film has, after developing and fixing, an absorbance at 540 nm of 0.1–0.2 (7).

4.3 RNA size markers

To determine the sizes of the protected RNA fragments, we prepare ^{35}S-labelled RNA markers of known lengths, synthesized off linearized plasmids by T7 RNA polymerase. The procedure is given in *Protocol 5*. These RNAs are then electrophoresed on the same gel as the T1-protected fragments. When comparing the size of the RNase T1-protected RNA fragments to the length of the hybridization probe, it is important to remember that, since RNase T1 cuts specifically at the 3′ side of G-residues, the enzyme will in most cases not trim precisely at the end of the DNA:RNA hybrid helix.

Protocol 5. T7 RNA polymerase transcription of molecular weight markers

1. Linearize plasmid DNA with an appropriate restriction endonuclease. Extract with phenol/chloroform, and precipitate with 95% ethanol (see

Protocol 2). Resuspend the DNA pellet in TE[a] to a final concentration of 0.5 μg/μl. (1 A_{260} unit = 50 μg/ml.)

2. For each 50 μl RNA synthesis reaction, mix 5 μl 10 × T7 buffer,[b] 2.5 μl 100 mM dithiothreitol, 0.5 μl 50 μg/μl BSA, 2 μl 10 mM each ATP, GTP, UTP, and CTP, 20 μCi [^{35}S]-UTP, 100–200 units T7 RNA polymerase, 0.5–1 μg linearized DNA and nuclease-free water (described in Section 2.1, part a) to bring the volume to 50 μl.
3. Incubate at 37 °C for 30 min.
4. Chill on ice. Extract one time with an equal volume of water-saturated phenol and one time with an equal volume of chloroform–isoamyl alcohol (see *Protocol 2*). Add 10 μl 10.5 M ammonium acetate and 150 μl 95% ethanol. Mix and store at −70 °C for 15 min or longer.
5. Centrifuge in a microcentrifuge for 20 min. Decant the supernatant and dry the pellet. Resuspend in 50–100 μl TE.[a] Store at −20 °C.
6. Just before electrophoresis, mix 1 μl ^{35}S-labelled RNA with 1 μl loading solution.[c]

[a] TE is 10 mM Tris–HCl (pH 7.4), 0.1 mM Na-EDTA.

[b] 10 × T7 buffer is 0.4 M Tris–HCl (pH 8.0), 0.15 M $MgCl_2$.

[c] Loading solution is 95% deionized formamide, 10 mM EDTA, 0.1% bromophenol blue, and 0.1% xylene cyanol.

4.4 Quantitation of [^{32}P] RNA

Protocol 6. Quantitation of [^{32}P]RNA

1. To quantitate the amounts of ^{32}P radioactivity, scan the bands on the autoradiogram with a densitometer (we use a Zeineh Soft Laser Densitometer, model SL504).
2. Quantitate the areas under each peak by planimetry (e.g. by using a Tektronix 4956 digitizer). Since there is rarely 1 : 1 proportionality between the amount of radioactivity in a band and the intensity of the band on the autoradiogram (7), it is necessary to construct a standard curve.
3. Blot serial dilutions of ^{32}P-labelled RNA in 20 × SSC (3 M NaCl, 0.3 M sodium citrate pH 7.0) on to strips of nitrocellulose using a slot blotter (e.g. Bethesda Research Lab.).
4. Place the nitrocellulose filter beside the dried-down polyacrylamide gel during exposure to the X-ray film.
5. After autoradiography, scan the bands of the radioactive standard solutions with the densitometer to provide data for constructing a standard curve.

4.5 Quantitation of [^{3}H] RNA

The amounts of [^{3}H] RNA are determined by scintillation counting.

Protocol 7. Scintillation counting

1. Before exposure of the fluor-impregnated, dried gel to the X-ray film, apply small dots of ^{35}S radioactivity in black ink along the edges of the gel.
2. After fluorography, line up the dots on the fluorogram with the ink spots on the gel and then mark the positions of bands on the gel by piercing a sharp needle (e.g. a dissecting needle) through the X-ray film into the gel around each band.
3. Cut out the bands and place them in scintillation vials.
4. Extract the radioactivity in each gel piece by adding 0.5 ml 50% hydrogen peroxide, capping the vials, and incubating 4–6 h at 75 °C.
5. Remove the hydrogen peroxide by adding 0.5 ml catalase (0.3 mg/ml in 25 mM Tris–HCl (pH 8.0)) and incubating at room temperature for 15 min.
6. Add scintillation cocktail, mix the samples thoroughly, and then count them in a scintillation counter.

---

5. Modifications

The basic techniques described above can be modified in several ways to extend their use.

5.1 Analysis of mRNA processing and decay

We have used the RNase T1 mapping technique for studies of mRNA decay and processing by inhibiting the initiation of new RNA synthesis with rifampicin. Cells are labelled with $^{32}PO_4$ for several minutes before addition of rifampicin, and aliquots of the culture are then removed before and at various times after addition of rifampicin for preparation of RNA. Accurate determination of mRNA half-life requires that initiation of mRNA synthesis be stopped quickly. Rifampicin enters most *E. coli* strains slowly and it typically takes several minutes before transcription initiation is completely stopped (8). However, in the membrane mutant *E. coli* AS19 rifampicin enters quickly, allowing complete inhibition of transcription initiation within seconds (8, 9). It should, however, be noted that transformation of strain AS19 with appropriate plasmids used for such studies requires modifications to the standard $CaCl_2$ protocol (8).

The background of ^{32}P-labelled DNA in the gels (see Section 2.3) becomes a serious problem with these studies of mRNA processing and decay because incorporation of $^{32}PO_4$ into DNA continues after RNA synthesis has been

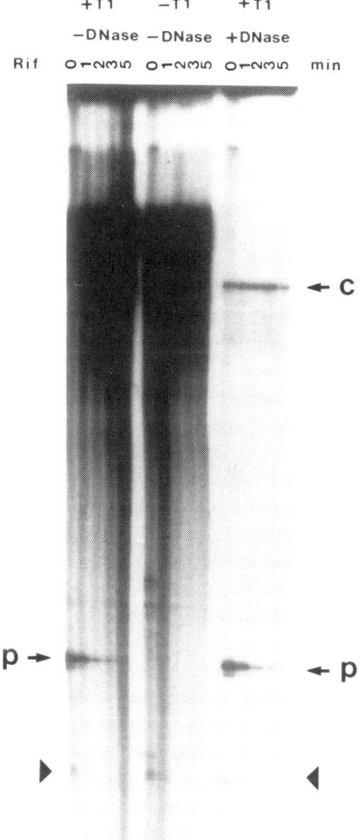

Figure 3. Hybrid selection of RNA synthesized after rifampicin treatment. Cells carrying an S10'/*lacZ'* plasmid similar to the plasmid shown in *Figure 2* were labelled with $^{32}PO_4$ for 5 min before inhibition of transcription initiation by addition of rifampicin. At the indicated times after rifampicin addition, samples were removed and processed as described in *Protocol 2*, part A. An aliquot of each sample was treated with DNase I as described in *Protocol 2*, part B. The RNAs were then hybridized to a probe specific for the beginning of the S10 operon, treated with RNase T1 (half of the minus DNase I RNA sample was not digested with the RNase), filtered, and electrophoresed on a denaturing gel. An autoradiogram of the gel is shown. The chromosomal-derived S10 operon transcript (c) can be distinguished from the plasmid-derived RNA (p) because the probe contains sequences from the S10 operon extending beyond the S10'/*lacZ'* junction on the plasmid (analogous with the example shown in *Figure 2*). The arrowheads show the attenuated transcript from both chromosome and plasmid. Note the effect of the DNase on the gel background.

inhibited by rifampicin. As shown in *Figure 3*, DNase treatment eliminates this problem.

5.2 Use of multiple probes in the same hybridization reaction

If synthesis of several different mRNA molecules is to be compared, the number of hybridization reactions makes the analysis very tedious. Multiple hybridization probes can be used in the same hybridization reaction, as long as probes with different size inserts are chosen so that the protected fragments can be separated by gel electrophoresis. This procedure also corrects for pipetting variabilty and differential recovery of mRNA fragments from the same RNA preparations protected by different probes. As illustrated in *Figure 4*, it is even possible to analyse radioactivity in separate segments of a single long transcript by using several non-overlapping probes in the same hybridization reaction. For such experiments, we use probes which do not contain immediately adjacent segments of the transcription unit to avoid possible interference between the hybridization on different probes of the same mRNA.

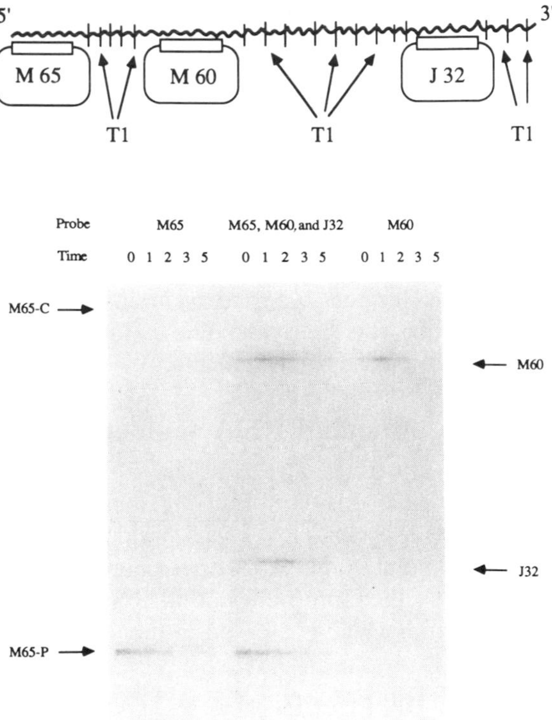

Figure 4. Simultaneous hybridization with multiple probes from a single transcription unit. Cells carrying a plasmid with an S10'/*lacZ'* fusion were labelled for 5 min with $^{32}PO_4$. Rifampicin was added to prevent further initiation of transcription, and samples were removed at the indicated times after rifampicin addition. RNA was extracted, treated with DNase I (see *Figure 3*), and hybridized either to a single DNA probe or to a mixture of three DNA probes from the S10'/*lacZ'* fusion gene (illustrated at the top of the figure). The RNase T1-protected fragments were then fractionated on a denaturing gel. The RNA fragments protected by each probe are indicated by labelled arrows. Note that probe M65 protects distinguishable chromosomal (M65-C) and plasmid (M65-P) RNAs (see *Figure 2* for an explanation).

6. Limitations and problems

6.1 Hybridization efficiencies

The hybridization efficiency varies at least several-fold for different probes. The hybridization efficiency for rRNA and tRNA is particularly low, presumably because intramolecular hybridization within these highly structured molecules competes with the bimolecular hybridization to the DNA probe. It is therefore only possible to compare the amounts of radioactivity from different samples hybridizing to the same probe, and not directly the amounts of radioactivity hybridizing to different probes. If the absolute amounts of radioactive RNA hybridizing to different probes must be compared, it is necessary to use internal standards labelled with an isotope different from the one used to label the *in vivo* synthesized RNA. Such standards can most easily be made *in vitro* using T7 or SP6 RNA polymerase (*Protocol 5*).

6.2 Incomplete trimming with RNase T1

For some hybridization probes, the RNase T1-protected RNA is of two or more different lengths and multiple bands are seen on the gel, even with increased amounts of RNase T1. Unfortunately, we have found only one solution to this problem: make another probe, either by using other restriction sites or by altering one or both endpoints by deleting a few bases with nuclease Ba131. Since RNase T1 cleaves only at G residues, we tried to eliminate the multiband problem by adding RNase A, which cuts at pyrimidines. However, RNase A did not generate well-defined protected RNA fragments.

6.3 Cleavage in the protected RNA segment

In a few cases we have observed cleavage by RNase T1 within the protected RNA sequence. We could distinguish this event from *in vivo* processing by showing that intact RNA transcripts made *in vitro* (*Protocol 5*) yield the same cleavage pattern. In all cases the cleavage point coincided with a potential stable hairpin in the hybridizing sequences. We therefore believe that RNA in the hybrid forms an internal stem-loop structure which is cleaved in the loop of the hairpin.

7. Comments and prospects

We have developed the hybrid selection technique described here to analyse the regulation of the *E. coli* ribosomal protein operons. This procedure has been especially useful in our analysis of the S10 ribosomal protein operon (*Figures 1*, *2*, and *3*), since standard RNA mapping techniques could not detect the unstable attenuated transcript. We believe the methodology outlined here will be of general use because it combines pulse labelling of RNA, allowing for measurements of RNA synthesis (rather than accumulation), with fine structure mapping of transcripts.

References

1. Hansen, U. and Sharp, P. A. (1983). *EMBO. J.*, **2**, 2293.
2. Lamond, A. I. and Travers, A. A. (1985). *Cell*, **40**, 319.
3. Shen, P., Zengel, J. M., and Lindahl, L. (1988). *Nucleic Acids Res.*, **16**, 8905.
4. Lindahl, L. and Zengel, J. M. (1988). In *Genetics of Translation* (ed. M. Tuite, M. Picard, and M. Bolotin-Fukuhara), p. 105. Springer-Verlag, Berlin.
5. Clark, D. J. and Maaløe, O. (1967). *J. Mol. Biol.*, **23**, 99.
6. Neidhardt, F. C., Bloch, P. L., and Smith, D. F. (1974). *J. Bacteriol.*, **119**, 736.
7. Laskey, R. A. (1984). *Radioisotope detection by fluorography and intensifying screens, Review 23*. Amersham Corp., Arlington Heights, IL.
8. McCormick, J. R. (1989). Ph.D. thesis, Univ. of Rochester. Rochester, NY.
9. Pato, M. L. and von Meyenburg, K. (1970). *Cold Spring Harbor Symp. Quant. Biol.*, **35**, 497.
10. Lindahl, L., Archer, R., and Zengel, J. M. (1983). *Cell*, **33**, 241.

11

Analysis of rRNA structure: experimental and theoretical considerations

JAN CHRISTIANSEN, JAN EGEBJERG, NIELS LARSEN, and ROGER A. GARRETT

1. Introduction

In the cell, RNAs are probably never naked but are always associated with various ligands. Ribosomal RNAs interact directly with ribosomal proteins, translational factors, and antibiotics, so it is of considerable structural and functional importance to elucidate the modes and sites of these interactions. As a first step towards this goal the secondary structures of the rRNAs have to be determined.

In this chapter, we present an experimental and a theoretical approach that are complementary, and together they provide a powerful approach to structure determination. Moreover, once the secondary structure is established, the experimental approach can be employed to include the analysis of ligand binding sites.

The following section contains detailed experimental protocols for the probing of RNA structure with chemicals and ribonucleases, and their analysis by reverse transcription. We regard this procedure as the most generally useful and expedient approach for any RNA molecule, regardless of size. However, alternative procedures will be outlined briefly towards the end of the section, without experimental details. Finally, there is a treatise on the application of computers to RNA structural analysis, where we will compare free-energy predictions with those originating from the compensating base-change approach.

2. Experimental approaches to higher-order RNA structures and ligand-binding sites

2.1. The principle behind reverse transcriptase analysis

Reverse transcriptase generates a complementary copy of DNA (cDNA) from an RNA template. When a base in RNA is chemically modified the polymerase will stop (or pause) 3′ to this position. Moreover, a break in the ribose-phosphate backbone, generated by a ribonuclease, will terminate the polymerase.

These properties of reverse transcriptase are exploited in the RNA structural analysis procedure, in which chemical and enzymatic probing are analysed by primer extension (1). The principle behind the method is outlined in *Figure 1*. Free or complexed RNA is modified or digested in a manner that only allows one 'hit' per molecule, or for large molecules about one 'hit' per 300 nucleotides, and the attacked position is identified by a stop in reverse transcription generated from a DNA primer. When the reverse transcripts are subjected to gel electrophoresis, the stops in cDNA synthesis are visualized as autoradiogram bands, each of which corresponds to an affected position on the RNA template. The reactivity of a position is reflected by the band intensity on the autoradiogram, and the exact position of a modification or digestion can be determined by co-electrophoresing dideoxy-sequencing reactions (2). By comparing band intensities from free and complexed RNA probed in this manner, it is often possible to deduce the attachment site(s) for a ligand, since interacting sites exhibit altered, and generally reduced, reactivities.

2.2 Renaturation of RNA and 'activation' of complexes

Before free and complexed RNA are subjected to probing with chemicals and ribonucleases it is essential to ensure that a conformationally homogeneous population of molecules is studied. This important point is often neglected, but it is worthwhile to spend the small effort involved in carrying out a renaturation or 'activation'.

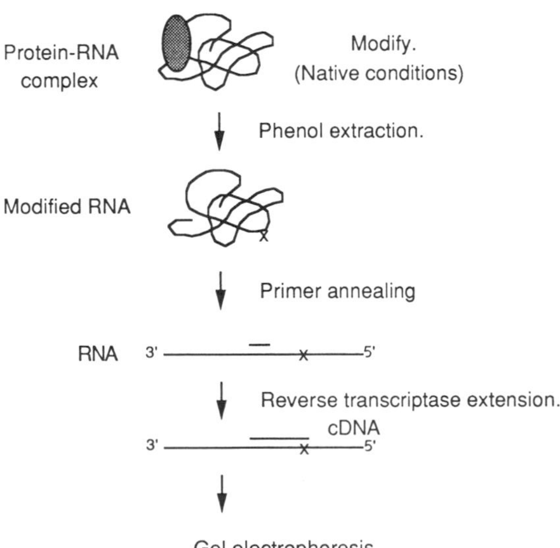

Figure 1. Schematic representation of the reverse transcriptase procedure used for chemical modification of a protein–RNA complex. The x marks a modified nucleotide.

Purified RNA has often been in contact with denaturants such as urea, SDS, EDTA, and guanidinium isothiocyanate, before it is deposited in the freezer. Therefore, a mixture of conformers will be present, so in order to obtain a 'native' conformation a heat treatment is necessary (3). The objective is to partially melt the RNA but to avoid a total denaturation that could trap the RNA in structures different from the thermodynamically most stable.

(a) Heat *E. coli* RNA at 50–60 °C for 10 min in an appropriate buffer (see Section 2.3) in the presence of at least 10 mM Mg^{2+} and 50 mM K^{+} (or NH_4^+) followed by slow cooling to room temperature. Renaturation of RNAs with a higher G+C content than those from *E. coli* require a higher temperature.

When it is desirable to obtain a specific protein–RNA complex by reconstitution from the components the optimal procedure will depend on the proteins involved, but it is critical to ensure specificity by using a high monovalent cation concentration. The risk of protein aggregation can be reduced by including a non-ionic detergent and remaining within the recommended temperature range.

(b) Cool purified RNA from step (a) to approximately 30 °C and adjust the KCl concentration to 300 mM. Add Nonidet NP-40 to 0.01% prior to a three-fold molar excess of *E. coli* ribosomal protein. Incubate for 40 min at 30–42 °C.

Complexes, that are isolated directly from cells, are usually in a 'native' conformation if Mg^{2+} and K^{+} are present throughout the purification. However, complexes are often stored frozen, so that an 'activation' step is highly recommended.

(c) Heat *E. coli* ribosomes or ribosomal subunits at 37–42 °C for 20 min in an appropriate buffer (see Section 2.3) containing 10 mM $MgCl_2$ and 100–270 mM KCl.

2.3 Chemicals and ribonucleases as probes

2.3.1 Chemicals

We routinely use five different chemical reagents as probes for RNA structure and complexes in the reverse transcriptase procedure. They are dimethyl sulphate (DMS), diethyl pyrocarbonate (DEP), Kethoxal, 1-cyclohexyl-3-[2-morpholinoethyl] carbodiimide metho-*p*-toluene sulphonate (CMCT), and 8-{[3-(4-methyl-1-piperazinyl)propyl]oxy}psoralen (4). *Figure 2* depicts the structures of the reagents and the base atoms which are attacked. The initial attacking atom of a particular probe is indicated by an asterisk. DMS, Kethoxal, and CMCT modify Watson–Crick pairing positions, so it is to be expected that these modifications cause reverse transcriptase to stop or pause. Although the psoralen reagent does not modify pairing positions, the cyclobutane adduct causes reverse transcriptase to stop due to a radical change of the uracil geometry (5). However,

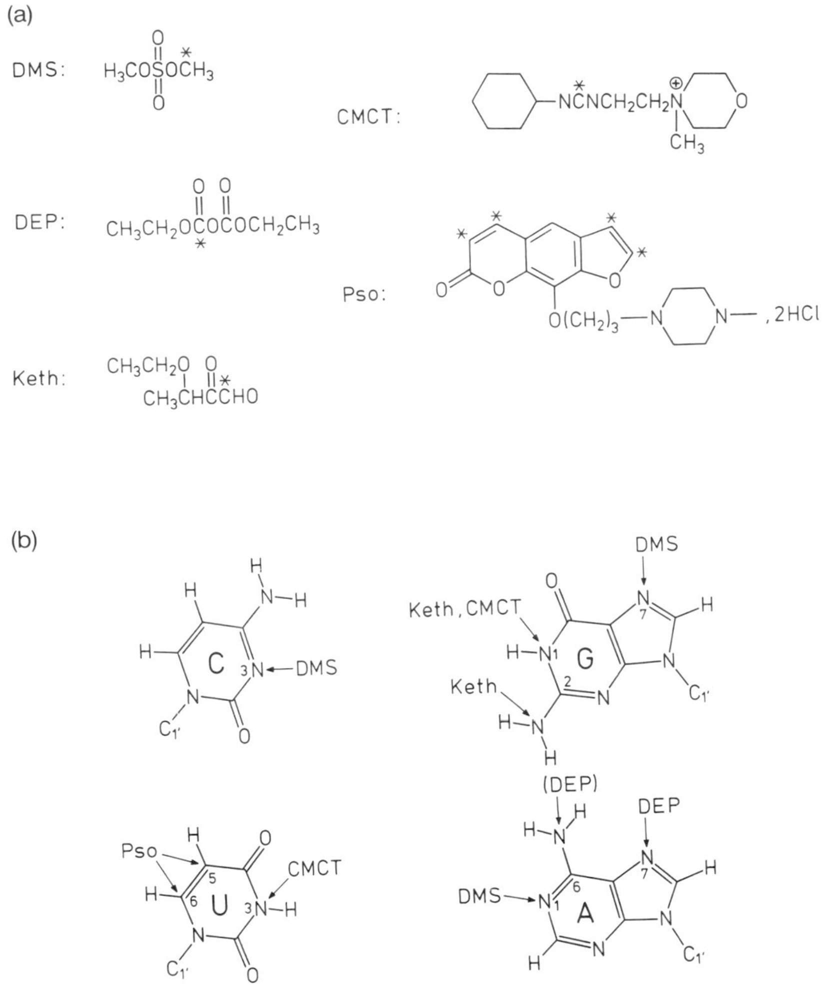

Figure 2. (a) Chemical formulae for the different modifying reagents. The reactive atoms are marked with asterisks. For psoralen, either the pyrone or the furan-ring reacts by creating a cyclobutane adduct with U(C5 and C6). (b) The modified atoms in the bases are marked by the appropriate reagent name. The order of reactivity of the different atoms with DMS is G(N7) > A(N1) > C(N3), and with CMCT is U(N3) > G(N1). DEP modification of A(N7) is a prerequisite for A(N6) modification (see text).

some latitude in base geometry is tolerated by the polymerase, because DMS-modified N-7 positions on guanine are not detected by reverse transcription (in order to monitor DMS-modified N-7 positions it is necessary to carry out a reduction with sodium borohydride followed by an acid-catalysed strand scission). This is worth keeping in mind when DMS is used to probe accessible N-1 positions on adenine and N-3 positions on cytosine, since the final gel electrophoretic analysis might indicate a limited extent of modification, and thus obscure extensive modification of guanine N-7 positions that could lead to secondary effects. In general, problems due to secondary effects are less pronounced with chemical probes than with ribonuclease probes where strand scission takes place. DEP-modified N-7 positions on adenine are detectable by reverse transcription. This may reflect either the bulkiness of the carbethoxy group or that the DEP causes spontaneous hydrolysis of the imidazole ring. However, another explanation is that subsequent to the adenine N-7 reaction the exocyclic N-6 position is modified and this blocks the reverse transcriptase.

DMS, DEP, Kethoxal, and CMCT all monitor unpaired bases and are therefore probes for RNA loop regions. RNA double-helical regions are often irregular, and such irregularities can also be detected by the chemical probes. Besides unpaired bases, DMS can be used to detect hydrogen-bonded adenines in the *syn* conformation which occur in G·A and Hoogsteen A·U pairings. Moreover, Kethoxal will often exhibit reactivity towards G-C pairs at the end of helices. In contrast to the above mentioned chemicals, the water-soluble psoralen derivative reacts strongly with uracils at or near internal loops, but not inside regular double helices and rarely in hairpin loops (6). The psoralen mono-addition is a two-step process: intercalation followed by near UV-induced cyclobutane formation, so the probe recognizes flexible RNA regions which can undergo the necessary axial rise per residue.

The detailed experimental protocols for employing the chemical reagents in studies of free and complexed RNA are provided in *Protocols 1* to *5*. Step 6 can be omitted if naked RNA is examined without a comparison with complexed RNA. Highly purified chemical probes should be used invariably, and in the studies of complexed RNA it is important to check the effect of the reagents on the stability of the complex, thus ensuring that extensive dissociation has not occurred. This is best done electrophoretically on SDS–polyacrylamide gels after separating protein–RNA complexes from free protein by either gel filtration for larger rRNAs (7) or gel electrophoresis for small rRNAs (8). In our experience, the main culprit in promoting dissociation is DEP which is an excellent protein modifying reagent. However, the use of DMS can also lead to complex breakdown, probably because of a pH decrease mediated by the invariable formation of some sulphuric acid. Therefore, we now employ Hepes as an universal buffer for chemical modifications since it has an optimal pK_a (7.5) and is insufficiently nucleophilic to interfere with the chemical reactions.

High concentrations of Mg^{2+} and K^+ are included to tighten the structure of free RNA and to provide specificity in studies of protein–RNA interactions. Protein–RNA complexes are most stable at 0 °C and it is preferable to perform reactions at this temperature; *Protocols 1* to *4* also provide alternative conditions at 30 °C since especially the Kethoxal reaction is sluggish at 0 °C. If it is desirable to use temperatures other than those given in *Protocols 1* to *4* a rough guide is to halve or double the reaction time for each increase or decrease of 10 °C, respectively.

Protocol 1. Modification with dimethylsulphate (DMS)

Caution: DMS (Aldrich–Gold brand label) is a strong methylating agent so all waste containing DMS should be inactivated in 10 M NaOH.

Steps for monitoring adenines and cytosines

1. Place 4 μg renatured RNA or complex in 200 μl 70 mM Hepes-KOH (pH 7.8), 10 mM $MgCl_2$, 270 mM KCl, 1 mM DTT on ice.
2. Add 1 μl of a freshly made 1:1 dilution of DMS in ethanol.
3. Incubate at 30 °C for 5 min (*or* at 0 °C for 20 min).
4. Terminate the methylation by adding 50 μl 1.0 M Tris–acetate (pH 7.5), 1.0 M 2-mercaptoethanol, 1.5 M sodium acetate, 0.1 mM EDTA and precipitate RNA with 750 μl ethanol (leave in a dry ice/ethanol bath for 30 min).
5. Centrifuge, and dissolve the pellet in 150 μl 0.3 M sodium acetate (pH 6.0).
6. Extract once with phenol, once with phenol/chloroform (1:1), and once with chloroform.
7. Precipitate with 2.5 volumes ethanol.
8. Centrifuge, wash with 80% ethanol, and dry the pellet in a vacuum desiccator.
9. Redissolve the dried pellet in 10 mM Tris–HCl (pH 7.5), 0.1 mM EDTA at a concentration of 0.5 pmol/μl. (1 A_{260} unit $\simeq$ 40 μg/ml.)

Additional steps for monitoring N-7 *positions on guanines*

9b. Redissolve the dried pellet from step 8 in 10 μl 1.0 M Tris–HCl (pH 8.2) containing 8 μg of methylated carrier RNA[a].

10. Add 10 μl of freshly prepared 0.2 M sodium borohydride and incubate at 0 °C for 30 min in the dark.
11. Terminate the reduction by adding 200 μl 0.6 M sodium acetate, 0.6 M acetic acid and precipitate RNA with 600 μl ethanol (leave in a dry ice/ethanol bath for 30 min).
12. Centrifuge, and reprecipitate from 200 μl 0.25 M sodium acetate (pH 6.0) with 500 μl ethanol.
13. Centrifuge, wash the pellet with 80% ethanol, and dry in a vacuum desiccator.
14. Induce strand scission by adding 20 μl aniline/acetic acid solution[b] and incubate for 20 min at 60 °C in the dark.
15. Lyophilize.
16. Add 20 μl water and lyophilize.
17. Repeat step 16.

18. Redissolve in 10 mM Tris–HCl (pH 7.5), 0.1 mM EDTA at a concentration of 0.5 pmol/μl (see step 9).

[a] Heavily methylated carrier RNA is prepared as follows: Dissolve 1 mg RNA carrier in 300 μl 50 mM sodium cacodylate (pH 5.5), 1 mM EDTA and add 5 μl DMS. Incubate for 5 min at 90 °C and chill on ice. Add 75 μl 1.0 M Tris–acetate (pH 7.5), 1.5 M sodium acetate, 1.0 M 2-mercaptoethanol and 900 μl ethanol. Centrifuge at 14 000 × g for 8 min and reprecipitate RNA from 0.3 M sodium acetate with ethanol. Wash the pellet with 80% ethanol, dry, and dissolve in water at a final concentration of 8 mg/ml.

[b] The aniline/acetic acid solution is prepared immediately before use. Add 30 μl redistilled aniline to 90 μl glacial acetic acid in 210 μl water.

Protocol 2. Modification with diethyl pyrocarbonate (DEP)

Caution: DEP (Eastman-Kodak) is a potent carbethoxylating agent so all waste containing DEP should be inactivated in 10 M NaOH.

1. Place 4 μg renatured RNA or complex in 200 μl 70 mM Hepes-KOH (pH 7.8), 10 mM $MgCl_2$, 270 mM KCl, 1 mM DTT on ice.

2. Add 5 μl DEP.

3. Incubate at 30 °C for 10 min (or at 0 °C for 30 min, shake after 15 min).

4. Terminate the carbethoxylation by adding 20 μl 3 M sodium acetate (pH 6.0) and precipitate the RNA with 660 μl ethanol (leave in a dry ice/ethanol bath for 30 min).

5–9. Follow steps 5–9 in *Protocol 1*.

Protocol 3. Modification with Kethoxal

Kethoxal (Serva) is normally supplied as a lyophilized compound that is not soluble in water. Dissolve Kethoxal in a small volume of ethanol and dilute with water until a 20% ethanol solution is obtained. The final concentration of Kethoxal should not exceed 40 mg/ml. Store this stock at 4 °C.

1. Place 4 μg renatured RNA or complex in 200 μl 70 mM Hepes–KOH (pH 7.8), 10 mM $MgCl_2$, 270 mM KCl, 1 mM DTT on ice.

2. Add 5 μl Kethoxal from the stock.

3. Incubate at 30 °C for 10 min (or use 20 μl Kethoxal stock at 0 °C for 60 min).

4. Terminate the reaction by adding 20 μl 3 M sodium acetate (pH 6.0), 10 μl 0.5 M boric acid[a], and 660 μl ethanol (leave in a dry ice/ethanol bath for 30 min).

5. Centrifuge, and dissolve the pellet in 150 μl 0.3 M sodium acetate, 50 mM boric acid[a] (pH 6.0).

6–8. Follow steps 6–8 in *Protocol 1*.

9. Redissolve the dried pellet in 50 mM potassium borate[a], 10 mM Tris–HCl, 0.1 mM EDTA (pH 7.5) at a concentration of 0.5 pmol/μl. (1 A_{260} unit $\simeq$ 40 μg/ml.)

[a] The presence of borate stabilizes the tricyclic adduct formed between Kethoxal and guanine.

Protocol 4. Modification with carbodiimide (CMCT)

CMCT (Sigma) is stored as a 42 mg/ml stock in 70 mM Hepes–KOH (pH 7.8), 10 mM $MgCl_2$, 270 mM KCl at -20 °C.[a]

1. Place 4 μg of renatured RNA or complex in 20 μl 70 mM Hepes–KOH (pH 7.8), 10 mM $MgCl_2$, 270 mM KCl, 1 mM DTT on ice.

2. Add 20 μl CMCT from the stock.

3. Incubate at 30 °C for 20 min (or at 0 °C for 90 min).

4. Terminate the reaction by adding 100 μl 0.3 M sodium acetate (pH 6.0) and 420 μl ethanol (leave in a dry ice/ethanol bath for 30 min).

5–9. Follow steps 5–9 in *Protocol 1*.

[a] The CMCT reaction is critically dependent on pH that should not be lower than 7.8.

Protocol 5. Modification with a water-soluble psoralen derivative

Many psoralens are only soluble in organic solvents such as DMSO, the presence of which destabilizes RNA complexes. However, psoralens derivatized in the 8-position with a positively charged group are water-soluble and they are useful photoaddition reagents. The derivative employed here[a] can be stored as a 20 mM stock in water.

1. Place 4 μg renatured RNA or complex in 20 μl 70 mM Hepes–KOH (pH 7.8), 10 mM $MgCl_2$, 270 mM KCl, 1 mM DTT on ice.

2. Add the psoralen derivative to a final concentration of 0.2 mM.

3. Irradiate with long-wave UV light (320–400 nm) through a filter, composed of Pyrex glass and acetone, for 15 min at 0 °C (the optimal time of irradiation depends on the luminosity of the chosen mercury lamp).

4. Add 50 μl 0.3 M sodium acetate (pH 6.0) and precipitate the RNA with 210 μl ethanol (leave in a dry ice/ethanol bath for 30 min).

5–9. Follow steps 5–9 in *Protocol 1*.

[a] 8-{[3-(4-methyl-1-piperazinyl)propyl]oxy}psoralen dihydrochloride can be obtained at a nominal fee from Professor O. Buchardt, Chemical Laboratory II, H. C. Ørsted Institute, DK-2100 Copenhagen Ø, Denmark.

2.3.2 Ribonucleases

Ribonuclease probes generate breaks in the ribose-phosphate backbone that lead to termination of reverse transcriptase. In studies of free and complexed RNA we routinely employ RNases T1 and T2 and the ribonuclease from cobra venom (CV) (9).

All the ribonucleases perform well in a ribosomal reconstitution buffer (30 mM Tris–HCl (pH 7.8), 20 mM $MgCl_2$, 300 mM KCl, 1 mM DTT; TMKD) at 0 °C. *Protocol 6* lists the conditions necessary for obtaining the appropriate partial digests for primer extension analysis. RNase T1 (G-specific) and T2 (low specificity with preference for A) exhibit strong reactivities with nucleotides at the apex of terminal loops while their reactivity with internal loops is substantially weaker. The mechanism of recognition of RNase T1 is known at atomic resolution, and the requirement for the less favourable *syn* conformation of guanosine explains the strong preference for terminal loops (10). In contrast, RNase CV cuts helical RNA regions with no apparent sequence specificity. Its mechanism of recognition remains unknown. The RNase CV employed in our studies was isolated and provided by Vassilenko and co-workers (9), and appears to attack double helices primarily, while the commercially available RNase V1 (Pharmacia) from cobra venom may have an additional strong specificity for helical single strands (11).

Protocol 6. Probing with ribonucleases T1, T2, and CV (or V1)

1. Place 4 μg renatured RNA or complex in a total volume of 20 μl 30 mM Tris–HCl (pH 7.8), 20 mM $MgCl_2$, 300 mM KCl, 1 mM DTT on ice.
2. Add 0.01 unit RNase T1 (Sankyo), *or* 0.05 unit RNase T2 (Sankyo), *or* 0.16 unit RNase CV (9) or V1 (Pharmacia).
3. Incubate at 0 °C for 30 min.
4. Stop the digestion by adding 130 μl 0.3 M sodium acetate (pH 6.0) and extract with 130 μl phenol.
5. Re-extract the aqueous phase with an additional 130 μl phenol.
6. Extract the aqueous phase with an equal volume of phenol/chloroform (1:1) followed by extraction with an equal volume of chloroform.
7. Precipitate the RNA with 2.5 volumes of ethanol.
8. Centrifuge, and wash the pellet in 80% ethanol.
9. Dry the pellet in a vacuum desiccator.
10. Redissolve in 10 mM Tris–HCl (pH 7.5), 0.1 mM EDTA at a concentration of 0.5 pmol/μl. (1 A_{260} unit $\simeq$ 40 μg/ml).

Ribonuclease probes have the advantage that they will only attack the RNA component in the complexes. However, commercial preparations of these enzymes often exhibit protease activity, so it is essential to establish the integrity of the protein component in a complex after ribonuclease treatment. This can best be done by purifying the complex and then electrophoresing in SDS–polyacrylamide gels (7, 8). The drawbacks with ribonuclease probes are their size and their propensity to cause secondary effects due to the primary event of strand scission. Regardless of the type of probe employed, it is important to use conditions given in *Protocol 6* that produce single cuts per molecule in order to minimize secondary effects. Secondary effects are a greater problem with single-strand-specific RNases than with RNase CV. A good rule of thumb is to obtain gel analysis tracks where more than 50% of the RNA molecules are unaffected, i.e. that at least 50% of the radioactive counts are in the band containing the intact RNA molecule.

We cannot recommend the use of single-strand-specific enzymes such as RNase S1 and the adenosine-specific RNase U2 for studying complexes due to their low pH optima. Moreover, owing to the high affinity of RNase A for pyrimidine–adenosine linkages, this enzyme tends also to cut in double-helical regions and is therefore a bad choice for establishing true single-stranded regions. Our experience with the cytidine-specific ribonuclease from chicken liver (CL3) is that it is also unsuitable due to its low activity in TMKD buffer. The purine-specific α-sarcin enzyme cuts both helices and loops and it is potentially useful as a universal probe with a minimum of 'blind spots' (i.e. regions of the molecule that are unreactive with all the probes) in studies of complexes (12). Its major drawback, however, is its inhibition by Mg^{2+} which is required for the stability of many complexes.

2.4 Primer extension

Analysis of the positions in the RNA attacked by chemical and enzymatic probes is performed by annealing a complementary DNA primer to the 3′ end of the molecule and extending with reverse transcriptase. If the target under study is large it is necessary to use additional primers typically spaced at a distance of about 300 nucleotides to cover the whole RNA molecule (see also Chapter 7). The spacing between primers is governed by the resolving power of the final gel electrophoretic analysis and does therefore depend on the choice of isotope (^{32}P generates broader autoradiographic images of bands than does ^{35}S). The primers are usually synthetic 17–20 mers of DNA that give sufficiently high priming specificity at primer:RNA ratios of about 1 (see Chapter 7 for further details). General considerations regarding the design of primers, such as G+C content, the number of G and C residues at the 3′ terminus, and the absence of competing sites in the target, all apply to the primer extension by reverse transcriptase.

The radioactive label can be introduced into the reverse transcript in two ways, namely by the use of a 5′ end-labelled primer, or by the use of an α-labelled deoxynucleoside triphosphate in combination with an unlabelled primer. The former procedure results in autoradiogram bands originating from fragments of identical specific activity and eliminates artefacts due to self-priming of the RNA template. However, since the reverse transcriptase is terminated at nicks in the template and slowed down by certain RNA structural features, as revealed in control tracks (untreated samples), the intensity of bands originating from long fragments does not represent the number of corresponding RNA templates. This situation is exacerbated at low primer : RNA ratios where preferential annealing to short templates occurs. Labelling of reverse transcripts with an α-labelled deoxynucleoside triphosphate compensates for premature termination of reverse transcriptase since large fragments have higher specific activity than shorter ones. However, this procedure can generate artefactual bands due to self-priming. Regardless of the labelling method employed, caution must be exercised in comparing band intensities within one track. Comparisons of intensities should, therefore, be restricted to bands of identical size in adjacent tracks. Fortunately, this is the main objective in studies of RNA complexes, where positions of protection and enhancement are examined.

Protocol 7 contains detailed procedures for primer extension analysis of modified or digested RNA using a 5′ ^{32}P-labelled primer, or labelled dATP and an unlabelled primer. The use of ^{35}S, in combination with electrophoresis in 80 cm long wedge-shaped (0.2 mm–0.7 mm thick) urea gels, allows the analysis of the modification or digestion pattern up to 500 nucleotides from the primer, thus enabling a molecule such as 16S rRNA to be analysed with three primers.

Protocol 7. Primer extension of modified or digested RNA

Annealing

1. Mix 0.5 pmol RNA and 0.5 pmol primer (if a 5′ ^{32}P-labelled primer is used, the specific activity should be about 20 Ci/mmol) in a total volume of 6 μl 10 mM Tris–HCl (pH 6.9), 40 mM KCl, 0.5 mM EDTA.
2. Heat for 30 sec at 95 °C, transfer to a 50 °C water bath, and leave for 20 min.
3. Spin down evaporated drops in a microfuge.

Extension from a 5′end-labelled primer

4. Add 4 μl of an extension mixture made up of 0.4 μl 25 × RT buffer[a], 0.8 μl 2.5 mM of each dNTP, 2.8 μl water and 1 unit AMV reverse transcriptase (Life Sciences) to the annealed primer/template solution.
5. Incubate at 37–42 °C for 30 min.
6. Stop the extension by adding 40 μl 0.3 M sodium acetate (pH 6.0) and precipitate RNA with 125 μl ethanol.
7. Centrifuge, wash the pellet in 80% ethanol, and dry in a vacuum desiccator.
8. Redissolve the dried pellet in loading solution[b] (an approximate volume is 1 μl per 100 c.p.s. as detected by a Mini-Monitor).

9. Denature samples by heating at 95 °C for 3 min and chill on ice.
10. Load approximately 1 μl on a sequencing gel.

Extension from an unlabelled primer

4. Dispense 3 μl of the annealed primer/template solution into a well in a microtitre plate (e.g., no. 163118, Nunc).
5. Add 2 μl of an extension mixture made up of 0.2 μl 25 × RT buffer[a], 0.2 μl α-[^{35}S]dATP (1000 Ci/mmol; 10 mCi/ml) *or* 0.2 μl α-[^{32}P]dATP (3000 Ci/mmol; 10 mCi/ml), 1.6 μl 0.75 mM dGTP, dCTP, and dTTP, and 1 unit AMV reverse transcriptase to the well.
6. Incubate at 37–42 °C for 15 min.
7. Chase with 2 μl 0.75 mM dATP for 15 min at 37–42 °C.
8. Stop the extension by adding 5 μl load solution.[b]
9. Denature samples by heating at 95 °C for 3 min and chill on ice.
10. Load 1 μl on a sequencing gel.

[a] 25 × RT buffer is 1.25 M Tris–HCl (pH 8.4), 250 mM $MgCl_2$, 50 mM DTT.

[b] Load solution is 80% deionized formamide, 10 mM NaOH, 1 mM EDTA, 0.05% xylene cyanol, 0.05% bromophenol blue.

Establishing which nucleotides have reacted in free RNA and complexes is usually straightforward if sequencing tracks are co-electrophoresed. However, the unequivocal identification of a modified site at a distance 500 nucleotides away from the primer depends heavily on the quality of the dideoxynucleotide sequencing tracks. We recommend the use of single-stranded DNA as the template for generating sequencing tracks by a polymerase such as the Klenow fragment, modified T7 DNA polymerase, or Taq DNA polymerase, instead of using untreated RNA as the template for reverse transcription. Sequencing tracks obtained by using dideoxynucleoside triphosphates are displaced by one nucleotide relative to the modified positions, since the last nucleotide incorporated by reverse transcriptase is complementary to the one on the 3′ side of the modified position in the RNA template.

A picture of a binding site emerges when the 'fragmentation' patterns of the free and complexed RNA are compared for the various probes. Usually, multiple protection effects occur with occasional enhancements. Interpretation of the protection of a ribonuclease cut is less straightforward than for chemical modification owing to the sheer bulk of the ribonucleases. However, free RNA is more flexible conformationally than complexed RNA and some protection effects will result from ligand-induced tightening of conformation; enhanced reactivities may reflect an increase in local conformational homogeneity. Most probes exhibit single-strand specificity and the data inevitably have a bias towards loop regions. It is important, therefore, not to dismiss double-helical regions that are 'blind spots' as necessarily being inaccessible.

2.5 An illustration of an attachment site for a protein

Figure 3 shows an autoradiogram of a reverse transcript that was produced using an unlabelled primer (complementary to positions 417–433 in *E. coli* 23S rRNA) and α-[^{35}S]dATP. The depicted region, between positions 284 and 392, contains an attachment site for the primary binding protein L24 (13). This was identified by probing free 23S rRNA and the L24–23S rRNA complex, in parallel, with chemicals (Kethoxal, DMS, DEP, CMCT) and ribonucleases (T1, T2, CV).

The importance of including untreated samples (preferably from both free RNA and complex) is illustrated in *Figure 3*. The predominant 'control cuts' occur at phosphodiester linkages U358–G359 and U365–C366 but weaker ones are also visible; they all appear as bands across the autoradiogram. The positions

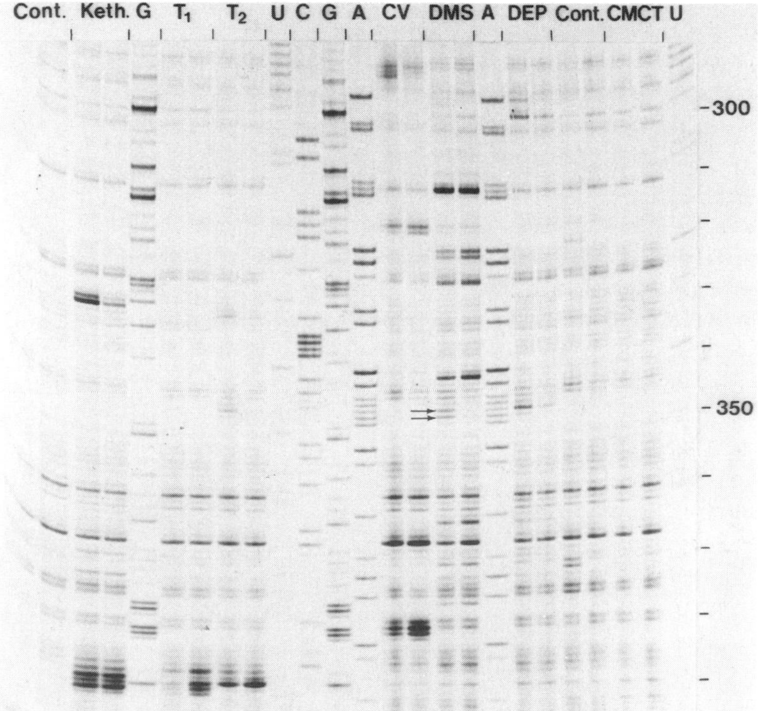

Figure 3. Reverse transcriptase results obtained with chemically and ribonuclease treated samples. A comparison of the data for free 23S rRNA and the L24–23S rRNA complex between nucleotides A_{284}–U_{392}. For each probed, and untreated sample (Cont), reverse transcripts are shown for both the free RNA (left side) and the complex (right side). Dideoxy-sequencing tracks (U,C,G,A) are included. As an example, two bands are indicated (by arrows) in the DMS-modified 23S RNA track that line up with A_{346} and A_{347} in the sequence track. Since each band produced by a modified nucleotide is displaced by one nucleotide from the corresponding sequence band, the modified nucleotides correspond to A_{345} and A_{346}, respectively. In the adjacent DMS track, originating from the modified L24–23S RNA complex, no bands occur at the same site and, therefore, the nucleotides are protected in the protein–RNA complex.

corresponding to 'control cuts' are eliminated from the analysis, thus reducing the information obtainable.Therefore, it is desirable to keep the number and the level of 'control cuts' at a minimum by not creating 'nicks' in the RNA through radiolysis or nuclease degradation during the preparation and analysis. However, 'control cuts' cannot be eliminated and some probably reflect the difficulty that the reverse transcriptase has in traversing certain RNA secondary/tertiary structural features.

In the present example, the aim is to map a protein-binding site on RNA rather than to deduce a putative secondary structure that is already known. Nevertheless, the probing data for free RNA, as seen on the autoradiogram, reinforce the secondary structural model in *Figure 4*. Such verification is not

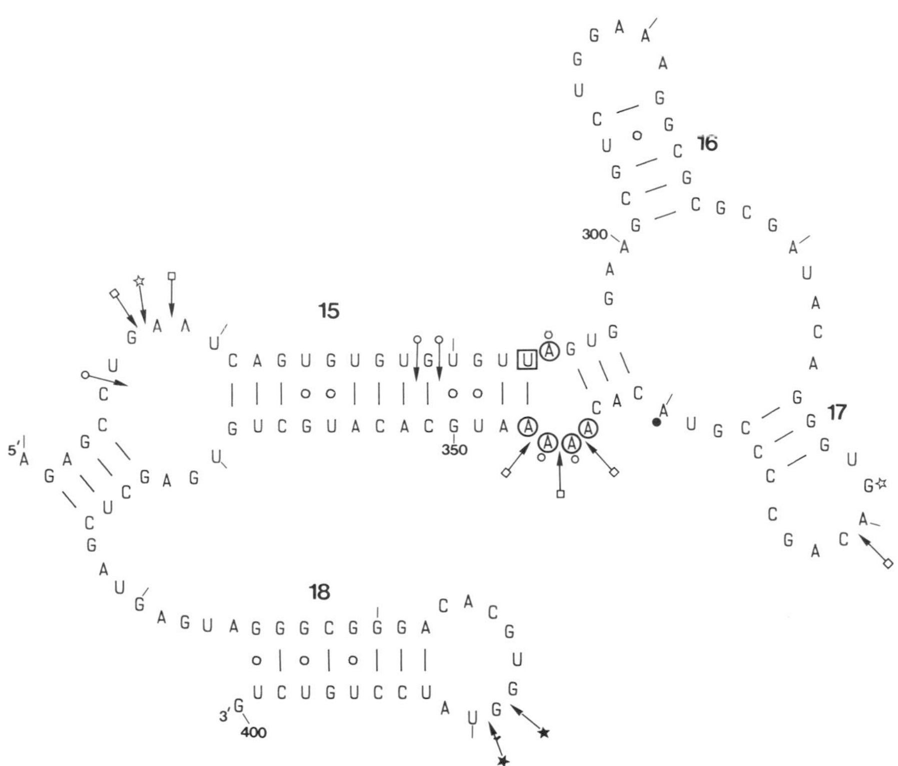

Figure 4. Protection and enhancement of ribonuclease and chemical reactivities caused by L24. The data are superimposed on the putative secondary structure of fragment A_{270}–G_{400} in domain I of *E. coli* 23S rRNA. Helix numbers are indicated by bold type. The open symbols indicate protection while filled symbols denote enhancement: (○) DMS; (◯) DEP; (□) Psoralen (pso); (☆) Kethoxal (keth); (☆→) Ribonuclease T1; (□→) T2; (○→) CV. The two DMS positions shown in *Figure 3* are indicated at A_{345} and A_{346} by small circles. Several chemical protection effects are clustered in the internal loop of helix 15 and they are indicative of a direct interaction site. The ribonuclease protections at nucleotides 275–278 probably reflect steric effects resulting from the large sizes of the enzymes.

always a matter of course. Occasionally the secondary structure deriving from phylogenetic evidence (see Section 3.2) disagrees with the thermodynamically preferred structure obtained by probing naked RNA (1).

Figure 4 illustrates the secondary structure of the part of 23S rRNA analysed on the autoradiogram. In order to identify an attachment site it is preferable to observe a clustering of protection effects for both chemical and enzymatic probes. This requirement is fulfilled by the internal loop in helix 15 which is, therefore, a strong candidate for such a binding site. Clustering of chemical protection effects alone is also indicative of an attachment site, but it is difficult to interpret ribonuclease protection effects. This is mainly due to the size of the enzymes. It is especially important to interpret apparent protection effects with caution when RNase CV is the probe, since this enzyme is very sensitive to small changes in helical geometry. In the same vein, the isolated effects seen in terminal loop 17 are probably the result of a local conformational change or a protein-induced tertiary interaction, rather than direct protein contacts.

Although the analysis of protected positions is the main objective of a typical experiment, enhanced or new reactivities can give clues to protein-induced conformational changes. In the experiment illustrated, terminal loop 18 exhibits enhanced reactivities towards RNase T1. Since the loop is displaced from the attachment site in the internal loop in helix 15, it is feasible that an additional attachment site for L24 is located elsewhere in the molecule, thereby changing the conformation of terminal loop 18. In fact, an additional site can be identified around position 500 of 23S rRNA (not shown on the autoradiogram and secondary structure), so the enhancements can in a crude way be rationalized in terms of a 'sandwich' effect; L24 binds to two distinct sites thus inducing conformational changes in the linker region (13).

2.6 Alternative probing and analysis strategies

In this chapter, we have provided detailed experimental protocols for carrying out structural analysis of free and complexed RNA by primer extension, because we regard this procedure as being the most useful that can be applied to any RNA molecule. Historically, other procedures have been employed in probing studies, and some are still useful for analysing RNA regions that are unsuitable for primer extension analysis. In the following, alternative strategies are briefly outlined with emphasis on where they can be used to advantage.

2.6.1 End-labelling methods

The methods are best suited to small RNAs (less than 300 nucleotides) that have at least one homogeneous end, and these methods will complement the primer extension procedure at the 3′ end where information is invariably lost due to primer annealing. However, the techniques have also been applied successfully to the analysis of protein–RNA fragment complexes isolated from large RNAs (14). Detailed experimental protocols for carrying out the end-labellings can be found in a previous volume in this series (15). A requirement for the end-labelling

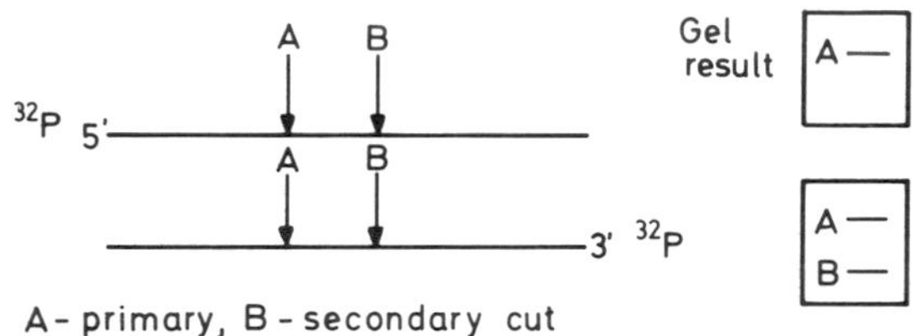

Figure 5. A primary cut (A) is distinguished from a secondary cut (B) by the presence of the former in autoradiograms of both 5′ and 3′ end-labelled RNA while the latter is always absent from one of the end-labelled samples. Thus, in the example shown, the secondary cut is only seen when it falls between the end label and the primary cut.

methods, when used with chemical reagents, is that the initial modification can be converted quantitatively into a strand scission. Therefore, purines can only be monitored with respect to their N-7 positions on 3′ end-labelled species. Most laboratories use only one type of end-labelling but it should be emphasized that the parallel use of 3′ and 5′ end-labelling is a powerful way of distinguishing between the primary and secondary effects which always occur in enzymatic probing experiments. This principle is illustrated diagrammatically in *Figure 5*.

2.6.2 Modification-selection

This procedure is an appropriate option if RNA–protein complexes dissociate during chemical treatment, provided a specific complex can be formed by reconstitution and the complex can be separated from free RNA. The RNA component is modified, complexed, and then the distribution of modified nucleotides between the free and the complexed RNA is analysed by primer extension or end-labelling (16). If RNA molecules exhibiting a particular modified base are excluded from the complex, then it is inferred that this nucleotide is involved, directly or indirectly, in binding. When renatured RNA is modified the structural interpretation is fairly straightforward, but difficulties arise when denatured RNA is modified and then renatured, since many modifications will be selected against simply because they inhibit renaturation of the RNA structure.

2.6.3 Hybridization method

This method involves hybridizing a restriction fragment of DNA to an RNA target after it has been modified in the free or complexed state (17). Non-hybridized RNA is then removed by ribonuclease treatment and the hybridized RNA is subsequently isolated, end-labelled, and analysed as described above (Section 2.6.1). The method can be applied to the analysis of internal RNA regions if these cannot be traversed by reverse transcriptase, since it is the only procedure whereby intact internal regions can be analysed by the end-labelling method. However, like the modification–selection method, it is only applicable to chemically modified samples, and it is also very laborious, although there are fewer problems with control bands than occur with the reverse transcriptase approach.

3. Theoretical approaches to higher-order RNA structures

In the following sections some guidelines are given for obtaining secondary structures from one or more homologous RNA sequences that involve methods of prediction and sequence comparison without mathematical formulation. Analysis of RNA sequences can be done either manually or automatically, but the speed of the computer may bring the 'impossible' within the reach of the possible. We cannot review here all the available software, but representative, affordable programs, and their functions, are listed in Table 1. Moreover, readers are referred to an earlier book in this series dealing with more general methods for sequence analysis using computers (18).

Table 1. Examples of inexpensive programs for RNA structure analysis.

Program	Source
Folding of a single sequence	(21), (24)
Dot matrix comparison of two sequences	(27)
Motif search	(30)
Automatic multiple sequence alignment	(28)
Sequence editing and automatic alignment	Thirup and Larsen, in preparation[a]
Search for compensating base changes	N. Larsen, unpublished[b]
Edit and plot of secondary structure	(18), N. Larsen, unpublished[b]
3D modelling	(40)

[a] Contact Søren Thirup, soren@kemi.aau.dk. Postal address: Biostrukturkemi, Kemisk Institut, Aarhus Universitet, DK-8000 Aarhus C, Denmark.

[b] The software is available from the author upon request provided the expenses are covered.

3.1 Structure prediction from a single RNA sequence

Single RNA sequences are available on magnetic tape (and single entries via electronic networks) at nominal cost from EMBL in Europe (19). GenBank (20) in the USA distributes a similar database on magnetic tape as well as on diskettes.

Two general approaches have been used for predicting RNA secondary structures. In the first, the overall free energy is minimized by adding contributions from each base pair, each bulged base, loops and so on. Free energy values were obtained from melting points of synthesized oligonucleotides of known sequence and structure, and are listed in tables which are also computer readable (e.g. 21). This approach is now commonly used, and it is available in many low-cost applications. Experimental data may be used for constraining folding, and recent developments include improved free energy parameters and computer time reduction. A second more empirical approach involves searching for the combination of non-exclusive helices with a maximum number of base pairing (22). Both approaches are reviewed in (18) and (23).

There are, however, major objections to both of these approaches: the computer time needed is still proportional to the third power of sequence length, which makes folding of more than about 1000 nucleotides very time consuming. More seriously, their predicting ability is poor. This is illustrated for *E. coli* 5S rRNA in *Figure 6*, where the prediction from Zuker's (24) most recent implementation of the free energy method (*Figure 6A*) is compared with its actual secondary structure (*Figure 6B*) derived from phylogenetic sequence comparisons (see Section 3.2) and experimental data (25). Although the number of base pairs in the two structures are similar (40 and 34 respectively), only helix I is common. Obviously, the structure is influenced by additional factors such as ions, solvents, and proteins, and their effects remain to be quantified. The performance does not improve for larger RNAs, as illustrated by *E. coli* 16S rRNA, where the 'thermodynamically most stable helix' does not form (26). Therefore, use of this type of program is not recommended as a basis for establishing models. However, a map of possible pairings could facilitate manual selection of a model that is in accordance with experimental data.

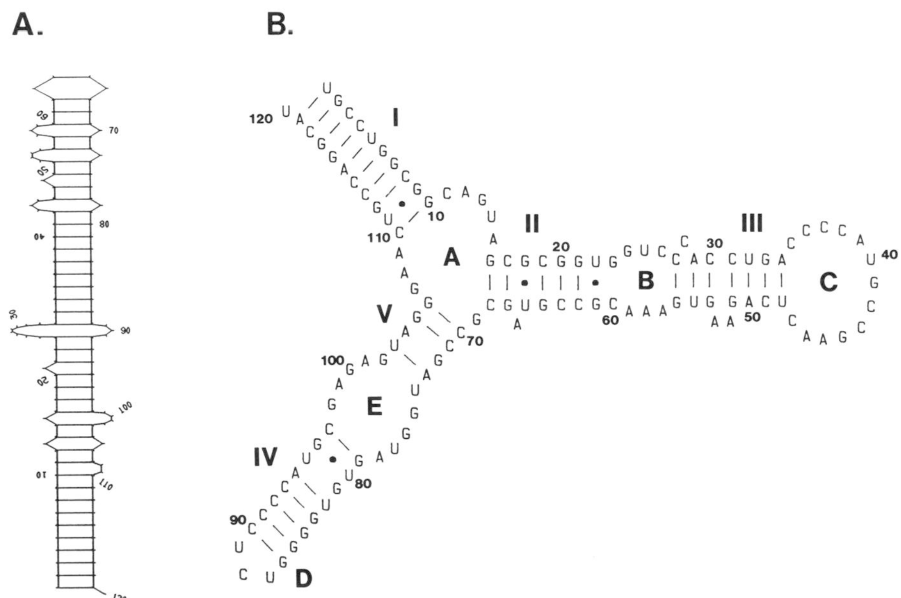

Figure 6. Alternative secondary structures for *E. coli* 5S rRNA. A. Output from Zuker's FOLD prediction program included in the UWGCG sequence analysis package (24) version 5.3 (released in July 1988). It was generated in 25 sec of MicroVax II computer time. B. Most recent structure based on comparative sequence analysis and experimental data (25).

3.2 Structure prediction by comparative sequence analysis

When more than one sequence is available, the phylogenetic sequence comparison approach can be used. The underlying assumption of this method is that molecules with different sequences, which perform similar functions, will also have similar higher-order structures. It requires the initial alignment of the available set of sequences. Homologous sequences can generally be aligned by sequence similarity, particularly if they have comparable lengths. The degree of similarity may be checked initially by quick alignment routines or dot matrix programs. The latter type of program graphically displays all significant similarities as diagonal lines in a rectangle with two sequences along the axes, as well as internal repeats (27).

3.2.1 Primary structure alignment

When significant primary structure similarity exists between several sequences, the sequences should be grouped in order of relatedness, if possible, since close relatives are easier to align, and they may reveal common inserts or gaps. Thus, regions of obvious sequence similarity among groups of relatives should be aligned first, then the groups themselves should be ordered together. In the more variable regions, sets of highly conserved nucleotides can often be identified and used as markers: for example, if five out of 10 sequences show the consensus sequence AA---U---GAA-AC--G, the remaining five will often exhibit a closely related pattern. Computer programs which exploit this phenomenon (called pattern matching) are available (e.g., 28); they are also fast enough in practice and are initially useful. A basically different class of alignment programs are based on the Needleman–Wunsch algorithm (29). These are not to be recommended, since they have inherent drawbacks in practice. Firstly, when sequences are long (>1000), or when three or more are aligned, almost all the available programs are too slow for practical use. Secondly, gaps are penalized, so that alignments with a number of smaller gaps are created (to maximize similarity) rather than single major insertions or deletions, which are more likely to have occurred at specific points during evolution. Especially in regions of little similarity, computer time is wasted, and incorrect alignments are frequently produced.

We favour strongly a manual refinement of the alignment at the primary structure level with the help of a dedicated sequence editor which can, for example, insert many gaps simultaneously. Recently, such tools have become available at low cost, with and without interplay between manual and automatic primary structure alignment. Finally, we emphasize the importance of taking the time necessary to prepare a correct alignment.

3.2.2 Secondary structure alignment

If a set of sequences share no obvious similarity at the primary structure level, a direct search for common secondary structural elements should be attempted, rather than attempting to align full-length sequences. This can be done by using a motif search program (30), where both primary and secondary structure patterns may be investigated in a quite flexible manner. Manual structure assembly, using exhaustive lists of potential helices for every sequence, is also possible. If regions of little similarity interrupt an otherwise well established alignment, common secondary structural elements should be used as additional alignment markers. Computer methods for aligning according to secondary structure alone are under development (23).

3.2.3 The use of sequence alignments in structure prediction

Historically, this procedure in its simplest form has revealed correct and fairly detailed structures, and it relies on evidence for compensatory base changes (CBC's). In 1969 an essentially correct tRNA structure including tertiary base pairs was predicted from just 14 sequences (35), and later minimal secondary structural models of 5S rRNA (25 and references therein), 16S rRNA (36 and references therein) and 23S rRNA (37 and references therein) were derived by independent groups. Tertiary interactions involving base pairing have also been proposed (36, 38). The structural predictions which were based primarily on alignment information required a computer program for analysis of groups of aligned nucleotides. Certain RNA alignments are now freely available: tRNAs (31) are distributed on tapes or electronic networks, 5S rRNAs (32) on diskettes, 16S rRNAs (34) on magnetic tape and diskettes, and small nuclear RNAs among others (33) at least as hardcopy.

3.2.4 The sequence comparison method

When a mutation in a Watson–Crick base pair creates a mismatch, which is then converted to another canonical base pair by a second mutation, the initial mismatch is said to be compensated for by the second change. Thus when sequences from different organisms are compared, and different base pairs are observed at equivalent positions, this is taken as evidence that they have arisen by the above process during evolution. Assuming that the RNA molecules are similar with respect to function and, therefore, to their higher-order structure, such patterns also strongly support the existence of the base pair. The more sequences and CBC's available, the stronger the positive evidence or phylogenetic proof. Non-Watson–Crick juxtapositions are regarded as negative evidence, or disproof, except that G–U (U–G) changes are considered neutral. All alignment positions are compared two by two to examine the phylogenetic support for all base pairing possibilities in the organism of interest. In each comparison, the most frequent Watson–Crick pair should be identified first, and the remaining Watson–Crick pairs then counted as CBCs. Then count non-Watson–Crick pairs, except G–U (U–G), as mismatches, and calculate the ratio between the positive and negative evidence. If this ratio is better than for example 2 : 1, then regard the two alignment positions as a possible base pair candidate.

3.2.5 Enhancement of the method using computers

The following application of the method requires computers. Firstly, prior to a CBC search, a table of similarity percentages is calculated from the best-aligned sequence regions, and this is used to re-group the molecules according to their degree of relatedness, so that the sequence of interest is at the top with its most distant relative at the bottom of the grouping. Each CBC providing positive evidence is then weighted, for example, by multiplying by its percentage similarity to the top sequence. In this way, positive phylogenetic evidence is taken as most significant when it derives from close relatives and, if the evidence is negative, there is no need to look at more distant sequences. Weighting and the use of the ratio of positive to negative evidence for base pairs have the advantage that interactions in highly conserved regions are not as readily confused with 'statistical noise'. A second application of the method involves informing the computer of where ambiguous alignments occur using a +/− ruler as shown in *Figure* 7, where part of the 5.8S rRNA (sequence positions 20–120) is aligned for 36 eukaryotic organisms (39). This ruler distinguishes well- and less well-aligned regions. The former are denoted with a + and are fixed. The latter are denoted with a − and can be displaced laterally by an amount specified by the program. This is analogous to what one would do manually but it ensures an exhaustive and flexible search. A third application involves a search for CBC's which can easily be extended to a search for any co-ordinated changes between non-Watson–Crick pairs, which may also be of structural importance.

After the computer analysis is complete, each potential base pair should be tested by the following criteria:

(a) The sequence alignment is unambiguous;

(b) The changes are strictly compensatory and the ratio of positive to negative evidence is favourable;

(c) The experimental evidence is not conflicting. Additional support for an interaction would derive from the occurrence of CBC's at adjacent positions.

Deriving secondary structures, as described above, leads to a model that includes interactions for which there is positive evidence. However, if unproven base pairs are included in a model they should be indicated as such. An example is given for the 5.8S rRNA alignments shown in *Figure* 7. This reveals three potential base pairing regions 5/5′, 6/6′, and 7/7′. Clearly, the ratio of positive to negative evidence is unfavourable for the shaded helices 5 and 7, but favourable for helix 6, and also for the interactions between nucleotide pairs indicated by diamonds and squares, respectively.

Secondary structure diagrams may be drawn manually, or edited and plotted by computer programs (see 18). Finally, if tertiary interactions constrain the degrees of freedom sufficiently, model building in three dimensions can be attempted, which may permit clearer interpretation of chemical or enzymatic probing data. Low-cost modelling programs for RNA are available (40).

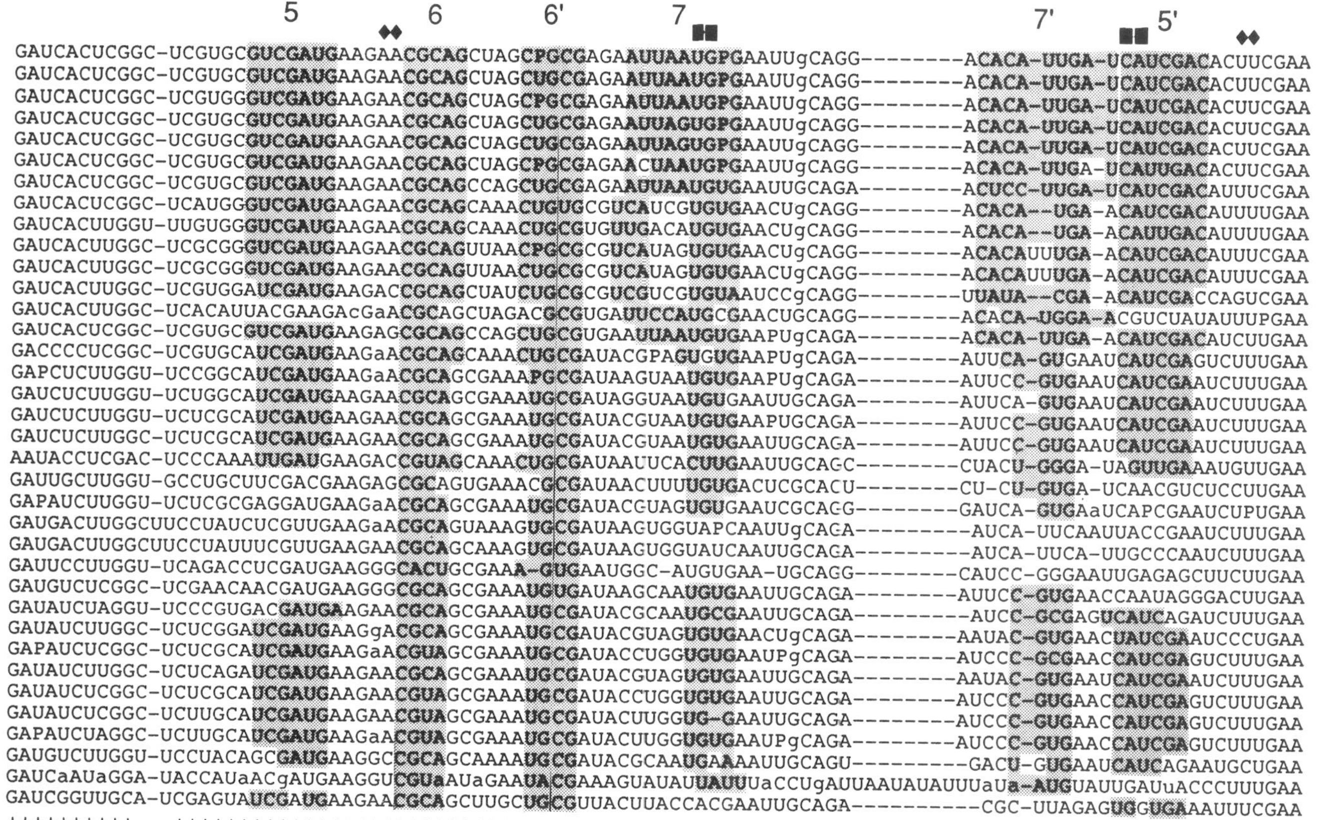

Figure 7. Section of a 5.8 S rRNA alignment (nucleotides 20–120, modified from reference 39). The regions shaded in grey denote areas that were base paired in secondary structure models presented with the sequences. Thus sequence 5 was paired with 5′, and so on. Examination of these putative interactions using the CBC criteria reveals that only helix 6/6′ [the helix numbering derives from the 5′-end of 23S rRNA (38)] is supported phylogenetically for all organisms. The +/− ruler is presented at the bottom; + denotes unambiguous alignment and − indicates regions where the alignment may be adjusted. Diamonds and squares indicate additional interactions that are supported by CBC's. Nucleotides denoted by small letters

3.3 Future developments

The capacity to predict a correct structure from a single sequence is an important goal, but the current methods need refining. A better understanding of the factors that influence RNA folding could help to achieve this. Comparative structure analysis by computer will become increasingly important in the future given the dramatically increasing availability of sequences, since the effectiveness of the approach increases with the number of sequences. From the rather simple core of the analysis described above, new and more sophisticated procedures are being developed. For example, extrapolations from tRNA crystal structures are possible. Given such detailed crystal structures and alignments, co-axially stacked helices, base triples, pseudoknots, or U-turn loops may be identified. The approach could also help to localize the regions within which functionally important movements occur. Ultimately, each step in such a procedure could be integrated by using parallel computers and high speed colour graphics, which would take the comparative approach into a new dimension.

3.4 Programs available

The guide to inexpensive programs presented in *Table 1* is by no means complete. Many of the programs occur in special issues of *Nucleic Acids Research*, **12**—1, **14**—1, and **16**—5, as well as in *Computer Applications in the Biosciences*, and in reference 18.

References

1. Moazed, D. Stern, S., and Noller, H. F. (1986). *J. Mol. Biol.*, **187**, 399.
2. Davies, R. W. (1982). In *Gel Electrophoresis of Nucleic Acids; a Practical Approach* (ed. D. Rickwood and B. D. Hames), p. 117. IRL Press Ltd., Oxford.
3. Ungewickell, E., Garrett, R. A., and Le Bret, M. (1977). *FEBS Lett.*, **84**, 37.
4. Hansen, J. B., Bjerring, P., Buchardt, O., Ebbesen, P., Kanstrup, A., Karup, G., Knudsen, P. H., Nielsen, P. E., Norden, B., and Ygge, B. (1985). *J. Med. Chem.*, **28**, 1001.
5. Kim, S.-H., Peckler, S., Graves, B., Kanne, D., Rapoport, M., and Hearst, J. E. (1982). *Cold Spring Harbor Symp. Quant. Biol.*, **47**, 361.
6. Christiansen, J. (1988). *Nucleic Acids Res.*, **16**, 7457.
7. Newberry, V., Brosius, J., and Garrett, R. A. (1978). *Nucleic Acids Res.*, **5**, 1753.
8. Newberry, V., Yaguchi, M., and Garrett, R. A. (1977). *Eur. J. Biochem.*, **76**, 51.
9. Vassilenko, S. and Babkina, V. (1965). *Biokhimiya (Moscow)*, **30**, 705.
10. Heinemann, U. and Saenger, W. (1982). *Nature*, **299**, 27.
11. Lowman, H. B. and Draper, D. E. (1986). *J. Biol. Chem.*, **261**, 5396.
12. Huber, P. and Wool, I. G. (1988). In *Methods in Enzymology*, Vol. 164 (ed. H. F. Noller and K. Moldave), p. 468. Academic Press, San Diego.
13. Egebjerg, J., Leffers, H., Christensen, A., Andersen, H., and Garrett, R. A. (1987). *J. Mol. Biol.*, **196**, 125.
14. Vester, B. and Garrett, R. A. (1984). *J. Mol. Biol.*, **179**, 431.

15. D'Alessio, J. M. (1982). In *Gel Electrophoresis of Nucleic Acids: a Practical Approach* (ed. D. Rickwood and B. D. Hames), p. 173. IRL Press Ltd., Oxford.
16. Peattie, D. A., Douthwaite, S., Garrett, R. A., and Noller, H. F. (1981). *Proc. Nat. Acad. Sci. USA*, **78**, 7331.
17. Van Stolk, B. J. and Noller, H. F. (1984). *J. Mol. Biol.*, **180**, 151.
18. Bishop, M. J. and Rawlings, C. J. (ed) (1987). *Nucleic and Protein Sequence Analysis—A Practical Approach.* IRL Press Ltd., Oxford.
19. Cameron, N. G. (1988). *Nucleic Acids Res.*, **16**, 1865.
20. Bilofsky, S. H. and Burks, C. (1988). *Nucleic Acids Res.*, **16**, 1861.
21. Zuker, M. and Stiegler, P. (1981). *Nucleic Acids Res.*, **9**, 133.
22. Nussinov, R., Pieczenik, G., Griggs, J. R., and Kleitman, D. J. (1978). *SIAM J. Appl. Math.*, **35**, 68.
23. Waterman, M. (1988). In *Methods in Enzymology*, Vol. 164 (ed. H. F. Noller and K. Moldave), p. 765. Academic Press, San Diego.
24. Devereux, J., Haeberli, P., and Smithies, O. (1984). *Nucleic Acids Res.*, **12**, 387.
25. Egebjerg, J., Christiansen, J., Brown, R. S., Larsen, N., and Garrett, R. A. (1989). *J. Mol. Biol.*, **206**, 651.
26. Noller, H. F. (1984). *Ann. Rev. Biochem.*, **53**, 119.
27. Staden, R. (1982). *Nucleic Acids Res.*, **10**, 2951.
28. Martinez, H. M. (1988). *Nucleic Acids Res.*, **16**, 1683.
29. Needleman, S. B. and Wunsch, J. (1970). *J. Mol. Biol.*, **48**, 44.
30. Staden, R. (1988). *Computer Appl. Biosci.*, **4**, 53.
31. Sprinzl, M., Moll, J., Meissner, F., and Hartmann, T. (1985). *Nucleic Acids Res.*, **13**, Suppl. 1.
32. Wolters, J. and Erdmann, V. A. (1988). *Nucleic Acids Res.*, **16**, Suppl. 1.
33. Reddy, R. (1988). *Nucleic Acids Res.*, **16**, Suppl. 71.
34. Dams, E., Hendriks, L., Van de Peer, Y., Neefs, J.-M., Smits, G., Vandenbempt, I., and De Wachter, R. (1988). *Nucleic Acids Res.*, **16**, Suppl. 87.
35. Levitt, M. (1969). *Nature*, **224**, 759.
36. Gutell, R. R., Weiser, B., Woese, C. R., and Noller, H. F. (1985). *Prog. Nucleic Acid Res. Mol. Biol.*, **32**, 155.
37. Gutell, R. R. and Fox, G. (1988). *Nucleic Acids Res.*, **16**, Sequences Suppl. 175.
38. Leffers, H., Kjems, J., Østergård, L., Larsen, N., and Garrett, R. A. (1987). *J. Mol. Biol.*, **195**, 43.
39. Erdmann, V. A. and Wolters, J. (1986). *Nucleic Acids Res.*, **14**, Suppl. 1.
40. Still, W. C., Richards, N. G. J., Guida, W. C., Lipton, M., Liskamp, R., Chang, G., and Hendrickson, T., *Macromodel v 2.0.* Department of Chemistry, Columbia University, New York, NY 10027.

12

Site-directed mutagenesis of *E. coli* rRNA

WILLIAM E. TAPPRICH, H. U. GÖRINGER, ELIZABETH A. DE STASIO, and ALBERT E. DAHLBERG

1. Introduction—mutagenesis of *E. coli* rRNA

Ribosomal RNA (rRNA) contributes both structural and functional activities to the ribosome. With the advances in biochemical and comparative phylogenetic analyses of rRNA it has been possible to define specific regions with particular structural and functional roles. It is now possible to investigate these regions of rRNA in greater detail by utilizing *in vitro* mutagenesis techniques.

Since *E. coli* contains seven operons coding for rRNA and direct mutagenesis of the genome would pose many problems, rRNA mutations are produced in plasmid-bourne copies of rDNA. Mutagenesis can be performed in several ways: deletions can be constructed at convenient restriction sites in the *rrn* operons, transition mutations can be made by bisulphite treatment of single stranded rDNA or specific nucleotides can be altered by oligonucleotide directed mutagenesis. Using these techniques it is possible to introduce mutations into any site in the 16S, 23S and 5S rRNAs of *E. coli*, to express and assemble the mutant rRNAs into ribosomes, and to study the structural and functional consequences of the mutations *in vivo* and *in vitro*. By using site-directed mutagenesis, the targets of study can be chosen such that important functional regions of the rRNA can be analysed in a base-by-base fashion. In this chapter we will describe methods for the construction, expression, and analysis of ribosomes containing mutations in the rRNA.

1.1 Enzymes

- DNA polymerase 1 Klenow fragment (New England Biolabs).
- Polynucleotide kinase (New England Biolabs).
- T4 DNA ligase (New England Biolabs).
- Lysozyme (Sigma).
- DNase 1 (RNase free) (Worthington Enzymes)
- AMV reverse transcriptase (Promega Biotech).

1.2 Reagents

- M13mp18, 19 replicative form (RF) DNA (Pharmacia-LKB).
- 2′-Deoxynucleoside 5′triphosphates (dATP, dCTP, dGTP, dTTP), and adenosine 5′-triphosphate, disodium salt (Sigma). Dissolve dNTP in sterile distilled water to make a 10 mM stock. Neutralize to pH 7 with NaOH. Freeze in small aliquots and store at −20 °C.
- Isopropyl β-D-thiogalactoside (IPTG) (Sigma). Dissolve in sterile distilled water to make a 100 mM stock. Store frozen at −20 °C.
- 5-bromo-4-chloro-3-indolyl-β-D-galactopyranoside (X-gal) (Sigma). Dissolve in dimethylformamide to make a 2% (w/v) stock solution. Store at −20 °C.
- Rifampicin (Sigma). Dissolve in DMSO to make a 10 mg/ml stock. Store at −20 °C.
- Carrier-free ortho[^{32}P]phosphate (New England Nuclear).
- [^{35}S]-dATP (1200 Ci/mmol) (New England Nuclear).
- Sucrose (RNase free) (ICN).
- Polyethylene Glycol (6000) (Sigma).

Protocol 1. Media and buffers

1. LB media (per litre): 10 g tryptone, 5 g yeast extract, 10 g NaCl.
2. YT media (per litre): 8 g tryptone, 5 g yeast extract, 5 g NaCl, 15 g agar (for plates), 7.5 g agar (for soft YT).
3. 2YT (per litre): 16 g tryptone, 10 g yeast extract, 5 g NaCl.
4. ZPM media: 50 mM Tris–HCl (pH 7.4), 2 mM Na-citrate, 2 mM KCl, 2 μM $FeCl_2$.
5. Phosphate-free casamino acids: dissolve 100 g of casamino acids in 500 ml of water. Add $MgCl_2$ to 50 mM. Adjust the pH to 8.4 with NH_4OH and place on ice for 2 h. Filter the solution and adjust the pH to 7.2 with HCl. Add water to make 1 litre and sterilize by autoclaving.
6. NaH_2PO_4/Na_2HPO_4 (pH 6.8): make 0.5 M stocks of NaH_2PO_4 and Na_2HPO_4. Titrate the NaH_2PO_4 to pH 6.8 with Na_2HPO_4. Dilute the pH 6.8, 0.5 M stock to the appropriate concentration with sterile distilled water.
7. 4 M sodium bisulphite solution: 136 mg $NaHSO_3$, 64 mg Na_2SO_3, 0.43 ml distilled water.
8. Polymerase reaction buffer: 7 mM Tris–HCl (pH 7.5), 7 mM $MgCl_2$, 50 mM NaCl, 1 mM DTT.
9. TE: 10 mM Tris–HCl (pH 7.6), 1 mM EDTA.
10. 10× kinase buffer: 1.0 M Tris–HCl (pH 8.0), 0.1 M $MgCl_2$, 0.1 M DTT.

11. 10× annealing buffer: 0.2 M Tris–HCl (pH 7.5), 0.1 M $MgCl_2$, 0.5 M NaCl, 0.01 M DTT.
12. Extension buffer: 0.2 M Tris–HCl (pH 7.5), 0.1 M $MgCl_2$, 0.1 M DTT.
13. Lysis buffer: 25 mM Tris–HCl (pH 7.6), 60 mM KCl, 10 mM $MgCl_2$, 20% (w/v) sucrose (RNase free), 150 μg/ml lysozyme.
14. Detergent buffer: 25 mM Tris–HCl (pH 8.0), 30 mM KCl, 0.2% (w/v) sodium deoxycholate (use a freshly made 1% stock solution), 0.6% (w/v) Brij 58 (6% stock solution can be stored at 4 °C).
15. 10× RT annealing buffer: 0.5 M Tris–HCl (pH 8.0), 1 M KCl, 0.2 M $MgCl_2$.
16. dNTP/ddNTP mix (N°): The following concentrations are given in μM and should be made up in 1× reaction buffer.

N° (dNTP/ddNTP)	A°	C°	G°	T°
ddATP	---	---	---	0.25
ddCTP	---	---	100	---
ddGTP	---	100	---	---
ddTTP	100	---	---	---
dCTP	100	100	100	100
dGTP	100	100	100	100
dTTP	100	100	100	100

17. 10× reaction buffer: 0.5 M Tris–HCl (pH 8.0), 0.5 M KCl, 0.1 M $MgCl_2$, 0.1 M DTT.
18. Chase solution: 50 mM Tris–HCl (pH 8.0), 2 mM DTT, 10% glycerol, 1.2 mM each dNTP.
19. Stop solution: 10 mM EDTA (pH 8.0), 0.2% (w/v) bromophenol blue, 0.2% (w/v) xylene cyanol in 95% deionized formamide.

Protocol 2. *E. coli* **strains**

1. HB101: $recA^-$, F^-, pro^-, leu^-, thi^-, $lacY^-$, str^r, $hsdM^-$, $hsdR^-$, $endol^-$, ara-14, galK2, xyl-5, mtl-1, sup44. Host for expressing rRNA from plasmids and for maxicell analysis.
2. DH1: F^-, recA1, endA1, gyrA96, thi-1, hsdR17, supE44, λ^-. Host for expressing rRNA from plasmids that does not contain a mutation in ribosomal protein S12.
3. K5637: F^-, $hsdR^-$, $hsdM^+$, c1857, ΔBam, cro27, Oam29. This strain contains the λ P_L respressor gene in the chromosome. It can be used to conditionally express rRNA from the plasmid pNO2680.
4. CSR603: recA1, uvrA6, phr-1, thr-1, leuB6, proA2, argE3, thi-1, ara-14,

lacY1, galK2, xyl-5, mtl-1, rpsL31, tsx-33, supE44, nalA98, $\lambda^- F^-$. Host for maxicell analysis.

5. BL21(DE3): $hsdM^-$, $hsdR^-$, rif^S, λ lysogen. Host for conditional expression of rRNA from pAR3056. Host for chemical maxicells using pAR3056.
6. CJ326: Commercially available $recA^-$, ung^-, dut^- strain for bisulphite mutagenesis and for preparing ss-template for oligonucleotide-directed mutagenesis. (Bio-Rad).
7. XL-1 Blue: Commercially available F^-, $recA^-$ strain for transfecting with M13. (Stratagene).

2. Construction of rRNA mutations

The first rRNA mutations constructed *in vitro* were deletions. These mutations were produced by restriction enzyme cleavage of a rRNA operon followed by limited digestion of the rRNA coding region with nuclease Bal 31 and religation of the plasmid prior to transformation (1). Deletion mutations were produced in regions of the rRNA important for rRNA processing (2), protein–rRNA interaction (3–5), and subunit assembly (2, 3, 6). Given the substantial structural perturbations often introduced into the rRNA by deletions, more recent studies of rRNA structure and function *in situ* have utilized base substitution mutations. Therefore in the following sections we will describe two methods for producing site specific substitutions: bisulphite mutagenesis and oligonucleotide-directed mutagenesis.

2.1 Directed mutagenesis using sodium bisulphite

Sodium bisulphite is a single strand specific mutagen that catalyses the deamination of unpaired cytosines, converting them to uridines in a multiple step reaction (*Figure 1*). The first reaction intermediate is 5,6-dihydrocytosine-6-sulphonate, which is converted to 5,6-dihydrouracil-6-sulphonate in a nucleophilic substitution reaction at the exocyclic amino group. The latter step is the rate-limiting reaction and requires sodium bisulphite concentration >1 M at pH values between 5 and 6. The subsequent elimination of the bisulphite moiety from the aromatic ring system requires alkaline pH conditions and results in the formation of uracil. Since the rate of the deamination is linear with time, and is first order with the concentration of the educt sodium bisulphite, the extent of the C to U conversion can be controlled by varying one of these two variables.

Bisulphite mutagenesis requires a 'window' of single stranded DNA in a circular DNA duplex. The conversion of a specific sequence in circular, duplex DNA to a single-stranded form can be achieved by several methodologies. One approach is to generate a single-strand break in the DNA phosphate backbone at a predetermined location, and then to extend this nick into a larger 'window' by controlled exonuclease treatment. To create a single strand nick it is possible to

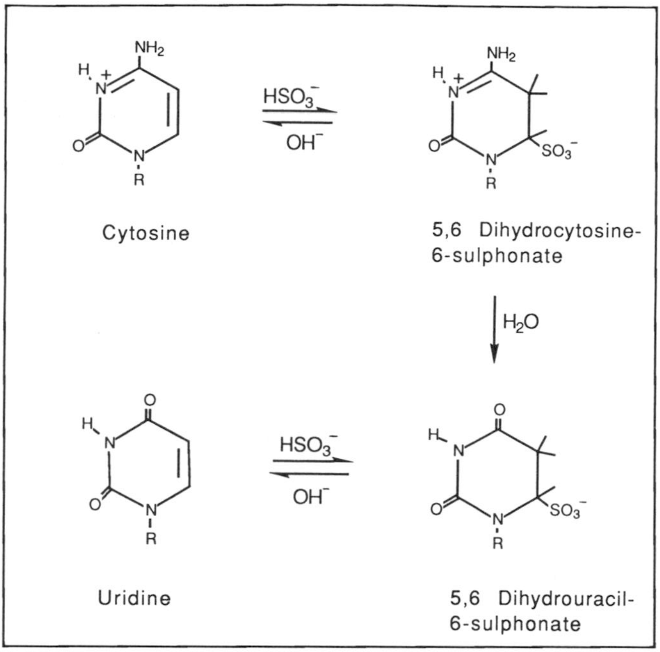

Figure 1. Reaction scheme for the deamination of cytosine to uracil.

use several type II restriction endonucleases, which can be induced to cleave only one strand within the target sequence. This is done by adding defined concentrations of ethidium bromide to the reaction mixture. The inhibition of the usual double-strand cleavage reaction is presumably the result of a partial unwinding of the DNA at the recognition sequence following the first strand cleavage, thereby creating new sites for the intercalating molecule. A thorough description of this approach is given by Shortle and Botstein (7).

A second method, detailed here, involves the generation of a heteroduplex by annealing two DNA molecules whose complementarity differs only at the target 'window' sequence. The two heteroduplex techniques outlined here were successfully used to create rRNA mutants in *E. coli* (8, 9). The gapped duplex method involves annealing the circularized plus strand of a recombinant M13 bacteriophage with a denatured, linear wild-type M13 DNA (10). Linearization of the parental M13 vector requires the same restriction endonucleases as used during the insertion of the target DNA fragment. Deletion loop mutagenesis uses denatured strands from two types of plasmid DNA. One plasmid contains the entire *rrn* operon and the other contains a characterized deletion mutation. The unhybridized ('non-deleted region') of the intact *rrn* operon loops out and is mutagenized. Since the restriction endonucleases used to linearize the wild-type plasmid are distinct from those used for the deletion plasmid, only the resulting heteroduplexes are circular, and therefore preferentially transform the host (11).

Protocol 3. Heteroduplex formation: gapped duplex method

1. Mix 5 μg of linearized M13 RF DNA with 2.5 μg of recombinant M13 single strand DNA (containing a fragment of an *rrn* operon) in 0.1 ml of 150 mM NaH_2PO_4/Na_2HPO_4(pH 6.8).
2. Boil for 3 min to denature.
3. To reanneal, incubate at 60 °C for 10 min then place the sample on ice.

Protocol 4. Bisulphite treatment

1. To the DNA sample on ice add 3 volumes of 4 M sodium bisulphite (pH 6.0) and 0.04 volumes of 50 mM hydroquinone (freshly prepared).
2. Incubate at 37 °C in the dark for the appropriate time (15 min–3 h, dependent upon the extent of modification desired).
3. To terminate the reaction, dialyse against:
   (a) 1000 volumes of 5 mM NaH_2PO_4/Na_2HPO_4 (pH 6.8), 3 mM hydroquinone (4 °C, 2 × 3 h).
   (b) 1000 volumes of 5 mM NaH_2PO_4/Na_2HPO_4 (pH 6.8) (4 °C, 1 × 1.5 h).
4. Add 1/10th volume of 1 M Tris base. Incubate for 12 h at 37 °C.
5. Recover the DNA by ethanol precipitation: Add 0.1 volume of 3 M sodium acetate and 2.5 volumes ethanol, leave at −70 °C for 30 min. Pellet the DNA in a microfuge for 10 min at 4 °C, carefully remove the supernatant and discard. Add 0.5 ml of cold (−20 °C) 70% ethanol to the tube and spin in a microfuge for 5 min. Carefully remove the supernatant and dry the pellet *in vacuo*.

Protocol 5. Filling-in the gapped, mutated, heteroduplex DNA

1. Dissolve the DNA pellet in 48 μl of Polymerase Reaction Buffer.
2. Add 1 μl of 2.5 mM dNTPs.
3. Start the reaction by adding 5 units (about 1 μl) of *E. coli* DNA polymerase (Klenow fragment).
4. Incubate at 37 °C for 60–90 min.
5. Transform competent *E. coli* (such as XL-1 Blue or JM109).
6. Identify mutant clones by dideoxy sequencing (*Figure 2*). The procedure for preparing single-stranded M13 DNA for sequencing is given in Section 2.2.2.

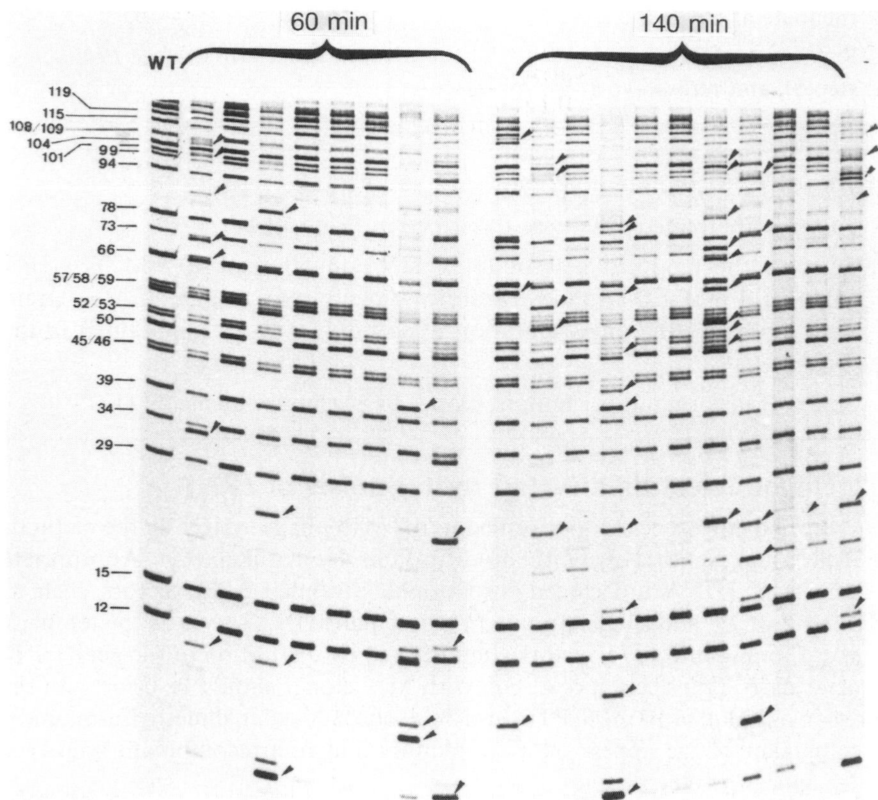

Figure 2. Identification of sodium bisulphite mutants generated by DNA sequence analysis. Autoradiograms of sequencing gels (only A specific tracks) from different 5S rDNA clones subjected to 60 and 140 min sodium bisulphite treatment. Mutagenized base positions are characterized by additional bands in comparison to the wild-type pattern (WT). The location of adenine nucleotides within the wild-type 5S RNA primary structure is shown on the left margin.

Protocol 6. Heteroduplex formation: deletion loop method

1. Linearize wild-type and deletion plasmid DNA (approximately 5 μg of each) with different single site restriction endonucleases.
2. Phenol extract the two samples and combine about 3.5 μg aliquots.
3. Ethanol precipitate the mixed sample and dry the DNA *in vacuo*.
4. To denature, redissolve the DNA in 20 μl of water and add 5 μl of 1 M NaOH.
5. Incubate at room temperature for 10 min.
6. To renature, add sequentially: 285 μl of water, 40 μl of 0.5 M Tris–HCl (pH 8.0) and 50 μl of 0.1 M HCl.

7. Incubate at 63 °C for 2–3 h.
8. Recover the DNA by repeated ethanol precipitation (3 times) (see *Protocol 4*, step 5), and redissolve in 0.1 ml of water.

Protocol 7. Bisulphite treatment

1. Samples are treated exactly as described in *Protocol 4*.
2. Dissolve the modified heteroduplex DNA in 50 μl of 10 mM Tris–HCl (pH 8.0), 1 mM EDTA, and transform directly into an ung^- *E. coli* strain such as CJ236 (the ung^- mutation allows dUTP to be maintained in the DNA).
3. Screen transformants for mutant clones by plasmid sequencing (12).

2.2 Oligonucleotide-directed mutagenesis of rRNA

Site-directed mutagenesis is performed in the M13 phage system by the methods of Zoller and Smith (13), with modifications by Kunkel (14). Appropriate fragments of rDNA are cloned into double stranded M13 vectors such as M13mp18 or 19, and single-stranded recombinant DNA serves as the template strand. Cloning into M13 vectors is monitored by disruption of the gene for β-galactosidase. Transfection of *E. coli* with M13 clones should be done with the addition of 100 μl of 10 mM IPTG and 50 μl of 2% X-gal in dimethylformamide. Recombinant phage appear as clear plaques and non-recombinant wild-type phage appear as blue plaques.

Single-stranded M13 template DNA is prepared from a dut^-, ung^- *E. coli* strain such as CJ236, to allow incorporation of uridine into the DNA. Uridine containing template strands are digested during transformation, thereby increasing the frequency of mutagenesis (14). An overnight culture of *E. coli* CJ236 should be diluted 1:100 into YT broth and grown to an A_{600} of 0.1 (the culture size depends on the amount of template required, 25 ml is adequate for mutagenesis, 250 ml or 500 ml cultures are used if large template stocks are needed). Titered phage are then added to yield a multiplicity of infection (phage per cell) of approximately 10:1. The infected culture is allowed to grow at 37 °C for 6 h and single-stranded DNA is prepared according to the procedure in *Protocol 8*.

Oligonucleotides of 15–20 bases in length should be synthesized or purchased such that the mismatched (mutagenic) nucleotide is positioned near the centre of the oligonucleotide. Separate oligonucleotides can be made for each nucleotide change, or a mixture of bases can be added at each of several mutagenic positions. Care must be taken that the oligonucleotide is not greater than 70% complementary to other sequences in the M13 or the cloned insert. We have found that the two-primer method of Zoller and Smith (13) is not necessary, and in fact the presence of the second primer can lower the frequency of mutagenesis.

Protocol 8. Isolation of single-stranded M13 template DNA

1. Remove bacterial cells by two centrifugations at $8000 \times g$ for 15 min each.
2. Add 1/3 volume of 20% polyethylene glycol (PEG6000) in 2.5 M NaCl to the phage-containing supernatant; place on ice for 30 min.
3. Centrifuge at $10\,000 \times g$ for 30 min. Discard the supernatant.
4. Resuspend the viral pellet in 1/100 of the original culture volume of TE [10 mM Tris–HCl (pH 7.6), 1 mM EDTA].
5. Extract twice with an equal volume of Tris-buffered phenol (pH 7.6).
6. Extract three times with an equal volume of 1:1 phenol:chloroform.
7. Add 0.1 volume of 3 M sodium acetate to the aqueous phase and then add $2\frac{1}{2}$ volumes of cold (−20 °C) 95% ethanol. Incubate at −70 °C for 15 min and pellet DNA at $12\,000 \times g$ for 15 min.
8. Resuspend the dry pellet in 1/1000 of the original culture volume of TE.

The template DNA preparation should be tested for self-priming activity by following the mutagenesis procedure below (*Protocols 10* to *12*) in the absence of added oligonucleotide primer. Self-priming leads to a reduction in the mutagenesis yield, making the recovery of mutants more difficult. As a general rule, the addition of oligonucleotide should increase the number of plaques at least two-fold.

Protocol 9. Phosphorylation of oligonucleotide

1. Mix: 3 μl of 10× Kinase Buffer, 2 μl of 10 mM ATP, 20 μl of oligonucleotide (20 pmole/μl), 3 μl of water and 2 μl of polynucleotide kinase (2 units/μl).
2. Incubate for 45 min at 37 °C; stop the reaction by heating at 65 °C for 10 min.

Protocol 10. Annealing reaction

For best results vary the molar ratio of template to oligonucleotide over the range of 0.1:1, 1:1, and 10:1.

1. Mix: 1 pmole of M13 template DNA, 10 pmole (1 pmol, 0.1 pmol) of phosphorylated oligonucleotide, 1 μl of 10× Annealing buffer and water to 10 μl.
2. Incubate at 65 °C for 5 min, then at room temperature for 5 min.

Protocol 11. Extension-ligation reaction

1. Mix: 2 μl of Extension Buffer, 4 μl of dNTPs (2.5 mM of each dNTP) 2 μl of 10 mM ATP, 2 μl of T4 DNA ligase (6 units) and 1 μl of *E. coli* DNA polymerase Klenow fragment (5 units).
2. Add 10 μl of the above to the annealing reaction mixture. Incubate at 15 °C overnight or at room temperature for 5 min, and 37 °C for 1 h.

The extension reaction can be treated *in vitro* with uracil-n- glycosylase (14) or simply used to transform an appropriate *E. coli* strain (such as XL-1 Blue). We do not recommend the use of JM101 for work with rDNA in M13 as recombination appears to be a problem. The $recA^-$ strain JM109, however, has been used successfully.

Protocol 12. Transformation

1. Add 10 μl of the extension–ligation reaction to 100 μl of competent *E. coli* (such as XL-1 Blue) (protocols for preparing competent cells are widely available, cells treated with $CaCl_2$ (15) or $RbCl_2$ (16) are both suitable).
2. Place on ice for 30 min, then incubate at 42 °C for 2 min.
3. Add 100 μl of log phase *E. coli* (XL-1 blue) and 2.5 ml of soft YT (maintained at 45 °C). Vortex gently.
4. Pour on to YT plates (at room temperature).
5. Incubate at 37 °C until plaques are visible (approximately 6 h).

2.2.1 Screening for mutants by sequencing

By using uracil-containing template DNA, the frequency of mutagenesis can range from 60–90%. Therefore the only screening necessary is sequencing. Single-strand DNA is prepared from phage plaques using a modified method of Zinder and Boeke (17).

Protocol 13. Preparation of single-strand DNA for sequencing

1. Core plaques with sterile microcapillaries.
2. Add one plaque to 1.5 ml of a 1:100 dilution of an overnight culture of wild type *E. coli* (XL-1 blue) in 2YT media. Incubate at 37 °C with good aeration (i.e. on a roller drum) for 6 h.
3. Pellet bacteria at 12 000 × *g* for 10 min. Respin the supernatant at 12 000 × *g* for 10 min (it is essential to remove all cells).
4. Add 750 μl of the supernatant to 300 μl of 20% PEG/2.5 M NaCl. Incubate on ice 30 min. Save the remaining supernatant as phage stock at −20 °C.

5. Pellet phage at 12 000 × *g* for 10 min. Carefully remove all PEG supernatant.
6. Resuspend phage pellets in 200 μl of water.
7. Extract once with Tris-buffered phenol (pH 7.6) and twice with phenol: chloroform (1:1).
8. Recover the DNA by ethanol precipitation (see *Protocol 4*, step 5).
9. Resuspend the DNA pellet in 30 μl of TE.

Sequencing is done using the dideoxy method of Sanger *et al.* (18). Protocols are available from several sources (New England Nuclear, BRL, and so on).

2.2.2 Preparation of replicative form (RF) of mutated M13 for cloning

Once a mutant sequence is identified, the phage stock saved above is used to prepare double-stranded DNA for cloning. The stock should be plaque-purified. A purified plaque is added to 10 ml of *E. coli* (XL-1 Blue) in 2 YT (A_{600} approximately 0.1) and the culture grown for 6 h at 37 °C. Phage are precipitated with 1/3 volume of PEG/NaCl on ice for 30 min, spun at 12 000 × *g*, and resuspended in 1 ml of TE. 500 μl of this stock (about 10^{12} phage per ml) is used to infect 1 litre of bacteria at $A_{600} = 0.1$. This is approximately a MOI of 10:1. This culture is shaken at 37 °C for 6 h. Double-stranded RF DNA can be prepared from the bacteria using any large-scale plasmid preparation such as the alkali lysis method (15).

2.2.3 Cloning into expression vectors

To construct a mutant expression plasmid, a sub-fragment of M13 RF DNA which contains the base substitution must be cloned into the plasmid vector (the plasmid vectors used for expressing mutant rRNA are described in detail in Section 3). Since the rRNA operon must be reconstructed exactly it is important that the same restriction endonucleases are used to generate the insert fragment derived from the M13 RF and the vector derived from the wild-type expression plasmid. For this reason the most convenient clonings utilize restriction endonucleases which cleave only one time in the expression vector. In regions of the operon where there are no single-cutting restriction sites, partial-restriction digests are used to generate the vector DNA. The cloning steps follow standard techniques for restriction enzyme digestion, ligation, and transformation (15, 16). After constructing a mutant expression vector it is extremely important to verify the mutation. This is done by sequencing the plasmid in the region of the operon containing the mutation (12).

3. Expression of rRNA mutants

Transcription, processing, and assembly of the mutant, plasmid-coded rRNA is carried out *in vivo* utilizing one of several vectors. A detailed description of these vectors (19), and a summary of mutations constructed in these systems (20), can be found elsewhere.

3.1 Vectors for rRNA mutants

3.1.1 pKK3535

Ideally, the expression of rRNA mutants is carried out from a plasmid coded operon which is regulated in the same manner as the chromosomal rRNA operons. The pBR322 derivative pKK3535 (21) contains the entire *rrnB* operon, including the natural promoters P1 and P2 and terminators T1 and T2 (*Figure 3*). The plasmid-bourne operons are under stringent control (22) and maintain a high-copy number (approximately 14 copies per cell). Plasmid-coded rRNA accounts for 40–70% of the rRNA in the cell, depending on the mutation (23–25). Many rRNA mutations expressed from pKK3535 drastically reduce the growth rate of the cell. In some cases the mutations cannot be cloned into the vector at all, presumably because the mutant rRNA product is lethal to the cell. For mutations that are unmanageably slow or lethal in pKK3535, alternative plasmids with inducible promoters must be used.

3.1.2 pNO2680

Several vectors are available for conditional expression of lethal rRNA mutations. The first to be constructed was pNO2680, in which the wild-type P1 and P2 promoters of pKK3535 are replaced by the λ P_L promoter (*Figure 4*) (26). This plasmid can be introduced into cells harbouring a gene for the temperature sensitive P_L repressor protein, either on a second plasmid, pCI857 (27) or in the host chromosome (strains such as K5637). At 30 °C the repressor is functional and plasmid-coded rRNA is not expressed. At 42 °C the repressor undergoes thermal denaturation and expression from P_L occurs. After 4 h of growth at 42 °C we find 50–60% of the cellular rRNA is contributed by the plasmid (23, and unpublished results).

3.1.3 pAR3056

A second conditional expression vector, pAR3056, has the ribosomal promoters P1 and P2 replaced by the T7 late-promoter (*Figure 5*) (28). In this system, a strain (BL21/DE3) harbouring a gene for T7 polymerase is encoded in the chromosome under the control of the *lac* UV5 promoter. With the addition of IPTG, T7 polymerase is produced and mutant rDNA is expressed. Lethal mutations can thus be propagated in this plasmid in any wild-type *E. coli* strain (lacking T7 polymerase) or in BL21/DE3 in the absence of IPTG. Unlike pNO2680, this system avoids possible complications associated with raising the temperature. However, expression of the *rrn* operon is limited to strains containing the T7 polymerase under *lac* UV5 promoter control.

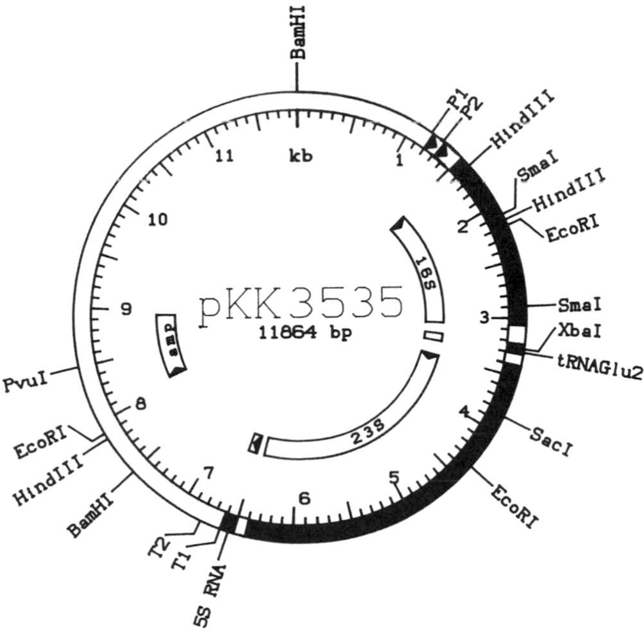

Figure 3. Plasmid pKK3535.

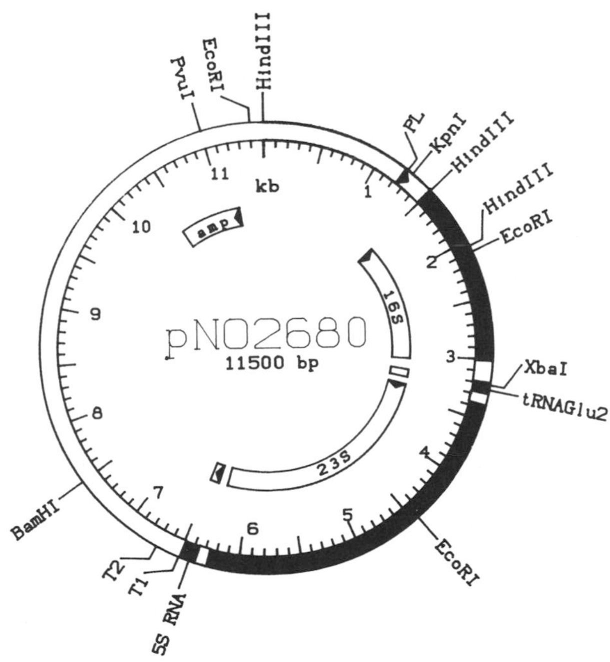

Figure 4. Plasmid pNO2680.

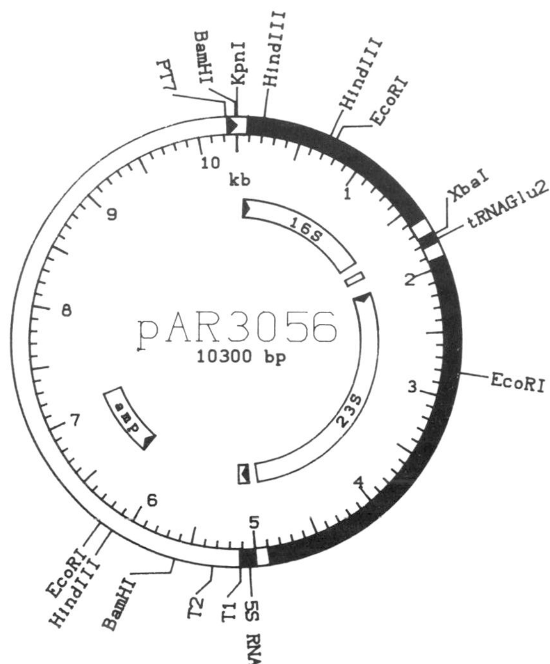

Figure 5. Plasmid pAR3056.

4. Analysis of rRNA mutants

The initial characterization of a rRNA mutation may occur during its construction and expression. Often the mutation confers an altered growth phenotype on the host cell or it is lethal when expressed in large quantities. The assays outlined in this section describe additional methods used to investigate specific structural or functional aspects of mutated rRNA.

4.1 Maxicell analysis

The processing and assembly characteristics of mutant rRNA can be analysed in maxicell systems. In these systems the plasmid-coded mutant rRNA is specifically labelled with ^{32}P following UV or chemical inactivation of the wild-type chromosomal rRNA operons (2, 29). Separation of the labelled RNA or ribosomal particles on agarose-acrylamide composite gels allows the processing and assembly fate of the mutant rRNA to be determined. An extensive description of the maxicell technique applied to rRNA can be found in Jemiolo *et al.* (29).

4.2 UV maxicells

The minimum host cell requirement for UV maxicells is that they be recA$^-$. Cell strains specifically designed for maxicell work, such as CSR603, have additional mutations in the uvr genes, decreasing the UV dose required to inactivate the chromosome and the phr genes, relieving the requirement that UV-treated cells be shielded from light. The maxicell conditions for both the $recA^-$, uvr^+ strain HB101 and CSR603 are described below.

Protocol 14. Growth, irradiation, and labelling

1. Suspend bacterial cells from one agar plate in 1.5 ml of LB and use to inoculate 25 ml of LB broth containing 200 μg/ml ampicillin. (A_{600} = 0.18–0.25). Shake the culture at 37 °C until the A_{600} reaches 0.5.
2. Transfer 12.5 ml of the culture to a sterile 100 mm × 15 mm plastic petri dish containing a 1.25 inch stir bar. Cover the dish with cardboard and place on a stir plate 25 cm below a Sylvania G15T8 germicidal lamp. With the culture gently stirring, remove the cardboard and expose the cells for the proper time (10–15 sec for CSR603, 75–90 sec for HB101).
3. Place 10 ml of the irradiated culture into a 125 ml flask (HB101 cultures must be shielded from the light) and shake at 37 °C for 6 h, adding 20 μl of 20 mg/ml D-cycloserine after 2 and 4 h.
4. Pellet the cells at 5000 r.p.m. for 2 min in an SS-34 or similar rotor at 4 °C.
5. Wash cells twice with 5 ml of cold ZPM supplemented with 0.2% phosphate-free casamino acids, 0.2% glucose, and 1 μg/ml thiamine. Resuspend the final pellet in 2 ml of the same media, transfer to a 30 ml Corex tube or 25 ml flask, and add: 20 μl of 20 mg/ml D-cycloserine, 10 μl of 1 M KH_2PO_4, and 50 μCi of carrier free ortho[^{32}P]phosphate.
6. Shake the cells at 37 °C for 10–16 h, then lyse as described in *Protocol 15*.

Protocol 15. Lysate preparation

1. Pellet the bacterial cells and wash in 1 ml of Lysis Buffer.
2. Resuspend the pellet in 25 μl of Lysis Buffer, then freeze-thaw 3 times.
3. Add 175 μl of Detergent Buffer and 2 μl of 10 mg/ml DNase 1 (RNase free). Place on ice for 10 min then spin in a microfuge for 10 min at 4 °C.
4. Carefully remove the supernatant, flash freeze, and store aliquots at −70 °C. Extract the rRNA from one aliquot (50 μl) using phenol and phenol-chloroform. Recover the extracted rRNA by ethanol precipitation. Analyse the rRNA as in Section 4.4.

4.3 Chemical maxicells

A second maxicell system takes advantage of the T7 expression system described earlier. Transcripts from the rRNA operon of plasmid pAR3056 are produced by inducing the T7 polymerase gene of the host strain BL21(DE3) with IPTG. By adding rifampicin to an induced culture the plasmid-coded rRNA can be specifically labelled. Rifampicin blocks transcription of chromosome-coded rRNA by inhibiting *E. coli* RNA polymerase, but T7 RNA polymerase is resistant so the plasmid-coded rRNA continues to be expressed. Surprisingly, the T7 transcripts are processed and assembled after rifampicin treatment (29).

Protocol 16. Cell growth and labelling

1. Grow a 10 ml culture of BL21(DE3) containing pAR3056 overnight to stationary phase at 37 °C in ZPM containing 200 μg/ml ampicillin and 10% LB. Dilute the culture 1:3 with ZPM containing 10% LB, 0.2% casamino acids, 0.2% glucose, and 200 μg/ml ampicillin. Grow at 37 °C with good aeration for 2.5 h (cells should be in late log phase, $A_{600} = 1.2$).
2. Add IPTG (to 0.5 mM) and continue growth for 45 min.
3. Add 450 μl of 10 mg/ml rifampicin (stock dissolved in DMSO and stored at −20 °C), grow 5 min, then add 60 μCi of carrier-free ortho[^{32}P]phosphate. Grow the culture for 20 min then rapidly chill by transferring to a centrifuge tube at −70 °C.
4. Lyse the cells as described in *Protocol 15*.

4.4 Composite gel analysis of maxicell products

Maxicell labelled rRNAs, the specific transcripts of plasmid bourne *rrn* operons, can be analysed in porous composite gels containing agarose and acrylamide. These gels are useful for separating precursor and mature forms of rRNAs as well as precursor and mature ribosomal subunits, 70S ribosomes, and polyribosomes. This technique makes possible the rapid determination of effects of rRNA mutations on processing and assembly into ribosomal subunits and formation of 70S ribosomes. *Figure 6a* shows the electrophoretic separation of ^{32}P-labeled, plasmid coded rRNAs in a slab gel containing 3% acrylamide and 0.5% agarose using a Tris–borate EDTA buffer. Note the separation of precursors (17S and p23S) from the mature 16S and 23S rRNAs. In *Figure 6b* cell lysates (containing ribosomal particles) are electrophoresed directly into a composite gel which contains a Tris, $MgCl_2$ and KCl buffer. The structure of the ribosomal particles is preserved in the presence of magnesium ions and the autoradiogram shows the separation of 30S, 50S, and 70S ribosomes in the first dimension. Duplicate samples, were soaked in SDS to deproteinize the ribosomes prior to electrophoresis in a second dimension, to identify the rRNAs associated with the respective subunits. Note the effect of a deletion mutation in 16S rRNA (Bgl II-66) which prevents both processing of 17S to 16S rRNA and assembly into mature 30S subunits. The mutant particles migrate faster than mature 30S subunits in the first dimension gel and, unlike wild-type (WT) subunits, they are not found in 70S

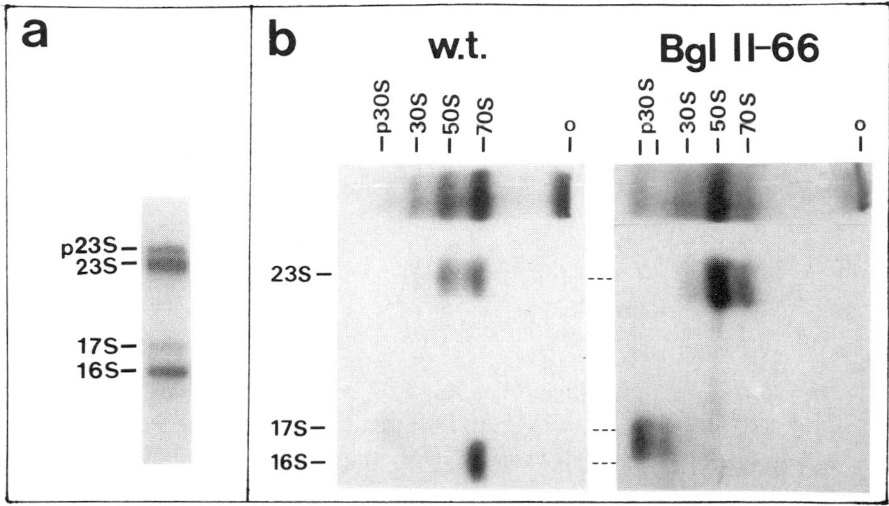

Figure 6. Gel analysis of maxicell labelled rRNA, ribosomal subunits, and ribosomes. Labelled rRNA was separated by electrophoresis in a 3% acrylamide, 0.5% agarose composite gel containing Tris-borate, EDTA buffer (a). Ribosomes and subunits were analysed on a two-dimensional gel (b). Lysates of wild-type (w.t.) and 16S rRNA mutants (Bgl II-66) plasmids were electrophoresed (right to left) into adjacent slots in a 3% acrylamide, 0.5% agarose composite gel containing 25 mM Tris–HCl (pH 8.0), 10 mM $MgCl_2$, 30 mM KCl buffer. Gel strips with duplicate samples were soaked in SDS, EDTA–Tris containing buffer, rotated 90°, fixed to the top of a new gel with agarose, and electrophoresed in the second dimension (top to bottom).

ribosomes. For a detailed description of the preparation of composite gels and their application to the characterization of rRNA and ribosomal particles, the reader is referred to a previous volume in the practical approach series (30).

4.5 rRNA sequence analysis with reverse transcriptase

The presence of a plasmid-coded, mutant rRNA in free subunits, 70S ribosomes, and polyribosomes, can be determined by reverse transcriptase sequence analysis. This method complements maxicell analysis and provides information about the distribution of mutant rRNA in ribosomal pools. The isolation of 70S ribosomes, ribosomal subunits, and polyribosomes, follows standard techniques (see Chapter 1 and Chapter 8). The sequencing reactions of the dideoxy chain termination method utilizing [^{35}S]-dATP are described here. A similar technique utilizing a ^{32}P end-labelled primer can also be used (31).

Protocol 17. Primer annealing reaction

1. Mix: 2 μl of rRNA template (approximately 5 pmol), 2 μl of primer (5–20 pmol), 1 μl of 10 × RT Annealing Buffer, and 5 μl of water.

2. Incubate at 95 °C for 2 min and then at room temperature for 20 min. The template: primer ratio can be reduced to 1:1 if false banding is a problem.

Protocol 18. Sequencing reaction

1. Make up an extension mix (for 4 reactions) from:
   - 1 μl of [^{35}S]-dATP
   - 1 μl of reverse transcriptase (7 units)
   - 1 μl of 10 × reaction buffer
   - 7 μl of water
2. Label four 0.5 ml microfuge tubes, G, A, T, and C. Add per tube: 5 μl of N° dNTP/ddNTP mix (i.e. dGTP/ddGTP mix to the tube labelled G) 2.5 μl of the primer annealing reaction and 2 μl of extension mix. Incubate at 42 °C for 20–25 min.
3. Add 2 μl of Chase Solution and 1 unit reverse transcriptase (additional enzyme is optional). Incubate at 42 °C for 15 min.
4. Add 5 μl of stop solution, heat samples for 2 min at 95 °C, and run on a sequencing gel.

4.6. Assays of structure and function

Choosing the appropriate assay for a particular rRNA mutation requires some prediction about the ribosomal function(s) affected by the mutation. Any assay which tests the structure or function of ribosomes *in vivo* or *in vitro* can be adapted to investigate mutant ribosomes. We refer the reader to the other chapters in this volume for the detailed protocols of ribosome assays.

Acknowledgements

We would like to thank Michael Stark, Richard Gourse, Rolf Steen, Christian Zwieb, William Jacob, and David Jemiolo for their contributions to the mutagenic systems. We also thank Catherine Prescott, Michael Rasmussen, Matthew Firpo, B. Kleuvers, M. Zacharias, and George Q. Pennabble for helpful contributions. This work was supported by the National Institutes of Health (GM19756 to A. E. D.)

References

1. Gourse, R., Stark, M, and Dahlberg, A. E. (1982). *J. Mol. Biol*, **159**, 397.
2. Stark, M., Gourse, R., and Dahlberg, A. E. (1982). *J. Mol. Biol.*, **159**, 417.
3. Stark, M., Gregory, R., Gourse, R., Thurlow, D., Zwieb, C., Zimmermann, R., and Dahlberg, A. E. (1984). *J. Mol. Biol.*, **178**, 303.

4. Gregory, R., Zeller, M., Thurlow, D., Gourse, R., Stark, M., Dahlberg, A. E., and Zimmermann, R. (1984). *J. Mol. Biol.*, **178**, 287.
5. Skinner, R., Stark, M., and Dahlberg, A. E. (1985). *EMBO J.*, **4**, 1605.
6. Zwieb, C., Jemiolo, D., Jacob, W., Wagner, R., and Dahlberg, A. E. (1986). *Mol. Gen. Genet.*, **203**, 256.
7. Shortle, D. and Botstein, D. (1983). In *Methods in Enzymology*, Vol. 100 (ed. R. Wu, L. Grossman, and K. Moldave), p. 457. Academic Press Inc, London.
8. Zwieb, C. and Dahlberg, A. E. (1984). *Nucleic Acids Res.*, **12**, 4361.
9. Göringer, H. U. and Wagner, R. (1988). In *Methods in Enzymology*, Vol. 164 (ed. H. Noller and K. Moldave), p. 721. Academic Press Inc, London.
10. Everett, R. D. and Chambon, P. (1982). *EMBO J.*, **1**, 433.
11. Kalderon, D., Oostra, B. A., Ely, B. K., and Smith, A. E. (1982). *Nucleic Acids Res.*, **10**, 5161.
12. Chen, E. Y. and Seeburg, P. H. (1985). *DNA*, **4**, 165.
13. Zoller, M. J. and Smith M. (1984). *DNA*, **3**, 479.
14. Kunkel, T. (1985). *Proc. Nat. Acad. Sci. USA*, **82**, 488.
15. Maniatis, T., Fritsch, E. F., and Sambrook, J. (1982). *Molecular Cloning: A Laboratory Manual.* Cold Spring Harbor Laboratory Press, NY.
16. Hanahan, D. (1985). In *DNA Cloning Volume 1—A Practical Approach* (ed. D. M. Glover), p. 109. IRL Press Ltd., Oxford.
17. Zinder, N. and Boeke, J. (1982). *Gene*, **19**, 1.
18. Sanger, F., Nicklen, S., and Coulsen, A. R. (1977). *Proc. Nat. Acad. Sci. USA*, **74**, 5463.
19. Steen, R., Jemiolo, D., Skinner, R., Dunn, J., and Dahlberg, A. E. (1986). *Prog. Nucleic Acids Res. Mol. Biol.*, **33**, 1.
20. De Stasio, E. A., Goringer, H. U., Tapprich, W. E., and Dahlberg, A. E. (1988). In *Genetics of Translation* (ed. M. F. Tuite, M. Picard, and M. Bolotin-Fukuhara), p. 17. Springer-Verlag, Berlin.
21. Brosius, J., Ullrich, A., Raker, M. A., Gray, A., Dull, T. J., Gutell, R. R., and Noller, H. F. (1981). *Plasmid*, **6**, 112.
22. Gourse, R. L., Stark, M. J. R., and Dahlberg, A. E. (1983). *Cell*, **32**, 1347.
23. Jacob, W. F., Santer, M., and Dahlberg, A. E. (1987). *Proc. Nat. Acad. Sci. USA*, **84**, 4757.
24. Thompson, J., Cundliffe, E., and Dahlberg, A. E. (1988). *J. Mol. Biol.*, **203**, 457.
25. Vester, B. and Garrett, R. A. (1988). *EMBO J.*, **7**, 3577.
26. Gourse, R. L., Takebe, Y. R., Sharrock, A., and Nomura, M. (1985). *Proc. Nat. Acad. Sci. USA*, **82**, 1069.
27. Remaut, E., Stanssens, P., and Fiers, W. (1981). *Gene*, **15**, 81.
28. Steen, R., Dahlberg, A. E., Lade, B., Studier, F., and Dunn, J. (1986). *EMBO J.*, **5**, 1099.
29. Jemiolo, D., Steen, R., Stark, M. J. R., and Dahlberg, A. E. (1988). In *Methods in Enzymology*, Vol. 164 (ed. H. Noller and K. Moldave), p. 691. Academic Press Inc, London.
30. Goodwin, G. H. and Dahlberg, A. (1982). In *Polyacrylamide Gel Electrophoresis of Nucleic Acids—A Practical Approach* (ed. D. Rickwood and B. D. Hames), p. 213. IRL Press Ltd., Oxford.
31. Moazed, D., Stern, S., and Noller, H. F. (1986). *J. Mol. Biol.*, **187**, 399.

13

Electron microscopy of ribosomes

MILOSLAV BOUBLIK

1. Introduction

Electron microscopy (EM) played a key role in the discovery of ribosomes. The ribosomes were first visualized by Palade (1) in liver cell tissue and named a 'small particulate component of the cytoplasm'. The term ribosome describing this intracellular organelle was coined later by Roberts (2). In the conventional transmission electron micrographs of thin sections of animal tissue, ribosomes appear as minute electron-dense granules (arrows in *Figure 1*), often attached to the membrane of endoplasmic reticulum. With the improvement of the isolation and purification procedures, and advancement of the EM techniques, free isolated ribosomes can now be visualized in much greater structural detail (inset in *Figure 1*).

Ribosomes are difficult objects for high-resolution EM imaging. They are small in size (with the largest dimensions between 200–300 Å), highly hydrated, low in contrast, and sensitive to radiation damage by the electron beam. They lack any symmetry or regular or repetitive structure which could be utilized for image enhancement by conventional techniques of image superposition and rotation. The complexity of the ribosome structure originates from the large number of ribosome constituents—proteins and RNAs. Although ribosomes were discovered in eukaryotic cells, most of the information on this organelle originates from the prokaryotic ribosome of *E. coli*, to date the only ribosome with a fully characterized composition (for reviews see 3, 4). It should be mentioned that some of the procedures used for electron microscopic imaging of ribosomes are routinely used for other biological specimens, and are exhaustively described in a series by Hayatt (5), and more recently, in the *Practical Approach* series, by Sommerville and Scheer (6). Ribosome related experimental procedures are published in (7).

From the rapidly growing number of highly specialized electron microscopic (frequently called electron spectroscopic) techniques [for a review see Hren *et al.* (8)], two are of dominant importance to high-resolution imaging of sensitive biological specimens such as ribosomes: conventional transmission electron microscopy (denoted mostly as CEM, TEM, CTEM, or simply EM) and scanning transmission electron microscopy (STEM) dedicated to biological

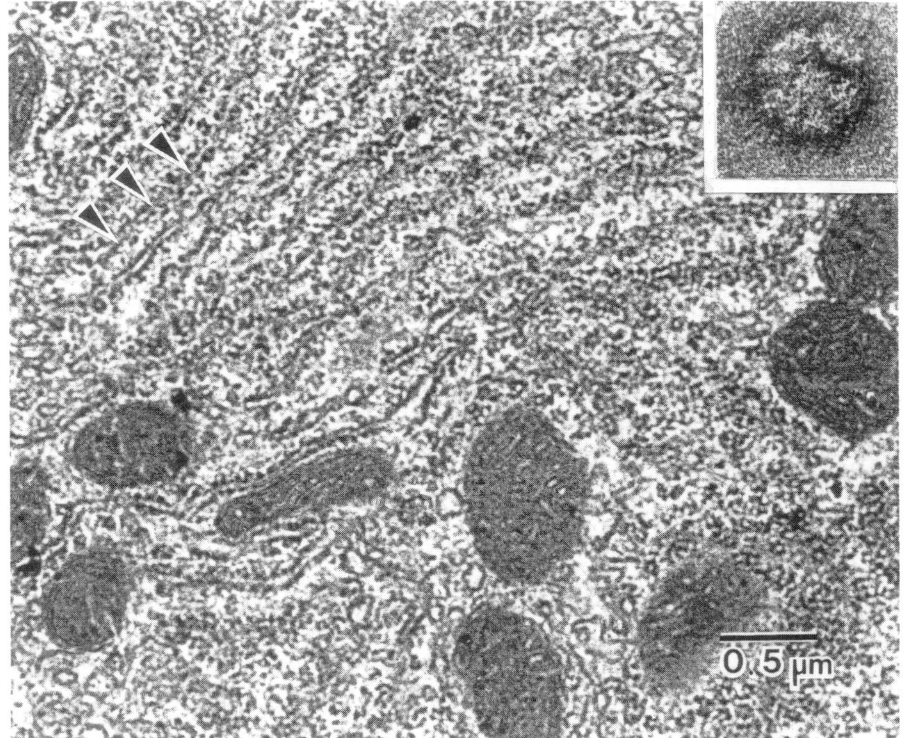

Figure 1. Electron micrograph (TEM) of a thin section of rat liver tissue. Arrow denotes ribosomes seen as small electron dense particles attached to the membrane of the endoplasmic reticulum. Bar represents 0.5 μm. Inset shows a high resolution electron micrograph of a free 80S monosome ($\times$ 610 000).

specimens (9). The aim of this chapter is to demonstrate the potential of TEM and STEM for elucidating the complexity of ribosomal structure. Although both techniques differ in instrumentation, specimen handling, and signal processing, they share the basic rule essential to high-resolution ($<$20 Å) imaging of ribosomes: full attention to detail from the preparation of specimen solutions through the final stage of photographic processing.

2. Preparative methods

2.1 Specimen preparation

2.1.1 Solutions

All solutions should be prepared from deionized water, processed through a Millipore Milli-Q Reagent Grade Water System, and distilled. Use of quartz-distilled water is recommended but is not always necessary. All chemicals used should be of analytical grade. Sterile conditions should be strictly observed with specimens of rRNAs in all preparative stages. Buffers should be similar to those used in biochemical assays, in order to preserve the native structure and activity of the specimen and facilitate the correlation of ribosome structure and function. For EM of ribosomes the most commonly used buffers contain 10–20 mM Hepes–KOH (or Tris–HCl), pH ~7.6, 50–100 mM KCl (or NH_4Cl), 2–20 mM Mg acetate and 0.5–2.0 mM dithiothreitol (DTT) or 2-mercaptoethanol. Magnesium ions are of particular importance; a minimal concentration of 2 mM Mg^{2+} is necessary for preserving the native structure of ribosomal subunits and their association into a monosome. Magnesium ions also affect stacking and stability of the nucleotide bases in rRNAs. Obviously, some experiments require special ionic conditions. For instance, total reconstitution of ribosomes from their constituent proteins and rRNAs has to be done in high Mg^{2+} (10–20 mM) and high salt conditions such as 20–30 mM Tris–HCl (pH 7.6), 300–400 mM KCl or (NH_4Cl), 0.5 mM DTT. Since the standard fixation procedure with 0.5% glutaraldehyde has no detectable effect on the ribosome appearance, apart from increasing their mass by about 10%, no fixatives are applied to stabilize the ribosome structure.

2.1.2 Ribosomes, ribosomal subunits, and their components

Prokaryotic (70S) (10) and eukaryotic (80S) (11) monosomes and their respective subunits are prepared by standard techniques (see Chapter 1). Whenever possible, the specimens for EM are immediately taken from the preparative sucrose density gradient fractions after dialysis into the 'ribosomal' buffer (usually 10 mM Hepes–KOH (pH 7.6), 60 mM NH_4Cl (or KCl), 2 mM Mg acetate). Ribosomal particles are best stored in small aliquots (~20–50 μl) at high concentration (~100–500 A_{260} units/ml) in 10 mM Hepes–KOH (pH 7.6), 60–100 mM NH_4Cl (or KCl), 10 mM Mg acetate, 10 mM 2-mercaptoethanol in a liquid nitrogen tank or at −70 °C. Comparatively good results are obtained by storing the ribosomes in the above buffer in the presence of 20–40% glycerol (or sucrose) at −20 °C.

Ribosomal RNAs are extracted from ribosomal subunits by standard phenol extraction and alcohol precipitation. Ribosomal proteins are isolated from the subunits and purified as described (12). See also Appendix 1.

2.1.3 Support films

Only pure carbon films (20–40 Å thick) are used as a specimen support because they are flat and stable in the electron beam. The films are obtained by carbon evaporation from a spectral grade carbon rod on to the surface of freshly cleaved mica or a rock salt crystal (NaCl) using the Balzers BAE 080T evaporator or a similar unit. The film thickness is measured with the quartz monitor of an EVM 052A control unit. However, with some experience it is possible to estimate the thickness of the carbon film from its colour against a white background. The films are floated off the mica on to the clean surface of deionized or quartz-distilled water (washed when floated off from a salt crystal) and picked up with a clean support grid (washed in acetone). The films, although prepared in the same way, can sometimes differ in their adsorption properties. Glow discharge, in our experiments, does not reduce the hydrophobicity of the carbon layer.

2.1.4 Staining

A variety of salt solutions of heavy metals has been tried for ribosome staining (6): uranyl acetate, uranyl formate, uranyl citrate, uranyl oxalate, uranyl sulphate, uranyl nitrate, ammonium molybdate, phosphotungstic acid (PTA), platinum–pyrimidine complexes (13), all in aqueous solutions at various pH, in concentration ranges between 0.05–2%.

Since the most satisfactory results are obtained with 0.5% aqueous uranyl acetate (unadjusted pH$\cong$4.2), this staining solution is being used almost exclusively. Specimen staining should be even and with minimum grain (without recrystallization of uranyl acetate) as demonstrated on 40S ribosomal subunits from HeLa cells (*Figure 2*). The intensity of staining and uniform distribution of the stain layer depends on the surface properties of the carbon supportive film. As

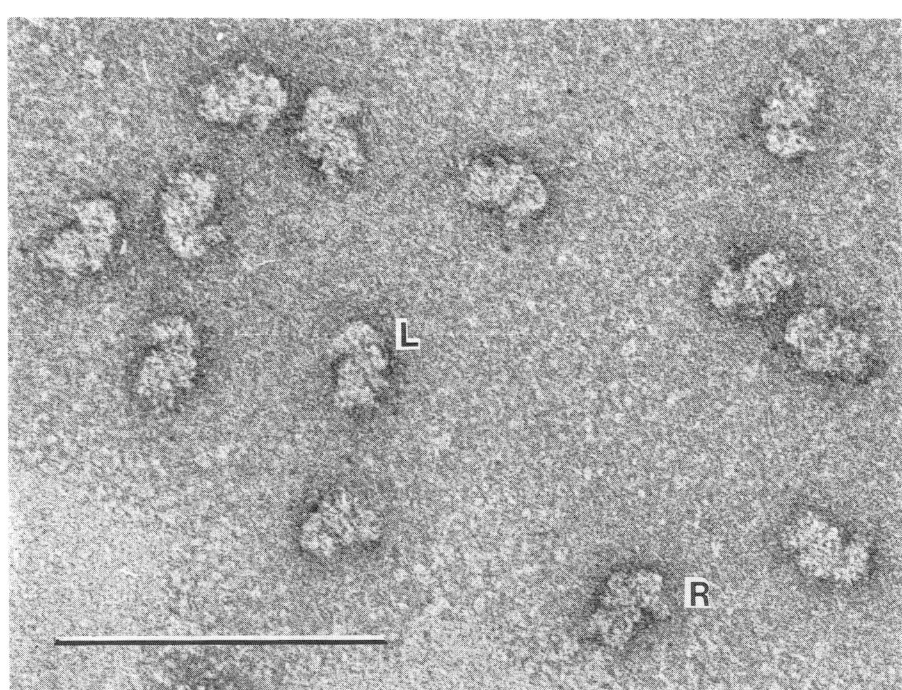

Figure 2. Electron micrograph (TEM) of 40S ribosomal subunits from Hela cells prepared by the single carbon film adsorption technique (Section 2.1.5) stained with 0.5% aqueous uranyl acetate and air-dried. Subunits show a strong preferential orientation for left (L) and right (R) 'side views'. Bar represents 0.1 μm.

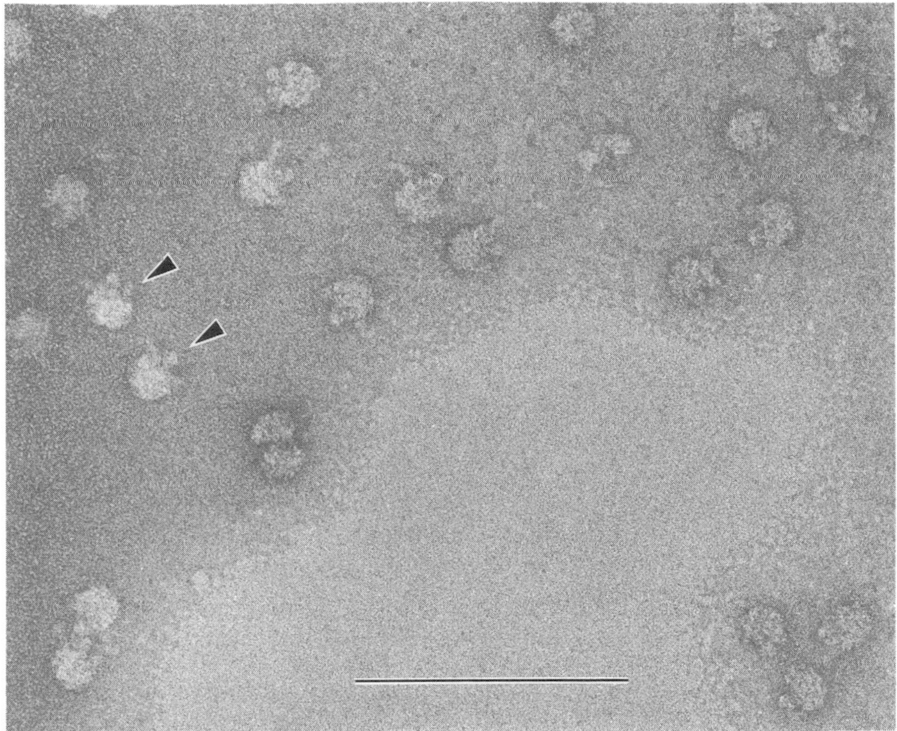

Figure 3. Demonstration of simultaneous positive and negative staining of specimen with 0.5% aqueous uranyl acetate. Arrows mark the negatively stained 50S *E. coli* ribosomal subunits. Notice the uneven distribution of the staining. Bar represents 0.1 μm.

mentioned above, while the thickness of the C films is reproducible, the extent of hydrophobicity and adsorption properties can vary from film to film. Uranyl acetate staining generally occurs as a positive contrasting, but may also (with the same specimen on the same grid) turn out to be negative as well (*Figure 3*).

2.1.5 Specimen deposition

Droplet technique

This technique, because of its simplicity, is generally used for 'quick' checking of the quality of the ribosome preparation. A solution of ribosomes in an appropriate buffer is deposited as a small droplet (volume $\cong$ 30 μl) on a sheet of paraffin film or other clean inert hydrophobic surface, e.g., a teflon plate. A standard 3.0 mm support grid (200–500 mesh) covered with a thin carbon support film (prepared as described in Section 2.1.3) is placed at room temperature on the top of the droplet for ribosome adsorption. Adsorption time depends on the concentration of ribosomal particles and is usually 1–2 min for a ribosome concentration of 0.3 A_{260} units. The time can be extended up to 20 min for specimens of substantially lower concentration (below 0.1 A_{260} unit), but extension of the adsorption time over 30 min does not increase the ribosome adsorption. The grid with the attached ribosomes is transferred on to a droplet of the staining solution (0.5% aqueous uranyl acetate), picked up after 1–2 min, blotted from the grid bottom side with filter paper, and air-dried.

Single carbon film adsorption technique

This technique of specimen deposition originally developed by Valentine and Green (14) became, in various modifications and under various names, the standard technique for high resolution imaging of ribosomes by TEM. In our version (15) a thin carbon support film (described in Section 2.1.3) is floated completely from a piece of mica ($\sim$4 mm $\times$ 4 mm) on to the surface of the ribosome solution in a small teflon well (*Figure 4*). In this way the 'adsorption' side of the carbon film is never exposed to the air. Since the freshly evaporated carbon film has very high surface activity, this precaution may prevent

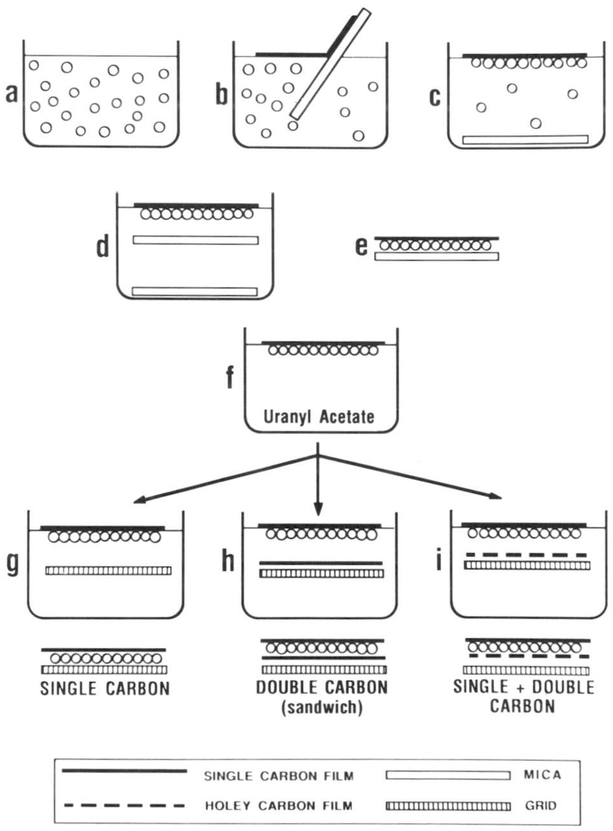

Figure 4. Cartoon of specimen deposition by modifications of single carbon film adsorption technique. (See Section 2.1.5 for details.)

adsorption of hydrophobic organic contaminants from the air. The solution volume in the well is about 400 μl, the concentration of ribosomes is usually between 0.05–0.1 A_{260} unit. Adsorption is done at ambient temperature for 1–5 min but can be extended up to 30 min, depending on the ribosome concentration. The carbon film with the adsorbed ribosomes is picked up from the bottom with a piece of plain mica and transferred into a contrasting solution of 0.5% aqueous uranyl acetate, uncorrected for pH. After 2–5 min in the staining solution the carbon film is picked up from the bottom with a clean 300–500 mesh tabbed copper grid (Ted Pella, Inc.) submerged under the floating carbon (*Figure 4g*). The specimen mount is blotted with filter paper and air-dried. In the microscopic stage (JEM 1200 EX) the thin carbon film faces the probe of electrons and protects the specimen against radiation damage.

Double carbon (sandwich) adsorption technique

This procedure is similar to that of the single carbon adsorption technique up to the step of staining [*Figure 4*, step (f)]. In the sandwich technique, however, the ribosomes adsorbed to the floating carbon film are picked up by a submerged copper grid coated with a thin carbon film (as in the droplet method). Thus, after lifting the grid, the specimen is 'embedded' in the staining solution between two thin carbon films (*Figure 4h*) and air-dried. Because of the evenly distributed staining solution around the ribosomal particles, this technique is preferable for obtaining electron micrographs for computer image analysis. On the other hand, one cannot exclude specimen flattening by cohesive forces in the carbon sandwich. The additional carbon can also adversely affect the resolution.

Because both single and double carbon film adsorption techniques have specific advantages, it is sometimes convenient to have images from both methods on the same grid. This can happen when one of the two carbon films is accidentally broken (*Figure 5*), or intentionally, when the floated carbon film is picked up with a grid covered with 'holey film' (*Figure 4i*). 'Holey' films can be prepared (16) or purchased from any of the major EM suppliers listed in reference 6.

3. Electron microscopy

3.1 Conventional transmission (TEM)

3.1.1 Techniques

A conventional electron microscope for high-resolution imaging of sensitive biological specimens should be equipped with an efficient oil-free ('dry') pumping system for attaining a good clean vacuum (10^{-5}–10^{-6} Torr for K-type filament, 10^{-6}–10^{-7} or better for LaB_6), an anti-contamination device (cold trap with liquid nitrogen), and a 'minimum dose' stage. For obtaining images for 3-D reconstruction, the microscope must be equipped with a wide angle range tilting

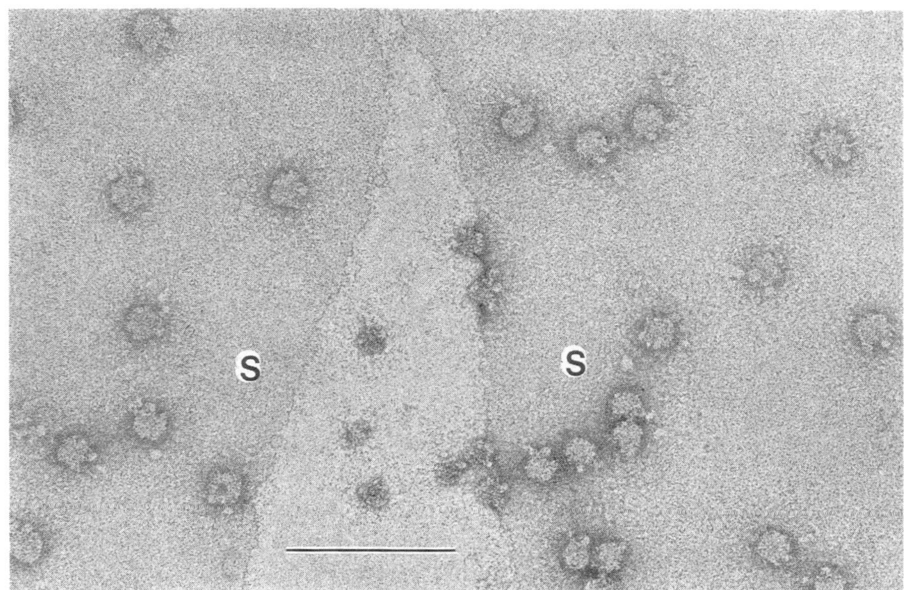

Figure 5. Demonstration of the effect of an additional carbon film (S) on the specimen appearance (50S ribosomal subunits from *E. coli*). Notice the distortion of the subunits due to lack of 'protective' staining (centre). Bar represents 0.1 μm.

stage ($\pm 50°$ or more). Resolution of such a microscope is better than 2 Å (for an ideal specimen such as carbon or gold lattice).

Stained, air-dried ribosomes prepared by one of the methods described in Section 2.1.5 are usually observed at direct magnification of 50 000–100 000 ×. Additional magnification is obtained by photographic enlargement (usually 2–6 times). From the range of accelerating voltages (60–120 keV), we consider 80 keV as a reasonable compromise between resolution, contrast, and radiation damage to the specimen.

Radiation damage in TEM occurs as a consequence of specimen exposure to the beam of electrons (inelastic scattering of the incident primary beam of electrons), particularly at higher magnifications. At the frequently used magnification of ~75 000 × the electron dose is in the range of 10^2–10^3 e/Å^2. Radiation damage in high-resolution EM is characterized by loss of fine (5–10 Å) structural features, loss of specimen crystalline structure, loss of mass, as well as by changes in the structure and composition of the specimen. Exposure to a high radiation dose results in recrystallization and migration of stain granules and creates artificial structures. Specimen radiation damage can be reduced by using minimal beam intensity, fast focusing, and short exposure time. The specimen can also be protected by lowering its temperature with an anti-contamination device and, in particular, by focusing away from the area of interest by using a 'minimum dose' attachment. Some specimen protection is also obtained by the deposit of staining solution and the thin carbon film on the top due to improved heat conductivity. Considerable specimen radiation damage is to be expected when taking images in a tilting series for 3-D image reconstruction.

3.1.2 Applications

Morphology

Electron micrographs from high resolution TEM, such as the example in *Figure 6*, are the most direct source of information for study of the morphology of ribosomal particles. The overall view (*Figure 6*, top) of 80S monosomes from slime mold (*Dictyostelium discoideum*) mainly reflects the quality of the ribosome specimen prepared according to Section 2.1.5, the degree of polydispersity in shape and size, and the extent of preferential orientation of the particles on the carbon support film. Contours and electron density pattern, the main criteria for the fine structure of ribosomes, are more clearly visible in the highly (~500 000×) enlarged electron micrographs in the gallery (*Figure 6*, bottom). These images are also used for determination of the particle characteristic dimensions: length, width, and axial ratio. The accuracy of these measurements is within ±10%. Images of particles selected from a variety of views can be utilized, after tilt angle assignment, for a 3-D reconstruction and model proposal.

Topographical mapping

The standard technique used to localize *in situ* individual ribosomal proteins, segments of rRNAs, and functional sites, is immuno electron microscopy (IEM). This technique has been recently reviewed in detail by Stöffler-Meilicke and Stöffler (17) and Glitz *et al.* (18).

In the original IEM technique (19), ribosomal particles were incubated with antibodies (IgG) specific for individual ribosomal proteins. The binding site of IgG marked the position of the antigen, a segment of the particular protein or rRNA against which the antibody was raised. The antibody can be resolved without additional markers because of its characteristic Y-shape. The example in *Figure 7a, b*, obtained by a combination of IEM and affinity labelling (20), demonstrates localization of the decoding region of the *E. coli* ribosome to the 'cleft' of the small subunit. Although IEM is responsible for rapid advancement of our understanding of the architecture of the ribosome (that of *E. coli* in particular), it is not the ultimate labelling solution. The major drawback, lack of specificity, has been eliminated by using monoclonal antibodies; however, the resolution of IEM is limited to 30–35 Å by the size of the IgG marker. Furthermore, the attachment site of the antibody (the Fab fragment) is not always clearly resolvable and can be hidden by the ribosomal particle. In some cases, the structure of the antibody is distorted and difficult to recognize. Thus, in IEM labelling, the structure of the antibody should always be checked first

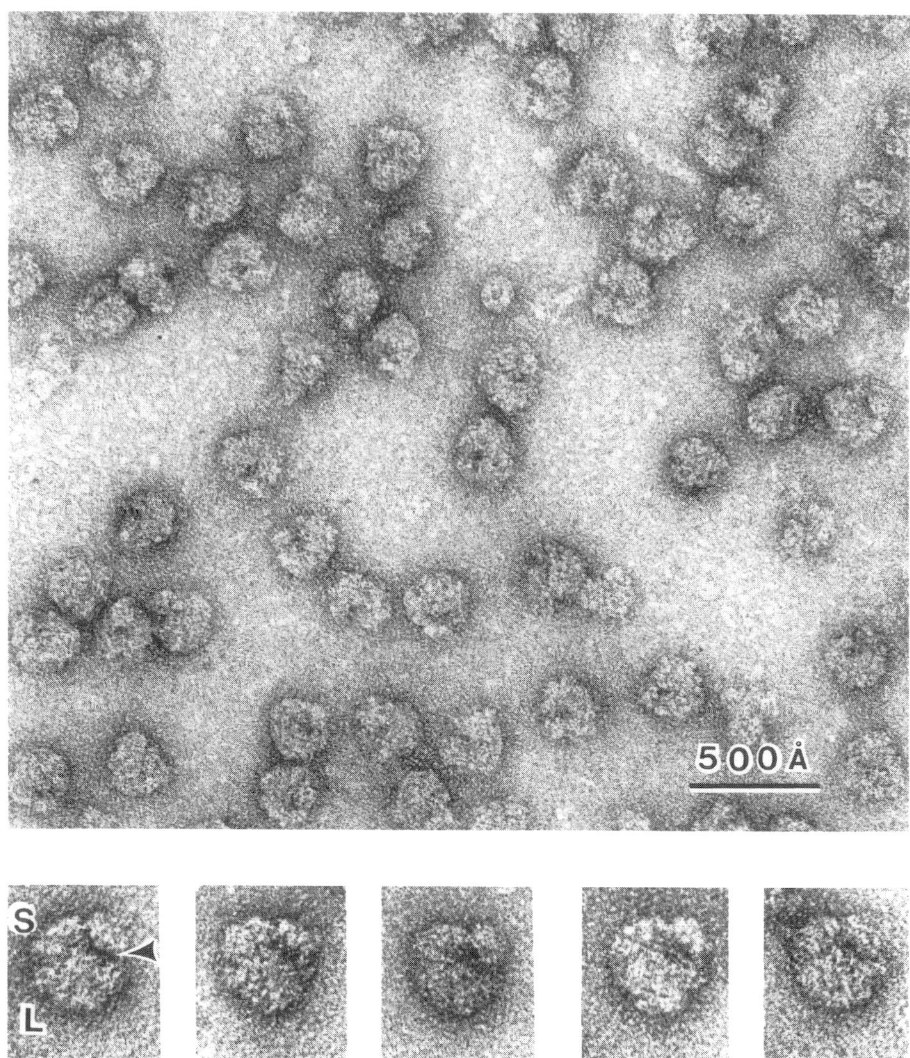

Figure 6. Electron micrographs (TEM) of the 80S monosomes from slime mold (*D. discoideum*) in an overall view (top), and in a gallery of selected views (bottom) showing the mutual orientation of the 40S small (S) and 60S large (L) subunits. Arrow points to the interface between the subunits (× 520 000).

(*Figure 7c*). Modifications using antibody probes to reagents applied in the site-specific modifications of rRNAs and to synthetic oligodeoxynucleotides complementing specific rRNA sequences (18) improve the specificity of labelling, but suffer from the same resolution restriction as the original IEM technique. The same criticism is valid for DNA hybridization electron microscopy using biotin-avidin complex (instead of IgG) as a topographical marker (21).

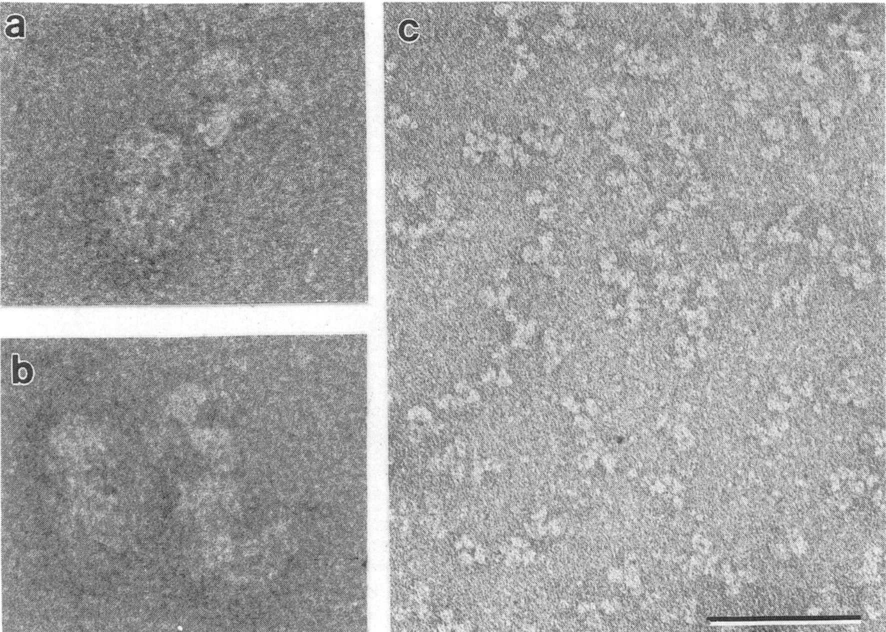

Figure 7. Electron micrographs (TEM) of (a) the 30S *E. coli* ribosomal subunit with anti-DNP antibody (IgG) attached to the 3′-end of the tRNA. Because of its divalency (b) IgG can be attached to two 30S subunits (×1 000 000). (c) Example of structurally preserved IgG for immuno EM labelling. Bar represents 0.05 μm.

3.1.3 Potential

The potential of TEM for structural studies on ribosomes is demonstrated in *Figure 8*. This figure summarizes data on 30S *E. coli* ribosomal subunits obtained by TEM and IEM in several laboratories (for references see figure legend): shape and dimensions of the subunit (shown in a 'side-view' projection of a proposed model of the 30S *E. coli* subunit based on TEM images), topographical mapping of ribosomal proteins, location of segments of 16S rRNA, functional sites (PM) and the decoding region (C1400, binding site for tRNA and mRNA).

3.2 Scanning transmission (STEM)

Application of dedicated STEM to the study of biological specimens in the early 1970s (22) marked a new trend in the EM imaging, characterized by the transition from a mere photographic record of the specimen to quantitative image analysis. Separation of components that affect resolution and contrast makes it possible to optimize detection of elastic electrons responsible for the contrast of the specimen. While conventional TEM operating in the dark-field mode utilizes only ~5% of the available elastically scattered electrons, the efficiency of dark-field STEM in collection of elastically scattered electrons is

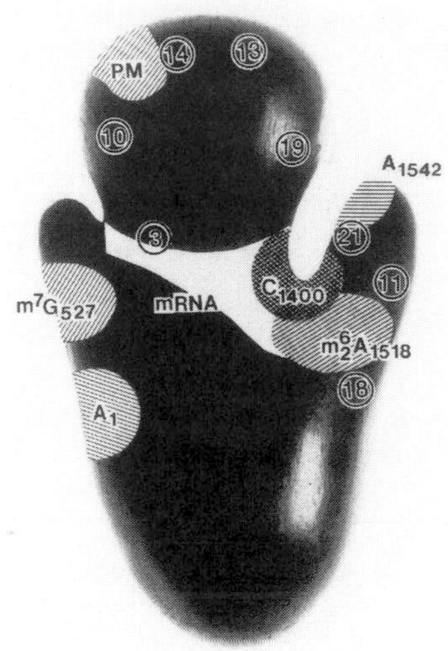

Figure 8. Demonstration of the potential of high-resolution TEM applied to the study of the 30S *E. coli* subunit. The model of the 30S subunit, according to Boublik *et al.* (40), shows topographically identified 16S rRNA sites: C_{1400}, the ribosomal decoding site (41); A_{1542}, the 3′-end of 16S ribosomal RNA (42); modified bases $m^6_2A_{1518}$ (43); m^7G_{527} (44); A_1, the 5′-end of 16S rRNA (45). Also shown are the puromycin cross-linking site, PM (46), the mRNA binding domain (47), and the locations of some ribosomal proteins (17).

almost 100%. The high contrast and superior signal/noise ratio associated with the dedicated STEM annular detector make it possible to visualize unstained freeze-dried ribosomal particles and rRNAs at extremely low radiation doses. Specimens prepared in this way are free of the main resolution-limiting conditions of TEM, staining, air-drying, and to a considerable extent, radiation damage. With the elimination of staining it becomes possible to relate image intensity to the local projected mass of the specimen, and thus to obtain (apart from the morphology) quantitative data on the molecular mass and mass distribution within a single macromolecule (23–26). Compared to conventional methods for mass determination, STEM has a unique option: selection and examination of any single molecule in the preparation. Centrifugation, chromatography, light scattering, small-angle X-ray, and neutron scattering methods measure averages over a total population of particles.

3.2.1 Techniques

Specimen preparation

Specimen preparation for STEM differs from the TEM technique mainly in elimination of staining, substitution of air-drying by freeze-drying, and application of an internal standard for quantitative analysis.

Protocol 1. Specimen preparation for STEM

1. A thin carbon film (usually 20–30 Å) evaporated on to a freshly cleaved surface of a rock salt (about 20 mm × 10 mm × 2 mm) is floated off the salt block on to a clean surface of distilled water, and picked up from the top with a titanium grid (2.3 mm diameter, 75 × 300 mesh) covered with a 'holey' carbon film. 'Holey' film in this application is rather thick (> 100 Å) and with relatively small holes (~ 1 μm in diameter).
2. The grid is exhaustively washed (10 times) with distilled water and partially blotted (with filter paper) between the washes, always keeping the carbon film wet (27).
3. Using an Eppendorf pipet tip, 3 μl (≃ 250 ng) of tobacco mosaic virus (TMV) in distilled water is gently injected into the remaining water layer. After 1 min adsorption at room temperature the grid is washed (up to ten times) with an appropriate buffer (Section 2.1.1) and partially blotted as above.
4. Ribosomal particles or rRNAs (individual ribosomal proteins are too small to be visualized by STEM in a meaningful way) at a final concentration of ~0.15 A_{260} unit in 5 μl of the respective buffer are injected below the surface of the buffer layer in a similar fashion as for TMV. Adsorption of ribosomal particles (monosomes, subunits, and nucleoprotein cores) is independent of the ionic strength of the buffer. However, adsorption of rRNAs to a 'wet' carbon film is drastically reduced in solutions with salt concentration below 1 μM. This effect can be overcome with a 10-fold increase in rRNA concentration. An alternative method to circumvent the problem of poor adsorption of rRNAs is pretreatment of the carbon film with polylysine (28), by injecting 5 μl of polylysine solution (10 μg/ml) into the water droplet. After 1 min adsorption the grid is washed eight times with water and air-dried. The deposition procedure continues by adding a droplet of water to the polylysine coated carbon film and injecting TMV as described above. Polylysine treatment reproducibly improves adsorption of rRNAs at least 10 times, which makes it possible to do the measurements with standard concentrations of rRNA solutions. A drawback of this procedure is slightly increased background (≤ 5%) and larger standard deviation of the mass measurement. For this reason, the first alternative (increasing concentration of rRNA) is preferred whenever possible.
5. After 1 min adsorption time for the specimen the grid is washed 6–10 times with a solution of 20–60 mM NH_4 acetate (volatile in vacuum) or 20 mM NH_4 acetate and 2 mM Mg acetate.

   The wash is necessary to remove salt from the buffer solution in order to prevent interference with the specimen mass measurement. The conformation of the adsorbed specimen does not appear to be significantly affected by the washing procedure. This has been documented by comparison of the values of radii of gyration (R_G] of *E. coli* rRNAs determined by STEM (29) and by hydrodynamic techniques (30) under various ionic strength conditions.

6. The grid, after the last wash, is blotted as much as possible (avoiding air-drying!) and rapidly frozen by dipping into liquid nitrogen slush (or liquid propane). Fast freezing is important to avoid specimen distortion by ice-crystal formation. The grid with the frozen specimen is transferred into the transport cartridge in a stainless steel chamber filled with liquid nitrogen. When most of the nitrogen is evaporated, the chamber is evacuated to less than 10^{-8} Torr. From the initial temperature of -150 °C the specimen is warmed up at a rate of 1 °C/min until the pressure in the freeze-drying system increases to 10^{-7} Torr. The temperature is then held constant at -95 °C to prevent the system pressure from exceeding 10^{-7} Torr. The completion of the freeze-drying cycle is indicated by a drop of pressure below 10^{-8} Torr and the resumption of the 1 °C/min warming. The freeze-drying ($\sim$15 h) is usually done overnight. The specimen, still under vacuum, is transferred into the STEM cold stage (-160 °C). Application of freeze-drying reduces specimen distortion by the surface tension forces characteristic of air-drying.

Dedicated STEM

The results presented in this chapter were obtained with dedicated STEM developed at the Brookhaven National Laboratory (9, 23). This high resolution (2–3 Å) STEM has a field emission gun that operates at 40 keV at 10^{-10}–10^{-12} A beam current. The vacuum in the gun chamber is maintained below 10^{-10} Torr. The objective lens (1 mm focal length) is cooled to about -170 °C by helium gas. The specimen is kept at -160 °C to reduce mass loss by electron irradiation. The electron dose depends on the magnification used, it is about 1 e/Å^2 at a magnificaiton of 50 000 × and about 30 e/Å^2 at 250 000 ×. Scans are controlled digitally. Detector signals are stored in a digital frame buffer (digital memory holding one TV frame) of a VAX 11-750 computer. This computer also supports additional terminals for image analysis (26). Digital frame buffers permit an image of 512 × 512 pixel resolution to be displayed at varying contrast, brightness, and colour. The operator may execute commands to change pan (image movement), zoom, and mark individual macromolecules (or their segments) with a trackball-controlled cursor. Display of co-ordinates and alphanumerics makes it possible to keep track of any selected particle. Image data are stored digitally on magnetic tape.

Most of the high-resolution transmission electron microscopes on the market have an optional STEM attachment. However, unless equipped with a field emission gun and computer interfaced with video graphics, their performance is not adequate for high resolution imaging of unstained biological macromolecules. At present, the availability of commercially manufactured dedicated STEMs is limited to HB 501 (VG Microscopes Ltd., England).

3.2.2 Applications

The combination of the unique features of dedicated STEM and improved preparation of the specimen (Section 3.2.1) has advanced EM studies of the structure of ribosomes and their components to a quantitative analytical level not attainable by TEM. Visualization of rRNAs under conditions in which they display native conformation, or conformation reflecting the ambient ionic conditions, has become possible only after application of STEM (29, 31, 32). The difference in the potential of TEM and STEM techniques for the structural studies becomes obvious from the comparison of 16S rRNA images by TEM (*Figure 9a*) and by STEM (*Figure 9b*). 16S rRNA (*Figure 9a*) spread on a water surface, and stretched under denaturing conditions using benzylalkylammonium chloride (BAC) according to reference 33, appears as a single filament about 5000 Å long. The length of the filament is in agreement with the number of 1542 nucleotides of 16S rRNA in the fully extended form. The length of 16S rRNA in the STEM image (*Figure 9b*) is ~1200 Å, about 1/4 of that in the fully extended state. The protrusion at about 2/5 of the RNA molecule is about 80 Å long and can be used as a marker for the polarity of the molecule. However, resolution of the 5′ and 3′ end would require an additional topographical marker. Differences in both the thickness and the length of the molecules in these images reflect the mode of their deposition. The difference in the thickness is due to the deposition of BAC and tungsten used for RNA shadowing, for obtaining contrast in the TEM. The difference in the length is due to the preservation of secondary structure of 16S RNA consisting of alternating intramolecular double stranded stems and single-strand loops. The extensive base-pairing is responsible for the 'shortening' of the native RNA molecule in the STEM image. The difference in the

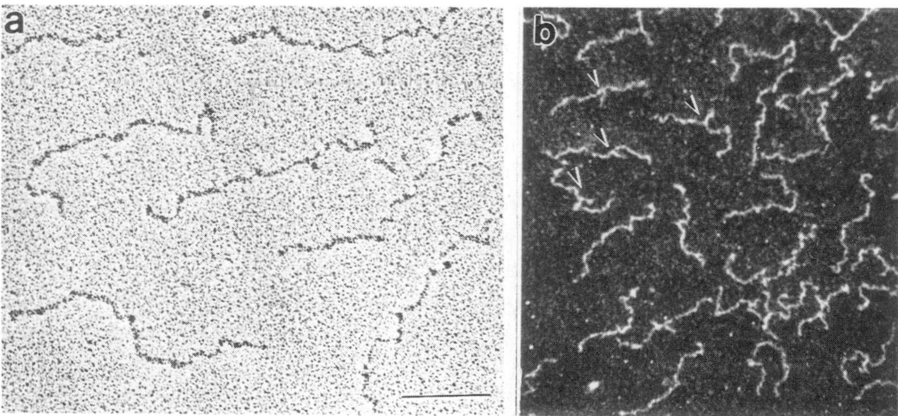

Figure 9. Electron micrographs of *E. coli* 16S rRNA obtained by (a) the conventional BAC-monolayer technique, air-dried, shadowed with PtPd. (b) STEM, unstained, freeze-dried. Arrows point to the protrusion at 2/5 of the RNA molecule. Deposition in both cases was from water. Bar represents 0.1 μm.

appearance of rRNA molecules visualized by TEM and STEM is even more striking with larger RNAs. While the TEM images of all RNAs prepared under denaturing (fully extended) conditions differ in their morphology only in their length, corresponding to their mass (not shown), STEM images of these RNAs show very different and very complex structures reflecting the ionic conditions of the environment (25). Structural information obtained from the TEM images is limited to the determination of the length of rRNA and assessment of the molecular mass from this parameter. By contrast, STEM images can be used for calculation of the molecular mass (M), assessment of polydispersity of the preparation, linear density (M/L), apparent number of strands in the rRNA segments, radius of gyration (R_G), interactions of rRNAs with proteins in the subunit assembly process, and topographical mapping of the distribution of the rRNA and protein moiety *in situ*.

Mass measurement (M)

Given the linearity of the STEM imaging process, direct measurement of particle mass is relatively straightforward, particularly for compact particles—ribosomal monosomes and subunits (23, 25, 26). Particles selected for mass measurement from the display on a TV monitor are marked with circles of a proper radius. The numbered circles in *Figure 10* (used as an example) mark individual 50S ribosomal subunits. The area within the circle is integrated and the intensity of scattering of the enclosed particle is corrected for the background. The resulting net intensity of scattered electrons is calibrated by a constant calculated for each specimen by using TMV as an internal mass standard. About 100 particles should be measured for obtaining statistically significant means, standard deviations (SD), and standard errors. Since the digitized images with x and y coordinates for each particle are stored in the computer memory, any selected particle can be retrieved for additional analysis. The results can be edited and plotted as a histogram. The accuracy of the mass measurement depends on the mass, size, and thickness of the particle, carbon support noise and counting statistics of the scattered electrons. Measured SD for compact macromolecules such as large ribosomal subunits and monosomes are below 5%, for extended filamentous molecules such as rRNA the SD$\cong$10%. Mass measurement of elongated molecules (*Figure 11*) is more elaborate as it requires cautious drawing of an arbitrary area around the molecule with a trackball-controlled cursor. Circle replacement by contour lines around the complex structure of rRNA includes substantially less background noise. Drawing of the contour lines which best fit the measured structure is usually done at a higher magnification (*Figure 11b*), which makes it possible to resolve and characterize structural segments of the molecule (*Figure 11c*).

Mass distribution within particles and radius of gyration (R_G)

Determination of mass distribution within compact particles is done by the circle routine and by the arbitrary area routine (26) for irregular and extended shapes.

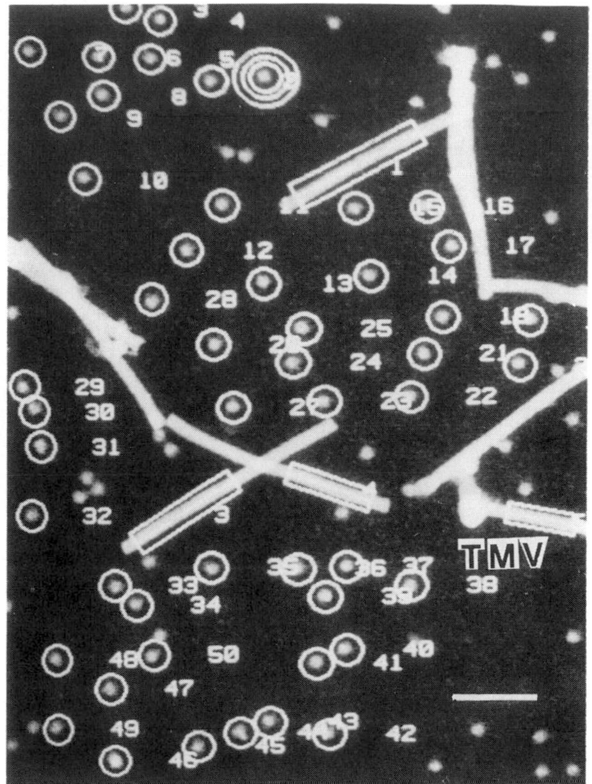

Figure 10. Molecular mass measurement by the circle sub-routine. The numbered circles enclose individual unstained, freeze-dried 50S *E. coli* ribosomal subunits selected for mass measurement. The first particle (without number) has an additional annulus around the circle which is used to calculate local background. Boxes denote the segments of tobacco mosaic virus (TMV) used as a standard for mass calibration. Bar represents 0.1 μm.

Since the shape of the particles is known, and since it is possible to determine the mass in each point of the particle (represented by a set of pixels the size of which is given by the applied magnification, e.g., 5 Å at 250 000 ×), one can determine the centre of gravity and calculate the radius of gyration from the equation

$$R_G = \left[\sum_i (I_i - I_b) \times R_i^2 \Big/ \sum_i (I_i - I_b) \right]^{1/2}$$

where I_i is the intensity of scattered electrons in pixel i, I_b is the intensity of the scattering from the support carbon film (background noise), and R_i is the distance from pixel i to the centre of gravity. The values of R_G for the ribosomal subunits and rRNAs obtained in this way are in very good agreement with the data of R_G from hydrodynamic techniques (for references see 29, 30, 34). The values of R_G are of particular importance to the studies of ribosomal particles

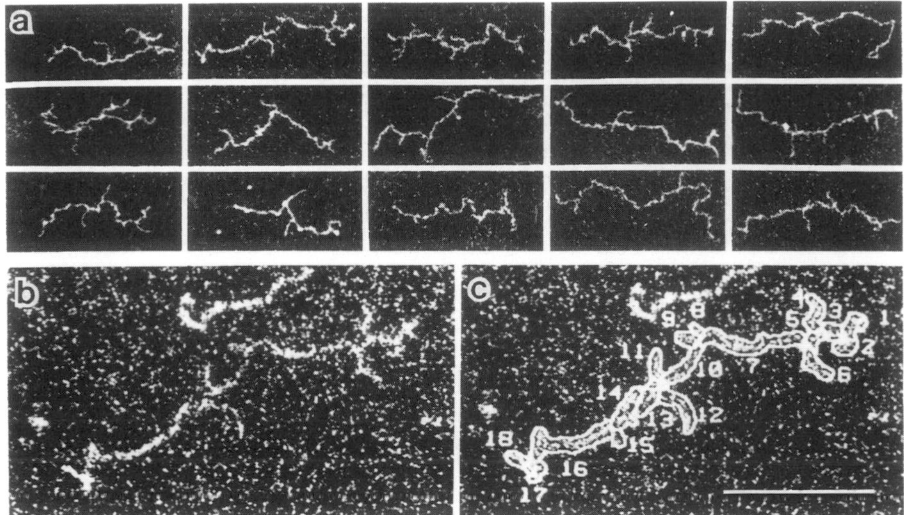

Figure 11. Mass distribution within the resolvable structural segments of an rRNA molecule. (a) 28S rRNA from BHK cells in water; (b) enlarged image of one molecule from (a); (c) the same molecule as in (b) divided into 18 arbitrary structural segments for mass and *M*/*L* determination. Bar represents 0.1 μm.

because they make it possible to compare morphological parameters of individual ribosomes and their components obtained under stringent conditions of electron microscopy (dehydration, radiation exposure) to those obtained from hydrodynamic and spectroscopic techniques in which the total specimen is studied in the fully hydrated state and in a variety of buffer conditions.

Mass per unit length (M/L) *and the apparent number of strands* (*n*)

Determination of these parameters can be applied only to highly extended structures. In the case of ribosomes it is limited to the rRNAs in relatively low ionic strength when the molecules are loosely folded and appear to be formed by a main backbone with several side branches and forks. The procedure is illustrated in *Figure 11* for 28S rRNA from baby hamster kidney (BHK) cells. The molecule is divided into 18 arbitrary segments (*Figure 11c*). The mass (*M*) within the enclosed area of each segment is calculated as above. The length (*L*) of each segment is measured directly on a highly enlarged electron micrograph of the molecule. These values are listed in *Table 1* together with the *M*/*L* ratio. The *M*/*L* values can be used for calculation of the apparent number of RNA strands (*n*) in the cross-section of each segment. The calculation is based on the value of 125 Da/Å for a single strand rRNA derived from 2.8 Å axial rise per residue for A-RNA double helices (35). The number of strands in this RNA varies from 1–5 (*Table 1*), the average being 4. This average value seems to be characteristic of all native rRNAs so far measured (16S, 18S, 23S, 28S) and it offers an explanation of

Table 1. Mass distribution within a 28S RNA molecule from the large ribosomal subunit from BHK cells

Segment	Mass (kDa)	Length (Å)	Mass/Length (Da/Å)	n^a
1	127	200	635	5
2	68	153	444	4
3	67	170	394	3
4	108	270	400	3
5	66	100	660	5
6	108	255	424	3
7	337	748	451	4
8	17	100	170	1
9	75	170	441	4
10	136	306	444	4
11	69	170	406	3
12	123	374	329	3
13	94	170	553	4
14	175	375	467	4
15	10	100	100	1
16	334	748	447	4
17	115	204	564	5
18	59	153	386	3
Σ	2088	4766		

[a] Number of strands based on a value of 125 Da/Å for a single strand of RNA derived from 2.8 Å axial rise per residue for A-RNA double helices (35).

why the total length of all the above rRNAs is only about 1/4 of that in the fully extended (denatured) state.

Monitoring the conformational changes in RNA

Conformation of ribosomes and subunits (as determined by EM) does not seem to be affected by moderate changes of ionic strength, except the well established effects of unfolding caused by low Mg^{2+} concentration (below 1 mM) and protein depletion by high salt. RNAs, on the other hand, undergo distinct conformational transitions from extended forms to tightly packed coils (*Figure 12*). The 28S rRNA molecules in distilled water (*Figure 12a*) appear extended ($R_G = 600 \pm 150$ Å). With increasing ionic strength (*Figure 12b, c*) these molecules become more compact, characterized by decreases of R_G values to $R_G = 240 \pm 30$ Å, and 160 ± 25 Å, respectively. However, the folding of 28S rRNA even in the reconstitution buffer (*Figure 12c*) does not reach the extent of the compactness and similarity to the shape of the native 60S ribosomal subunit (*Figure 12d*), characterized by $R_G = 87 \pm 8$ Å.

Interactions of rRNAs with ribosomal proteins

Protein-free deposition of rRNAs by the wet-film technique under non-denaturing conditions (27) makes it possible to visualize and analyse rRNA-protein interactions (32) and, ultimately, the process of complete ribosome

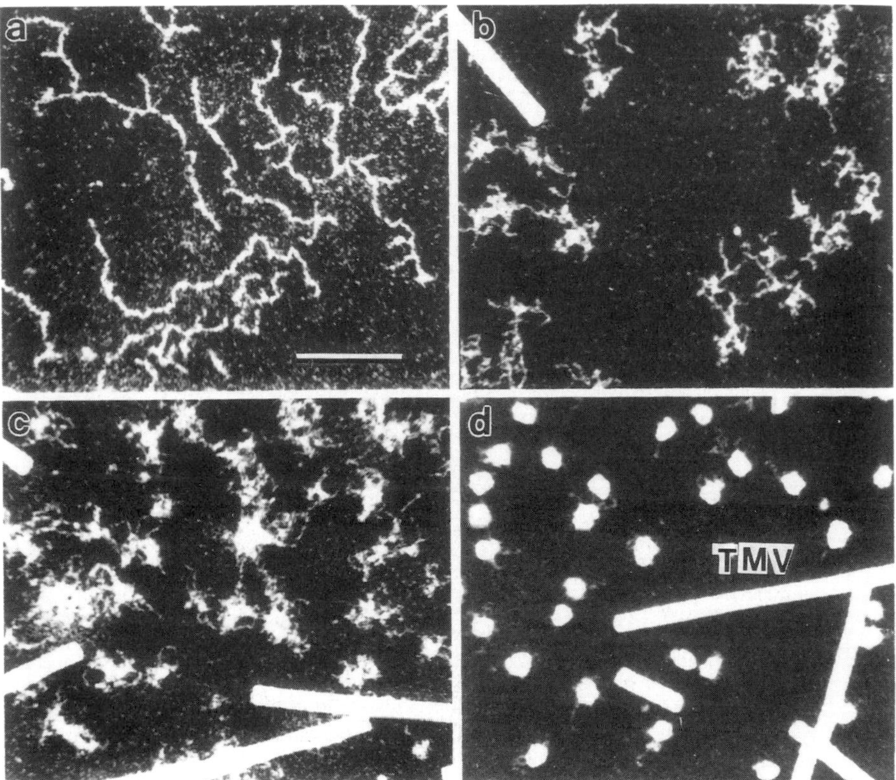

Figure 12. STEM images of unstained freeze-dried 28S rRNA molecules from BHK cells (a) in water, (b) in 60 mM ammonium acetate (pH 7.0), (c) in 30 mM Tris–HCl (pH 7.6), 360 mM KCl, 20 mM $MgCl_2$, and (d) corresponding 60S ribosomal subunits in 60 mM NH_4 acetate (pH 7.0), and 2 mM Mg acetate. TMV was used as an internal reference for mass measurements. Bar represents 0.1 μm.

assembly. The major criteria for monitoring the interactions are differences in mass, shape, and mass distribution (ΔR_G).

Visualization of rRNA–protein interactions is demonstrated using, as an example, 16S rRNA from *E. coli* associated with the six primary binding proteins—S4, S8, S15, S20, S17, and S7 (*Figure 13*). The complex of 16S rRNA with these proteins was prepared by incubation (at 42 °C for 1 h) of 16S rRNA in reconstitution buffer (30 mM Tris–HCl pH 7.6, 330 mM KCl, 20 mM Mg acetate, 1 mM DTT) with a three-fold molar excess of gradually added proteins S4, S8, S15, S20, S17, and S7. The reconstituted RNA–protein complex was separated on 15–30% sucrose gradients at 93 000 × *g* for 19 h in an SW28 rotor. The peak fractions were dialysed exhaustively against 10 mM Hepes–KOH (pH 7.5), 60 mM KCl, 2 mM Mg acetate, 1 mM DTT, and split into small

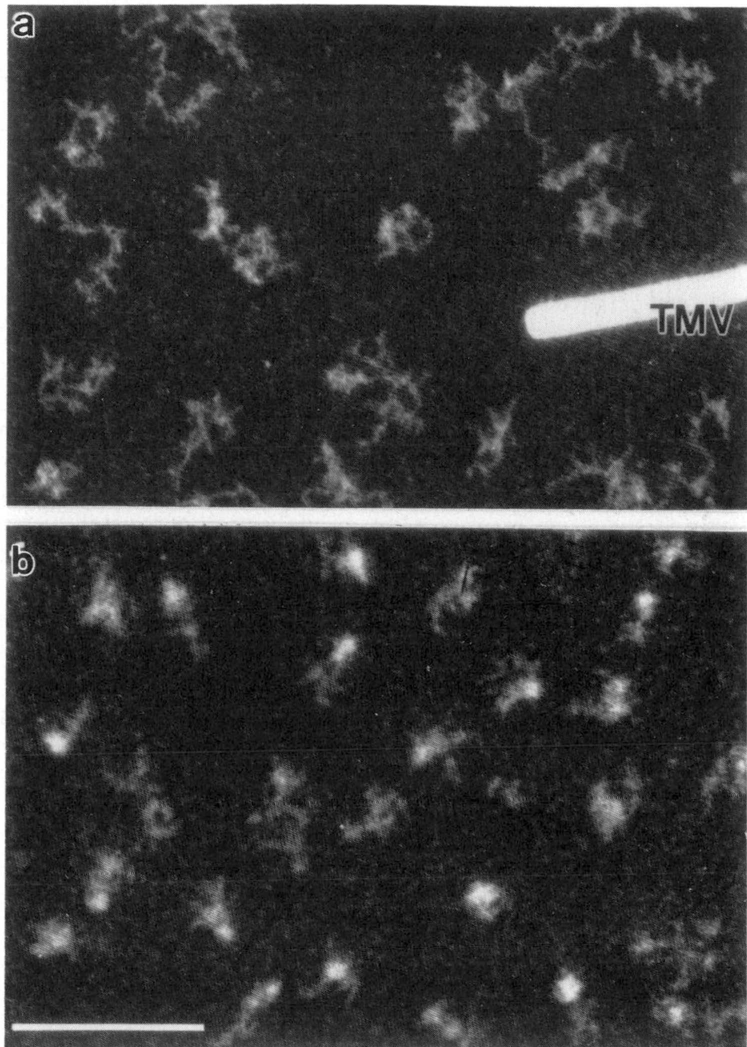

Figure 13. STEM images of unstained freeze-dried (a) 16S *E. coli* rRNA deposited from distilled water; (b) complex of 16S rRNA with ribosomal proteins S4, S8, S15, S20, S17, and S7 deposited from 10 mM Hepes–KOH (pH 7.4), 60 mM KCl, 2.0 mM Mg acetate, 1 mM DTT. Bar represents 0.1 μm.

aliquots used for measurement (32). Protein-free 16S rRNA (*Figure 13a*) in the above buffer is characterized by a loose coil with a mass of 551 ± 22 kDa and $R_G = 114 \pm 20$ Å. Complexes of 16S rRNA with the six bound proteins (*Figure 13b*) differ from 16S rRNA by the corresponding increase in mass ($M = 625 \pm 25$ kDa), by shape and mass distribution characterized by $R_G = 108 \pm 11$ Å.

4. Future trends

In future studies of ribosomes the emphasis will be on increasing resolution of EM imaging and quantitative analysis of the EM data. This trend should be pursued at the level of instrumentation, specimen preparation and preservation, and computer image processing.

In instrumentation, the resolution should be improved by reducing the radiation dose and by broadening the spectrum of high resolution electron spectroscopic techniques, e.g., electron energy loss spectroscopy (EELS) providing highly specific and quantitative information on mass and element distribution on the ribosome (36).

At the specimen level, the major effort will be directed towards preservation of the native structure of ribosomes and their components. The most promising and equally challenging approach seems to be imaging of frozen hydrated specimens embedded in thin films of vitreous ice. In topographical mapping of ribosomal proteins, rRNAs, and functional domains, substantial progress both in resolution and specificity is expected from application of 'cluster' compounds, small organic complexes with regularly spaced heavy atoms (Au, Pt, W) attached to specific sites on the ribosome (37).

Continuous progress in the development of sophisticated computer programs for analysis of high-resolution electron micrographs of biological specimens is reflected in three-dimensional reconstruction of ribosomal particles from their TEM images from a wide range of projections (38). Success of X-ray crystallography, the only direct technique introduced recently for the determination of the complete 3-D structure of the ribosome (39), depends primarily on availability of large well-ordered crystals of ribosomes or ribosomal subunits. Use of synchrotron radiation and prevention of crystal radiation damage by low temperatures (−180 °C) are important improvements in resolution of this demanding technique.

Application of the above directions should clarify the existing controversies in the structure of ribosomes, both prokaryotic and eukaryotic, and provide a deeper insight into the involvement of ribosomes in the process of protein synthesis. A detailed knowledge of ribosome architecture will be instrumental for establishing objective criteria for using ribosomes and their components as phylogenetic probes.

Acknowledgements

The author gratefully acknowledges stimulating discussions on EM imaging with Prof. A. K. Kleinschmidt in the very early phase of the project on structure–function relationships of ribosomes, extensive collaborative contributions of Drs. E. Spiess, N. Robakis, G. T. Oostergetel, V. Mandiyan, S. Tumminia, and the excellent assistance of W. Hellmann from my laboratory. Major progress in the field was possible by collaboration on image analysis with Drs. J. Frank, T.

Wagenknecht, M. Radermacher, and A. Verschoor from the Wadsworth Laboratory in Albany. Expertise of Drs. J. S. Wall and J. F. Hainfield from the Brookhaven STEM Biotechnology Resource was crucial for the advancement of electron microscopic imaging of ribosomes and their components to a quantitative analytical base.

References

1. Palade, G. E. (1955). *J. Biophys. Biochem. Cytol.*, **1**, 59.
2. Roberts, R. B. (1958). In *Microsomal Particles and Protein Synthesis* (ed. R. B. Roberts), p. 8. Pergamon Press, New York.
3. Hardesty, B. and Kramer, G. (1986). *Structure, Function, and Genetics of Ribosomes*. Springer-Verlag, New York.
4. Boublik, M. (1987). In *International Review of Cytology Supplement*, Vol. 17 (ed. G. H. Bourne), p. 357. Academic Press, San Diego.
5. Hayatt, M. A. (1970). *Principles and Techniques of Electron Microscopy*. Van Nostrand Reinhold Company, New York.
6. Sommerville, J. and Scheer, V. (1987). *Electron Microscopy in Molecular Biology—A Practical Approach*. IRL Press, Oxford.
7. Noller, H. F. and Moldave, K. (ed) (1988). *Methods in Enzymology*, Vol. 164. Academic Press, San Diego.
8. Hren, S. S., Goldstein, J. J., and Joy, D. C. (1979). *Introduction to Analytical Electron Microscopy*. Plenum, New York.
9. Wall, J. S. (1979). In *Introduction to Analytical Electron Microscopy* (ed. J. J. Hren, J. J. Goldstein, and D. C. Joy), p. 333. Plenum, New York.
10. Noll, M., Hapke, B., Schreier, M. H., and Noll, H. (1973). *J. Mol. Biol.*, **75**, 281.
11. Brown, G. E., Kolb, A. J., and Stanley, W. M., Jr. (1974). In *Methods in Enzymology*, Vol. 30 (ed. K. Moldave and L. Grossmann), p. 368. Academic Press, New York.
12. Zimmermann, R. A. (1979). In *Methods in Enzymology*, Vol. 59 (ed. K. Moldave and L. Grossman), p. 551. Academic Press, New York.
13. Aggarwal, S. K. (1976). *J. Histochem. Cytochem.*, **24**, 984.
14. Valentine, R. C. and Green, M. (1967). *J. Mol. Biol.*, **27**, 615.
15. Boublik, M., Hellmann, W., and Kleinschmidt, A. K. (1977). *Cytobiologie*, **14**, 293.
16. Fukami, A. and Köichi, A. (1965). *J. Electron Microscopy*, **14**, 112.
17. Stöffler-Meilicke, M. and Stöffler, G. (1988). In *Methods in Enzymology*, Vol. 164 (ed. H. F. Noller and K. Moldave), p. 503. Academic Press, San Diego.
18. Glitz, D. G., Cann, P. A., Lasater, L. S., and McKuskie Olson, H. (1988). In *Methods in Enzymology*, Vol. 164 (ed. H. F. Noller and K. Moldave), p. 493. Academic Press, San Diego.
19. Wabl, M. R. (1974). *J. Mol. Biol.*, **84**, 241.
20. Keren-Zur, M., Boublik, M., and Ofengand, J. (1979). *Proc. Nat. Acad. Sci. USA*, **76**, 1054.
21. Oakes, M. I., Clark, M. W., Henderson, E., and Lake, J. A. (1986). *Proc. Nat. Acad. Sci. USA*, **83**, 275.
22. Crewe, A. V. and Wall, J. (1970). *J. Mol. Biol.*, **48**, 375.
23. Wall, J. S. and Hainfeld, J. F. (1986). *Ann. Rev. Biophys, Biophys. Chem.*, **15**, 355.
24. Boublik, M., Oostergetel, G. T., Wall, J. S., Hainfeld, J. F., Radermacher, M., Wagenknecht, T., Verschoor, A. and Frank, J. (1986). In *Structure, Function and*

Genetics of Ribosomes (ed. B. Hardesty and G. Kramer), p. 68. Springer-Verlag, New York.
25. Boublik, M., Oostergetel, G. T., Mandiyan, V., Hainfeld, J. F., and Wall, J. S. (1988). In *Methods in Enzymology*, Vol. 164 (ed. H. F. Noller and K. Moldave), p. 49. Academic Press, San Diego.
26. Hainfeld, J. F., Wall, J. S., and Desmond, E. (1982). *Ultramicroscopy*, **8**, 263.
27. Wall, J. S., Hainfeld, J. F., and Chung, K. D. (1985). In *Proc. 43rd annual meeting of the electron microscopy society of America* (ed. G. W. Bailey), p. 716. San Francisco Press Inc., San Francisco.
28. Simon, M. N., Shiue, G. G., Wall, J. S., and Flory, P. J. (1988). In *Proc. 46th annual meeting of the electron microscopy society of America* (ed. G. W. Bailey), p. 416. San Francisco Press Inc., San Francisco.
29. Mandiyan, V., Hainfeld, J. F., Wall, J. S., and Boublik, M. (1988). *FEBS Lett.*, **236**, 340.
30. Vasiliev, V. P., Serdyuk, I. N., Gudkov, A. T., and Spirin, A. S. (1986). In *Structure, Function, and Genetics of Ribosomes* (ed. B. Hardesty and G. Kramer), p. 128. Springer-Verlag, New York.
31. Oostergetel, G. W., Wall, J. S., Hainfeld, J. F., and Boublik, M. (1985). *Proc. Nat. Acad. Sci. USA*, **82**, 5598.
32. Mandiyan, V., Wall, J. S., Hainfeld, J. F., and Boublik, M. (1989). *J. Mol. Biol.*, in press.
33. Sogo, J. M., Rodeno, P., Koller, T., Vinuela, E., and Salas, M. (1979). *Nucleic Acids Res.*, **7**, 107.
34. Kearney, H. R. and Moore, P. B. (1983). *J. Mol. Biol.*, **170**, 381.
35. Saenger, W. (1984). In *Principles of Nucleic Acid Structure* (ed. C. R. Cantor), p. 242. Springer-Verlag, New York.
36. Boublik, M., Oostergetel, G. T., Joy, D. C., Wall, J. S., Hainfeld, J. F., Frankland, B., and Ottensmeyer, P. F. (1986). *Ann. NY Acad. Sci.*, **463**, 168.
37. Wall, J. S., Hainfeld, J. F., Bartlett, P. A., and Singer, S. J. (1982). *Ultramicroscopy*, **8**, 397.
38. Frank, S., Radermacher, M., Wagenknecht, T., and Verschoor, A. (1988). In *Methods in Enzymology*, Vol 164 (ed. H. F. Noller and K. Moldave), p. 3. Academic Press, San Diego.
39. Yonath, A. and Wittmann, H. G. (1988). In *Methods in Enzymology*, Vol. 164 (ed. H. F. Noller and K. Moldave), p. 95. Academic Press, San Diego.
40. Boublik, M., Robakis, N., Hellmann, W., and Wall. J. S. (1982). *Eur. J. Cell Biol.*, **27**, 177.
41. Gornicki, P., Nurse, K., Hellmann, W., Boublik, M., and Ofengand, J. (1984). *J. Biol. Chem.*, **259**, 10493.
42. Olson, H. M. and Glitz, D. G. (1979). *Proc. Nat. Acad. Sci. USA*, **76**, 3769.
43. Politz, S. M. and Glitz, D. G. (1977). *Proc. Nat. Acad. Sci. USA*, **74**, 1468.
44. Trempe, M. R., Ohgi, K., and Glitz, D. G. (1982). *J. Biol. Chem.*, **257**, 9822.
45. Mochalova, L. V., Shatsky, I. N., Bogdanov, A. A., and Vasiliev, V. D. (1982). *J. Mol. Biol.*, **159**, 637.
46. Olson, H. M., Grant, P. G., Glitz, D. G., and Cooperman, B. S. (1980). *Proc. Nat. Acad. Sci. USA*, **77**, 890.
47. Stöffler, G. and Stöffler-Meilicke, M. (1981). In *International Cell Biology 1980/81* (ed. H. G. Schweiger), p. 93. Springer-Verlag, New York.

A1

Preparation of rRNA and r-proteins

1. rRNA

Two methods are described which have been used to prepare rRNA from prokaryotic ribosomes and subunits. A phenol extraction method is outlined in *Protocol 1* and an acetic acid-urea extraction method is presented in *Protocol 2*. The latter method, adapted from reference 1, is documented as producing rRNA which can bind more ribosomal proteins (under reconstitution conditions) than that prepared by phenol extraction (1). RNA prepared by the acetic acid-urea method is therefore typically used for protein-rRNA binding studies (e.g. see reference 2). It is also possible to prepare rRNA by loading ribosomal subunits directly on to 7.5–30% (w/v) sucrose gradients made up in 10 mM Tris–HCl (pH 7.8), 2 mM EDTA, 0.1% SDS and 4 mM 2-mercaptoethanol, followed by centrifugation at 82 000 *g* for 19 h at 10°C (e.g. 25 000 r.p.m. in a Beckman SW27 rotor). The ribosomal proteins remain at the top of the gradient. The rRNA may then be recovered by fractionation of the gradient, monitoring absorbance at 260 nm, pooling of appropriate fractions and precipitation with ethanol (see *Protocol 1* for details).

Protocol 1. rRNA preparation by phenol extraction[a]

1. Take ~50 A_{260} units/ml of 30S, 50S, or 70S ribosomal particles (see Chapters 1 and 8) in Buffer A[b]. Bring the solution to 1% (w/v) with respect to NaCl and then add SDS to 0.1–0.5% (w/v).
2. Add an equal volume of water-saturated phenol (highest quality phenol should be used, e.g. from BRL). Vortex for 5 min at room temperature then centrifuge at 14 000 *g* for 10 min at 4–6 °C.
3. Recover the upper aqueous phase and place on one side in ice. Re-extract the lower phenol phase with one-half volume of Buffer B[c] and centrifuge again as in step 2 to separate the phases.
4. Pool the two aqueous phases and then extract twice, each time with an equal volume of phenol followed by clarification of the phases as before.
5. Extract the final aqueous phase twice, each time with an equal volume of chloroform to remove traces of phenol.
6. Precipitate the RNA from the final aqueous phase with the addition of 2–2½

volumes of ethanol. Leave at −20 °C for 1 h. Recover the precipitated RNA by centrifugation at 14 000 *g* for 15 min at 4 °C.

7. Carefully decant the supernatant and then dry the pellet *in vacuo*. Resuspend the RNA in a small volume of 0.3 M sodium acetate (or 0.6 M potassium acetate) and re-precipitate with $2\frac{1}{2}$ volumes of ethanol. Leave at −20 °C for 1 h (minimum) or alternatively place into a dry ice/methanol bath for 10 min. Recover the RNA by centrifugation as in step 6.
8. Repeat step 7 twice. Rinse the final pellet of RNA with 70% ethanol, re-centrifuge following the rinse, decant the ethanol, and dry the pellet *in vacuo*.
9. Redissolve the RNA in water or Buffer C[d] (or other suitable buffer) at ~5 mg/ml. Store at −70 °C[e].
10. Total RNA from 70S ribosomes may be fractionated on sucrose gradients as described in *Protocol 2*. Two cycles of sucrose gradient purification may be necessary.

[a] When preparing intact rRNA it is important to follow the precautions outlined in Chapters 1 and 8.
[b] Buffer A: 10 mM Tris–HCl (pH 7.5 at 20 °C), 30 mM NH_4Cl, 10 mM $MgCl_2$, 6 mM 2-mercaptoethanol.
[c] Buffer B: 10 mM Tris-HCl (pH 7.6), 1 mM $MgCl_2$.
[d] Buffer C: 10 mM Tris–HCl (pH 7.6), 20 mM $MgCl_2$.
[e] The purity and integrity of rRNA may be determined on suitable polyacrylamide gels (3, 4).

Protocol 2. Preparation of RNA: acetic acid–urea method[a]

1. Resuspend crude or salt-washed ribosomes or subunits (see Chapter 1) in either Buffer A[b] or B[c]. Mix with an equal volume of 8 M urea (ultrapure, RNase free) and stir for 15 min at 4–8 °C.
2. Add solid magnesium acetate to raise the concentration to 0.8 M final and continue stirring for 15 min until dissolved.
3. Add 3 volumes of glacial acetic acid and stir gently for 1–2 h on ice. (The magnesium ion concentration is now 0.2 M.)
4. Recover the precipitated RNA by centrifugation at 10–14 000 *g* for 20 min (e.g. in a Beckman JA-21 or Sorvall SS34 rotor). Decant the supernatant[d].
5. Resuspend the RNA pellet in 1 ml of Buffer C[e]. Slowly add 0.214 g solid magnesium acetate and stir gently for 15 min. Add 4 ml glacial acetic acid (final [Mg^{2+}] again 0.2 M). Mix with stirring for 2 h in the cold.
6. Recover the RNA as in step 4. Resuspend the RNA and wash at least two times in 25 ml of Buffer D[f] followed by recovery by centrifugation as before. This washing procedure should be carried out until the pH of the resuspended RNA solution measures 8.0.
7. Resuspend the final RNA pellet in water or 30 mM Tris–HCl (pH 8.0). Heating at 37 °C for about 5–10 min may be necessary for the RNA to fully

dissolve. Store in aliquots at −70 °C (1 A_{260} unit is equivalent to about 27 pmol or 60 μg of total 70S rRNA when estimated in water).

8. Total 70S rRNA may be fractionated by density gradient centrifugation on linear 15–30% (w/v) sucrose gradients made up in Buffer E[g]. Centrifuge at 82 000 *g* for 20 h at 10–15 °C (e.g. 25 000 r.p.m. in an SW 27 rotor). Fractionate the gradients, monitoring absorbance at 260 nm. Recover 16S and 23S rRNA from appropriately pooled fractions by ethanol precipitation (see *Protocol 1*). Redissolve the RNA in water and reprecipitate, if necessary, following the addition of one tenth volume of 3 M sodium acetate and $2\frac{1}{2}$ volumes of ethanol.

[a] See note a in *Protocol 1*.
[b] Buffer A: 10 mM Tris–HCl (pH 7.5 at 20 °C), 50 mM NH_4Cl, 10 mM $MgCl_2$, 3 mM 2-mercaptoethanol.
[c] Buffer B: 30 mM Tricine (pH 8.0), 20 mM magnesium acetate.
[d] The supernatant containing total 70S or total subunit proteins may be reserved for further analysis or fractionation. Store frozen at −20 °C.
[e] Buffer C: 30 mM Tris–HCl (pH 8.0), 0.4 M KCl, 20 mM magnesium acetate, 1 mM DTT.
[f] Buffer D: 30 mM Tris–HCl (pH 8.0), 20 mM $MgCl_2$.
[g] Buffer E: 10 mM Tris–HCl (pH 7.5 at 20 °C), 100 mM LiCl, 10 mM Na_2 EDTA, 0.2% SDS.

2. Ribosomal proteins

A rapid method available for the preparation of total ribosomal proteins for use in two-dimensional gel electrophoretic analysis, which has found much favour, is presented in *Protocol 3*. It is based on references 5 and 6. This method is suitable for the preparation of ribosomal proteins from most sources. The method is also suitable for extraction of ribosomal proteins from whole cells, e.g. from yeast cells lysed with water and from extracts of *E. coli* or other bacterial cells lysed by freezing and thawing.

Alternative methods for extraction of ribosomal proteins from eukaryotic ribosomal subunits include extraction with LiCl and urea, with hydrochloric acid and by methods involving digestion of ribosomes and ribosomal subunits by RNase. Details of these methods may be found elsewhere (7, 8).

Protocol 3. Isolation of total ribosomal proteins

1. To a suspension of ribosomes (e.g. 200–1500 A_{260} units) add, whilst stirring vigorously, 1/10th volume of 1 M $MgCl_2$ (or acetate) and two volumes of ice cold glacial acetic acid in rapid succession.
2. Continue stirring at 4 °C for 45 min.
3. Precipitate the rRNA by centrifugation at 20 000 *g* for 10 min. Keep the supernatant on ice. Re-extract the pellet with 67% acetic acid and 0.1 M $MgCl_2$ (or acetate) for 20–30 min. Precipitate the RNA as above.
4. Pool the two supernatants. Precipitate the ribosomal proteins by adding 5–10

volumes of ice cold acetone. Leave the solution at −20 °C for 2 h or overnight.

5. Recover the precipitated proteins by centrifugation at 10 000 *g* for 30 min. Wash the resultant pellet twice with acetone to remove residual acetic acid and then dry under vacuum, preferably at 0 °C.
6. Store the pellets dry (at −70 °C) or dissolve them in a suitable two-dimensional gel sample buffer. Process accordingly (see Chapter 1, Section 4 for suggestions).

2.1 Individual ribosomal proteins

Four major methods have been employed for the isolation and purification of all the individual ribosomal proteins from *E. coli.*

(a) Acetic acid extraction of ribosomal proteins from subunits, followed by separation by cellulose ion exchange chromatography and gel filtration on Sephadex in the presence of 6 M urea (9, 10).

(b) Stepwise washing of ribosomal subunits with increasing concentrations of LiCl to remove specific subsets of proteins (as described by Nierhaus in *Chapter 8*). These steps are followed by purification of the proteins by column chromatography on carboxymethyl-Sephadex and then gel filtration on Sephadex G100. High ionic strength buffers are employed throughout the purification and no urea is used. This provides a gentle procedure for the isolation of the proteins avoiding the use of the denaturants acetic acid and urea (11).

(c) Washing of 50S subunits with increasing salt concentrations (NH_4Cl and LiCl) followed by purification of the proteins by gel filtration on Sephadex G100 and then ion-exchange chromatography on carboxymethyl-cellulose in buffers containing 6 M urea (12). Following purification proteins are dialysed extensively against renaturing buffers. (See *Chapter 8*, *Protocol 7*.)

(d) High performance liquid chromatography (HPLC) provides the most up-to-date means of preparing ribosomal proteins. It offers at least equivalent resolution and reproducibility and offers several advantages over conventional chromatographic procedures, including rapidity, good recovery yields and sensitivity. Starting materials usually include the NH_4Cl/ethanol split protein fractions or the LiCl-derived split proteins of ribosomal subunits (see *Protocols 3, 4, and 8* and Figure 3 in *Chapter 8* of this volume) or total 30S or 50S ribosomal subunit proteins (TP30 or TP50, respectively) etc. Two types of HPLC have been employed routinely, namely Reversed Phase high performance liquid chromatography (RP-HPLC) and Ion-Exchange high performance liquid chromatography (IE-HPLC]. The former method offers higher sensitivity and resolution over the latter but only with small total protein sample sizes. IE-HPLC offers much higher capacity. RP-HPLC has allowed the resolution of the 21 *E. coli* 30S ribosomal subunit proteins into 18 peaks and the resolution of the 33 *E. coli* 50S ribosomal subunit proteins into 28 peaks. IE-HPLC resolves the 21 small subunit proteins into 18 peaks and the 33 large subunit proteins into 25 peaks, respectively (13).

The description of these modern procedures is beyond the scope of this outline and the reader is referred to a series of articles in Section VII of reference 13. Further basic information concerning the HPLC purification of proteins can also be found in another book in the *Practical Approach* series (14).

References

1. Hochkeppel, H.-K., Spicer, E., and Craven, G. R. (1976). *J. Mol. Biol.*, **101**, 155.
2. Beauclerk, A. A. D. and Cundliffe, E. (1988). *EMBO J.*, **7**, 3589.
3. Lehrach, H., Diamond, D., Wozney, J. M., and Boedtker, H. (1977). *Biochemistry*, **16**, 4743.
4. Rickwood, D. and Hames, B. D. (ed.) (1982). *Gel Electrophoresis of Nucleic Acids: A Practical Approach*. IRL Press, Oxford.
5. Hardy, S. J. S., Kurland, C. G., Voynow, P., and Mora, G. (1969). *Biochemistry*, **8**, 2897.
6. Barritault, D., Expert-Bezançon, A., Guérin, M., and Hayes, D. (1976). *Eur. J. Biochem.*, **63**, 131.
7. Ogata, K. and Terao, K. (1979). In *Methods in Enzymology* (ed. K. Moldave and L. Grossman), Vol. 59, p. 502. Academic Press, NY.
8. Sherton, C. C. and Wool, I. G. (1974). *Mol. Gen. Genet.*, **135**, 97.
9. Giri, L., Hill, W. E., Wittmann, H. G., and Wittmann-Liebold, B. (1984). In *Advances in Protein Chemistry*, Vol. 36 (ed. C. B. Anfinsen, J. T. Edsall, and F. M. Richards), p. 1. Academic Press, Orlando.
10. Zimmermann, R. A. (1979). In *Methods in Enzymology* (ed. K. Moldave and L. Grossman), Vol. 59, p. 551. Academic Press, NY.
11. Dijk, J. and Littlechild, J. (1979). *ibid*, p. 481.
12. Wystup, G., Teraoka, H., Schulze, H., Hampl, H., and Nierhaus, K. H. (1979). *Eur. J. Biochem.*, **100**, 101.
13. Noller, H. F. and Moldave, K. (ed.) (1988). *Methods in Enzymology*, Vol. 164, Section VII. Academic Press, San Diego.
14. Oliver, R. W. A. (ed.) (1988). *HPLC of Macromolecules: A Practical Approach*. IRL Press, Oxford.

A2

Suppliers of specialist items

Items referred to in the text can be purchased from the following companies. In general, the address of the head office is listed, from which the names and addresses of more local suppliers can be obtained.

Aldrich Chemical Co., 940 West Saint Paul Avenue, Milwaukee, WI 53233, USA.

Amersham International plc, White Lion Road, Amersham, Buckinghamshire, HP7 9LL, UK.

Amicon Division, W. R. Grace and Co., 24 Cherry Hill Drive, Danvers, MA 01923, USA.

Applied Biosystems, 850 Lincoln Centre Drive, Foster City, CA 94404, USA.

Balzers Union AG, Postfach 75, FL-9496, Balzers, Fürstentum, Liechtenstein.

Beckman Instruments Inc, Spinco Division, 1050 Page Mill Road, Palo Alto, CA 94304, USA.

Bethesda Research Laboratories Inc., PO Box 6009, Gaithersburg, MD 20877, USA.

Bio 101 Inc., PO Box 2284, La Jolla, CA 92038-2284, USA.

Bio-Rad Laboratories, 1414 Harbour Way South, Richmond, CA 94804, USA.

Boehringer Mannheim GmbH Biochemica, Sandhofer Str. 116, Postfach 310120, D-6800 Mannheim 31, FRG.

Calbiochem. Corp., PO Box 12087, San Diego, CA 92112-4180, USA.

Collaborative Research Inc., Two Oak Park, Bedford, MA 01730, USA.

Desaga, 6900 Heidelberg, FRG.

Difco Laboratories Ltd., PO Box 14B, Central Avenue, East Moseley, Surrey, KT8 0SE, UK.

DuPont Company Biotechnology Systems, Barley Mill Plaza, Chandler Mill Building, Wilmington, DE 19898, USA.

Eastman–Kodak Co., 343 State Street, Rochester, NY 14650, USA.

EGA Chemie, 7924 Steinheim/Albuch, FRG.

Electro-Nucleonics Inc., Cell Science Lab., 12050 Tech Road, Silver Spring, MD 20904, USA.

Eppendorf Inc., 45635 Northport Loop East, Fremont, CA 94538, USA.

GIBCO Life Technologies Inc., 3175 Staley Road, Grand Island, NY 14072, USA.

Grain Processing Corp., 1600 Oregon Street, Muscatine, IA 52761, USA.

ICN Biochemicals, PO Box 28050, Cleveland, OH 44128, USA.

ISCO Inc., PO Box 5347, 4700 Superior Street, Lincoln, NE 68504, USA.

Life Sciences Inc., 2900 72nd Street North, Saint Petersburg, FL 33710, USA.
LKB-Produkter AB, Box 305, S-16126 Bromma, Sweden.
Macherey and Nagel, Neumann-Neander Str. 6-8, 5160 Düren, FRG.
E. Merck, Frankfurter Str. 250, Postfach 4119, D-6100 Darmstadt 1, FRG.
Miles Laboratories Inc., Research Products Division, PO Box 2000, 1127 Myrtle Street, Elkhart, IN 46514, USA.
Millipore Corporation, 80 Ashby Road, Bedford, MA 01730, USA.
Mini-Instruments Ltd, (for mini-monitors) Burnham-on-Crouch, Essex CM0 8RN, UK.
New England Nuclear, 549 Albany Street, Boston, MA 02118, USA.
Oxoid Ltd, Wade Road, Basingstoke, Hants, RG24 0PW, UK.
Pierce Biochemicals, PO Box 117, Rockford, IL 61105, USA.
Promega Corporation, 2800 South Fish Hatchery Road, Madison, WI 53711-5305, USA.
Pharmacia-LKB Biotechnology AB, S75182 Uppsala, Sweden.
San Francisco Bay Brand Inc., Newark, CA 94560, USA.
Savant Instruments Inc., 110–103 Bi-County Blvd, Farmingdale, NY 11735, USA.
Schleicher and Schuell GmbH, Postfach 4, D-3354 Dassel, FRG.
Serva Feinbiochemica GmbH, PO Box 105260, Carl-Benz-Str. 7, D-6900 Heidelberg 1, FRG.
Sigma Chemical Co., PO Box 14508, St Louis, MO 63178, USA.
SLM Aminco Instruments, (Aminco French Presses) 810 West Anthony Drive, Urbana, IL 61801, USA.
Sorvall, see DuPont.
Stratagene, 11099 North Torrey Pines Road, La Jolla, CA 92037, USA.
Sylvania, Graf Zeppelin Str. 9-11, 8520 Erlangen, FRG.
Ted Pella Inc., PO Box 510, Tustin, CA 92681, USA.
Tektronix Inc., (Zeineh densitometers) PO Box 1700, Beaverton, OR 97075 USA.
United States Biochemical Corp., Box 22400, Cleveland, OH 44122, USA.
V.G. Microscopes Ltd, Charlwoods Road, East Grinstead, West Sussex RH19 2JQ, UK.
Waters Chromatography Division, Millipore Corp., 34 Maple Street, Milford, MA 01757, USA.
Whatman Ltd, Springfield Mill, Maidstone, Kent ME14 2LE, UK.
Worthington Biochemical Corp., Halls Mill Road, Freehold, NJ 00728, USA.

Index

Index